T0296073

LONDON MATHEMATICAL SOCIETY LECTURE NOTE SERIES

Managing Editor: Professor M. Reid, Mathematics Institute,
University of Warwick, Coventry CV4 7AL, United Kingdom

The titles below are available from booksellers, or from Cambridge University Press at
www.cambridge.org/mathematics

287 Topics on Riemann surfaces and Fuchsian groups, E. BUJALANCE, A.F. COSTA & E. MARTÍNEZ (eds)
288 Surveys in combinatorics, 2001, J.W.P. HIRSCHFELD (ed)
289 Aspects of Sobolev-type inequalities, L. SALOFF-COSTE
290 Quantum groups and Lie theory, A. PRESSLEY (ed)
291 Tits buildings and the model theory of groups, K. TENT (ed)
292 A quantum groups primer, S. MAJID
293 Second order partial differential equations in Hilbert spaces, G. DA PRATO & J. ZABCZYK
294 Introduction to operator space theory, G. PISIER
295 Geometry and integrability, L. MASON & Y. NUTKU (eds)
296 Lectures on invariant theory, I. DOLGACHEV
297 The homotopy category of simply connected 4-manifolds, H.-J. BAUES
298 Higher operads, higher categories, T. LEINSTER (ed)
299 Kleinian groups and hyperbolic 3-manifolds, Y. KOMORI, V. MARKOVIC & C. SERIES (eds)
300 Introduction to Möbius differential geometry, U. HERTRICH-JEROMIN
301 Stable modules and the D(2)-problem, F.E.A. JOHNSON
302 Discrete and continuous nonlinear Schrödinger systems, M.J. ABLOWITZ, B. PRINARI & A.D. TRUBATCH
303 Number theory and algebraic geometry, M. REID & A. SKOROBOGATOV (eds)
304 Groups St Andrews 2001 in Oxford I, C.M. CAMPBELL, E.F. ROBERTSON & G.C. SMITH (eds)
305 Groups St Andrews 2001 in Oxford II, C.M. CAMPBELL, E.F. ROBERTSON & G.C. SMITH (eds)
306 Geometric mechanics and symmetry, J. MONTALDI & T. RATIU (eds)
307 Surveys in combinatorics 2003, C.D. WENSLEY (ed.)
308 Topology, geometry and quantum field theory, U.L. TILLMANN (ed)
309 Corings and comodules, T. BRZEZINSKI & R. WISBAUER
310 Topics in dynamics and ergodic theory, S. BEZUGLYI & S. KOLYADA (eds)
311 Groups: topological, combinatorial and arithmetic aspects, T.W. MÜLLER (ed)
312 Foundations of computational mathematics, Minneapolis 2002, F. CUCKER *et al* (eds)
313 Transcendental aspects of algebraic cycles, S. MÜLLER-STACH & C. PETERS (eds)
314 Spectral generalizations of line graphs, D. CVETKOVIĆ, P. ROWLINSON & S. SIMIĆ
315 Structured ring spectra, A. BAKER & B. RICHTER (eds)
316 Linear logic in computer science, T. EHRHARD, P. RUET, J.-Y. GIRARD & P. SCOTT (eds)
317 Advances in elliptic curve cryptography, I.F. BLAKE, G. SEROUSSI & N.P. SMART (eds)
318 Perturbation of the boundary in boundary-value problems of partial differential equations, D. HENRY
319 Double affine Hecke algebras, I. CHEREDNIK
320 L-functions and Galois representations, D. BURNS, K. BUZZARD & J. NEKOVÁŘ (eds)
321 Surveys in modern mathematics, V. PRASOLOV & Y. ILYASHENKO (eds)
322 Recent perspectives in random matrix theory and number theory, F. MEZZADRI & N.C. SNAITH (eds)
323 Poisson geometry, deformation quantisation and group representations, S. GUTT *et al* (eds)
324 Singularities and computer algebra, C. LOSSEN & G. PFISTER (eds)
325 Lectures on the Ricci flow, P. TOPPING
326 Modular representations of finite groups of Lie type, J.E. HUMPHREYS
327 Surveys in combinatorics 2005, B.S. WEBB (ed)
328 Fundamentals of hyperbolic manifolds, R. CANARY, D. EPSTEIN & A. MARDEN (eds)
329 Spaces of Kleinian groups, Y. MINSKY, M. SAKUMA & C. SERIES (eds)
330 Noncommutative localization in algebra and topology, A. RANICKI (ed)
331 Foundations of computational mathematics, Santander 2005, L.M PARDO, A. PINKUS, E. SÜLI & M.J. TODD (eds)
332 Handbook of tilting theory, L. ANGELERI HÜGEL, D. HAPPEL & H. KRAUSE (eds)
333 Synthetic differential geometry (2nd Edition), A. KOCK
334 The Navier–Stokes equations, N. RILEY & P. DRAZIN
335 Lectures on the combinatorics of free probability, A. NICA & R. SPEICHER
336 Integral closure of ideals, rings, and modules, I. SWANSON & C. HUNEKE
337 Methods in Banach space theory, J.M.F. CASTILLO & W.B. JOHNSON (eds)
338 Surveys in geometry and number theory, N. YOUNG (ed)
339 Groups St Andrews 2005 I, C.M. CAMPBELL, M.R. QUICK, E.F. ROBERTSON & G.C. SMITH (eds)
340 Groups St Andrews 2005 II, C.M. CAMPBELL, M.R. QUICK, E.F. ROBERTSON & G.C. SMITH (eds)
341 Ranks of elliptic curves and random matrix theory, J.B. CONREY, D.W. FARMER, F. MEZZADRI & N.C. SNAITH (eds)
342 Elliptic cohomology, H.R. MILLER & D.C. RAVENEL (eds)
343 Algebraic cycles and motives I, J. NAGEL & C. PETERS (eds)
344 Algebraic cycles and motives II, J. NAGEL & C. PETERS (eds)
345 Algebraic and analytic geometry, A. NEEMAN
346 Surveys in combinatorics 2007, A. HILTON & J. TALBOT (eds)
347 Surveys in contemporary mathematics, N. YOUNG & Y. CHOI (eds)

London Mathematical Society Lecture Note Series: 409

Surveys in Combinatorics 2013

Edited by

SIMON R. BLACKBURN
Royal Holloway, University of London

STEFANIE GERKE
Royal Holloway, University of London

MARK WILDON
Royal Holloway, University of London

CAMBRIDGE
UNIVERSITY PRESS

University Printing House, Cambridge CB2 8BS, United Kingdom

One Liberty Plaza, 20th Floor, New York, NY 10006, USA

477 Williamstown Road, Port Melbourne, VIC 3207, Australia

4843/24, 2nd Floor, Ansari Road, Daryaganj, Delhi - 110002, India

79 Anson Road, #06-04/06, Singapore 079906

Cambridge University Press is part of the University of Cambridge.

It furthers the University's mission by disseminating knowledge in the pursuit of education, learning and research at the highest international levels of excellence.

www.cambridge.org
Information on this title: www.cambridge.org/9781107651951

© Cambridge University Press 2013

First published 2013

A catalogue record for this publication is available from the British Library

ISBN 978-1-107-65195-1 Paperback

Cambridge University Press has no responsibility for the persistence or accuracy of URLs for external or third-party internet websites referred to in this publication, and does not guarantee that any content on such websites is, or will remain, accurate or appropriate.

Contents

Preface

The Twenty-Fourth British Combinatorial Conference was organised by Royal Holloway, University of London. It was held in Egham, Surrey in July 2013. The British Combinatorial Committee had invited nine distinguished combinatorialists to give survey lectures in areas of their expertise, and this volume contains the survey articles on which these lectures were based.

In compiling this volume we are indebted to the authors for preparing their articles so accurately and professionally, and to the referees for their rapid responses and keen eye for detail. We would also like to thank Roger Astley and Sam Harrison at Cambridge University Press for their advice and assistance.

Finally, without the previous efforts of editors of earlier *Surveys* and the guidance of the British Combinatorial Committee, the preparation of this volume would have been daunting: we would like to express our thanks for their support.

Simon R. Blackburn, Stefanie Gerke and Mark Wildon

Royal Holloway, University of London

January 2013

Graph removal lemmas

David Conlon[1] and Jacob Fox[2]

Abstract

The graph removal lemma states that any graph on n vertices with $o(n^h)$ copies of a fixed graph H on h vertices may be made H-free by removing $o(n^2)$ edges. Despite its innocent appearance, this lemma and its extensions have several important consequences in number theory, discrete geometry, graph theory and computer science. In this survey we discuss these lemmas, focusing in particular on recent improvements to their quantitative aspects.

1 Introduction

The triangle removal lemma states that for every $\varepsilon > 0$ there exists $\delta > 0$ such that any graph on n vertices with at most δn^3 triangles may be made triangle-free by removing at most εn^2 edges. This result, proved by Ruzsa and Szemerédi [94] in 1976, was originally stated in rather different language.

The original formulation was in terms of the $(6,3)$-problem.[3] This asks for the maximum number of edges $f^{(3)}(n, 6, 3)$ in a 3-uniform hypergraph on n vertices such that no 6 vertices contain 3 edges. Answering a question of Brown, Erdős and Sós [19], Ruzsa and Szemerédi showed that $f^{(3)}(n, 6, 3) = o(n^2)$. Their proof used several iterations of an early version of Szemerédi's regularity lemma [111].

This result, developed by Szemerédi in his proof of the Erdős-Turán conjecture on arithmetic progressions in dense sets [110], states that every graph may be partitioned into a small number of vertex sets so that the graph between almost every pair of vertex sets is random-like. Though this result now occupies a central position in graph theory, its importance only emerged over time. The resolution of the $(6,3)$-problem was one of the first indications of its strength.

---

[1] Supported by a Royal Society University Research Fellowship.

[2] Supported by a Simons Fellowship and NSF Grant DMS-1069197.

[3] The two results are not exactly equivalent, though the triangle removal lemma may be proved by their method. A weak form of the triangle removal lemma, already sufficient for proving Roth's theorem, is equivalent to the Ruzsa-Szemerédi theorem. This weaker form states that any graph on n vertices in which every edge is contained in exactly one triangle has $o(n^2)$ edges. This is also equivalent to another attractive formulation, known as the induced matching theorem. This states that any graph on n vertices which is the union of at most n induced matchings has $o(n^2)$ edges.

The Ruzsa-Szemerédi theorem was generalized by Erdős, Frankl and Rödl [32], who showed that $f^{(r)}(n, 3r-3, 3) = o(n^2)$, where $f^{(r)}(n, 3r-3, 3)$ is the maximum number of edges in an r-uniform hypergraph such that no $3r - 3$ vertices contain 3 edges. One of the tools used by Erdős, Frankl and Rödl in their proof was a striking result stating that if a graph on n vertices contains no copy of a graph H then it may be made K_r-free, where $r = \chi(H)$ is the chromatic number of H, by removing $o(n^2)$ edges. The proof of this result used the modern formulation of Szemerédi's regularity lemma and is already very close, both in proof and statement, to the following generalization of the triangle removal lemma, known as the graph removal lemma.[4] This was first stated explicitly in the literature by Alon, Duke, Lefmann, Rödl and Yuster [4] and by Füredi [47] in 1994.[5]

Theorem 1.1 *For any graph H on h vertices and any $\varepsilon > 0$, there exists $\delta > 0$ such that any graph on n vertices which contains at most δn^h copies of H may be made H-free by removing at most εn^2 edges.*

It was already observed by Ruzsa and Szemerédi that the $(6, 3)$-problem (and, thereby, the triangle removal lemma) is related to Roth's theorem on arithmetic progressions [92]. This theorem states that for any $\delta > 0$ there exists an n_0 such that if $n \geq n_0$, then any subset of the set $[n] := \{1, 2, \ldots, n\}$ of size at least δn contains an arithmetic progression of length 3. Letting $r_3(n)$ be the largest integer such that there exists a subset of the set $\{1, 2, \ldots, n\}$ of size $r_3(n)$ containing no arithmetic progression of length 3, this is equivalent to saying that $r_3(n) = o(n)$. Ruzsa and Szemerédi observed that $f^{(3)}(n, 6, 3) = \Omega(r_3(n)n)$. In particular, since $f^{(3)}(n, 6, 3) = o(n^2)$, this implies that $r_3(n) = o(n)$, yielding a proof of Roth's theorem.

It was further noted by Solymosi [105] that the Ruzsa-Szemerédi theorem yields a stronger result of Ajtai and Szemerédi [1]. This result states that for any $\delta > 0$ there exists an n_0 such that if $n \geq n_0$ then any subset of the set $[n] \times [n]$ of size at least δn^2 contains a set of the form $\{(a, b), (a + d, b), (a, b + d)\}$ with $d > 0$. That is, dense subsets of the

---

[4]The phrase 'removal lemma' is a comparatively recent coinage. It seems to have come into vogue in about 2005 when the hypergraph removal lemma was first proved (see, for example, [68, 79, 107, 113]).

[5]This was also the first time that the triangle removal lemma was stated explicitly, though the weaker version concerning graphs where every edge is contained in exactly one triangle had already appeared in the literature. The Ruzsa-Szemerédi theorem was usually [40, 41, 46] phrased in the following suggestive form: if a 3-uniform hypergraph is linear, that is, no two edges intersect on more than a single vertex, and triangle-free, then it has $o(n^2)$ edges. A more explicit formulation may be found in [23].

2-dimensional grid contain axis-parallel isosceles triangles. Roth's theorem is a simple corollary of this statement.

Roth's theorem is the first case of a famous result known as Szemerédi's theorem. This result, to which we alluded earlier, states that for any natural number $k \geq 3$ and any $\delta > 0$ there exists n_0 such that if $n \geq n_0$ then any subset of the set $[n]$ of size at least δn contains an arithmetic progression of length k. This was first proved by Szemerédi [110] in the early seventies using combinatorial techniques and since then several further proofs have emerged. The most important of these are that by Furstenberg [48, 50] using ergodic theory and that by Gowers [54, 55], who found a way to extend Roth's original Fourier analytic argument to general k. Both of these methods have been highly influential.

Yet another proof technique was suggested by Frankl and Rödl [42]. They showed that Szemerédi's theorem would follow from the following generalization of Theorem 1.1, referred to as the hypergraph removal lemma. They proved this theorem for the specific case of $K_4^{(3)}$, the complete 3-uniform hypergraph with 4 vertices. This was then extended to all 3-uniform hypergraphs in [78] and to $K_5^{(4)}$ in [90]. Finally, it was proved for all hypergraphs by Gowers [56, 57] and, independently, by Nagle, Rödl, Schacht and Skokan [79, 89]. Both proofs rely on extending Szemerédi's regularity lemma to hypergraphs in an appropriate fashion.

Theorem 1.2 *For any k-uniform hypergraph $\mathcal{H}$ on h vertices and any $\varepsilon > 0$, there exists $\delta > 0$ such that any k-uniform hypergraph on n vertices which contains at most δn^h copies of $\mathcal{H}$ may be made $\mathcal{H}$-free by removing at most εn^k edges.*

As well as reproving Szemerédi's theorem, the hypergraph removal lemma allows one to reprove the multidimensional Szemerédi theorem. This theorem, originally proved by Furstenberg and Katznelson [49], states that for any natural number r, any finite subset S of $\mathbb{Z}^r$ and any $\delta > 0$ there exists n_0 such that if $n \geq n_0$ then any subset of $[n]^r$ of size at least δn^r contains a subset of the form $a \cdot S + d$ with $a > 0$, that is, a dilated and translated copy of S. That it follows from the hypergraph removal lemma was first observed by Solymosi [106]. This was the first non-ergodic proof of this theorem. A new proof of the special case $S = \{(0,0), (1,0), (0,1)\}$, corresponding to the Ajtai-Szemerédi theorem, was given by Shkredov [103] using a Fourier analytic argument. Recently, a combinatorial proof of the density Hales-Jewett theorem, which is an extension of the multidimensional Szemerédi theorem, was discovered as part of the polymath project [82].

As well as its implications in number theory, the removal lemma and its extensions are central to the area of computer science known as property testing. In this area, one would like to find fast algorithms to distinguish between objects which satisfy a certain property and objects which are far from satisfying that property. This field of study was initiated by Rubinfield and Sudan [93] and, subsequently, Goldreich, Goldwasser and Ron [52] started the investigation of such property testers for combinatorial objects. Graph property testing has attracted a particular degree of interest.

A classic example of property testing is to decide whether a given graph G is ε-far from being triangle-free, that is, whether at least εn^2 edges will have to removed in order to make it triangle-free. The triangle removal lemma tells us that if G is ε-far from being triangle free then it must contain at least δn^3 triangles for some $\delta > 0$ depending only on ε. This furnishes a simple probabilistic algorithm for deciding whether G is ε-far from being triangle-free. We choose $t = 2\delta^{-1}$ triples of points from the vertices of G uniformly at random. If G is ε-far from being triangle-free then the probability that none of these randomly chosen triples is a triangle is $(1 - \delta)^t < e^{-t\delta} < \frac{1}{3}$. That is, if G is ε-far from being triangle-free, we will find a triangle with probability at least $\frac{2}{3}$, whereas if G is triangle-free, we will clearly find no triangles. The graph removal lemma may be used to derive a similar test for deciding whether G is ε-far from being H-free for any fixed graph H.

In property testing, it is often of interest to decide not only whether a graph is far from being H-free but also whether it is far from being induced H-free. A subgraph H' of a graph G is said to be an induced copy of H if there is a one-to-one map $f : V(H) \to V(H')$ such that $(f(u), f(v))$ is an edge of H' if and only if (u, v) is an edge of H. A graph G is said to be induced H-free if it contains no induced copies of H and ε-far from being induced H-free if we have to add and/or delete at least εn^2 edges to make it induced H-free. Note that it is not enough to delete edges since, for example, if H is the empty graph on two vertices and G is the complete graph minus an edge, then G contains only one induced copy of H, but one cannot simply delete edges from G to make it induced H-free.

By proving an appropriate strengthening of the regularity lemma, Alon, Fischer, Krivelevich and Szegedy [6] showed how to modify the graph removal lemma to this setting. This result, which allows one to test for induced H-freeness, is known as the induced removal lemma.

Theorem 1.3 *For any graph H on h vertices and any $\varepsilon > 0$, there exists a $\delta > 0$ such that any graph on n vertices which contains at most δn^h induced copies of H may be made induced H-free by adding and/or deleting at most εn^2 edges.*

A substantial generalization of this result, known as the infinite removal lemma, was proved by Alon and Shapira [12] (see also [76]). They showed that for each (possibly infinite) family $\mathcal{H}$ of graphs and $\varepsilon > 0$ there is $\delta = \delta_{\mathcal{H}}(\varepsilon) > 0$ and $t = t_{\mathcal{H}}(\varepsilon)$ such that if a graph G on n vertices contains at most δn^h induced copies of H for every graph H in $\mathcal{H}$ on $h \leq t$ vertices, then G may be made induced H-free, for every $H \in \mathcal{H}$, by adding and/or deleting at most εn^2 edges. They then used this result to show that every hereditary graph property is testable, where a graph property is hereditary if it is closed under removal of vertices. These results were extended to 3-uniform hypergraphs by Avart, Rödl and Schacht [14] and to k-uniform hypergraphs by Rödl and Schacht [87].

In this survey we will focus on recent developments, particularly with regard to the quantitative aspects of the removal lemma. In particular, we will discuss recent improvements on the bounds for the graph removal lemma, Theorem 1.1, and the induced graph removal lemma, Theorem 1.3, each of which bypasses a natural impediment.

The usual proof of the graph removal lemma makes use of the regularity lemma and gives bounds for the removal lemma which are of tower-type in ε. To be more specific, let $T(1) = 2$ and, for each $i \geq 1$, $T(i+1) = 2^{T(i)}$. The bounds that come out of applying the regularity lemma to the removal lemma then say that if $\delta^{-1} = T(\varepsilon^{-c_H})$, then any graph on n vertices with at most δn^h copies of a graph H on h vertices may be made H-free by removing at most εn^2 edges. Moreover, this tower-type dependency is inherent in any proof employing regularity. This follows from an important result of Gowers [53] (see also [24]) which states that the bounds that arise in the regularity lemma are necessarily of tower type. We will discuss this in more detail in Section 2.1 below.

Despite this obstacle, the following improvement was made by Fox [38].

Theorem 1.4 *For any graph H on h vertices, there exists a constant a_H such that if $\delta^{-1} = T(a_H \log \varepsilon^{-1})$ then any graph on n vertices which contains at most δn^h copies of H may be made H-free by removing at most εn^2 edges.*

As is implicit in the bounds, the proof of this theorem does not make an explicit appeal to Szemerédi's regularity lemma. However, many of the ideas used are similar to ideas used in the proof of the regularity lemma. The chief difference lies in the fact that the conditions of the removal lemma (containing few copies of a given graph H) allow us to say more about the structure of these partitions. A simplified proof of this theorem will be the main topic of Section 2.2.

Though still of tower-type, Theorem 1.4 improves substantially on the previous bound. However, it remains very far from the best known lower bound on δ^{-1}. The observation of Ruzsa and Szemerédi [94] that $f^{(3)}(n, 6, 3) = \Omega(r_3(n)n)$ allows one to transfer lower bounds for $r_3(n)$ to a corresponding lower bound for the triangle removal lemma. The best construction of a set containing no arithmetic progression of length 3 is due to Behrend [16] and gives a subset of $[n]$ with density $e^{-c\sqrt{\log n}}$. Transferring this to the graph setting yields a graph containing $\varepsilon^{c\log\varepsilon^{-1}}n^3$ triangles which cannot be made triangle-free by removing fewer than εn^2 edges. This quasi-polynomial lower bound, $\delta^{-1} \geq \varepsilon^{-c\log\varepsilon^{-1}}$, remains the best known.[6]

The standard proof of the induced removal lemma uses the strong regularity lemma of Alon, Fischer, Krivelevich and Szegedy [6]. We will speak at length about this result in Section 3.1. Here it will suffice to say that, like the ordinary regularity lemma, the bounds which an application of this theorem gives for the induced removal lemma are necessarily very large. Let $W(1) = 2$ and, for $i \geq 1$, $W(i + 1) = T(W(i))$. This is known as the wowzer function and its values dwarf those of the usual tower function.[7] By using the strong regularity lemma, the standard proof shows that we may take $\delta^{-1} = W(a_H\varepsilon^{-c})$ in the induced removal lemma, Theorem 1.3. Moreover, as with the ordinary removal lemma, such a bound is inherent in the application of the strong regularity lemma. This follows from recent results of Conlon and Fox [24] and, independently, Kalyanasundaram and Shapira [62] showing that the bounds arising in strong regularity are necessarily of wowzer type.

In the other direction, Conlon and Fox [24] showed how to bypass this obstacle and prove that the bounds for δ^{-1} are at worst a tower in a power of ε^{-1}.

Theorem 1.5 *There exists a constant $c > 0$ such that, for any graph H on h vertices, there exists a constant a_H such that if $\delta^{-1} = T(a_H\varepsilon^{-c})$ then any graph on n vertices which contains at most δn^h induced copies of H may be made induced H-free by adding and/or deleting at most εn^2 edges.*

---

[6]It is worth noting that the best known upper bound for Roth's theorem, due to Sanders [96], is considerably better than the best upper bound for $r_3(n)$ that follows from triangle removal. This upper bound is $r_3(n) = O\left(\frac{(\log\log n)^5}{\log n}n\right)$. A recent result of Schoen and Shkredov [100], building on further work of Sanders [97], shows that any subset of $[n]$ of density $e^{-c\left(\frac{\log n}{\log\log n}\right)^{1/6}}$ contains a solution to the equation $x_1+\cdots+x_5 = 5x_6$. Since arithmetic progressions correspond to solutions of $x_1 + x_2 = 2x_3$, this suggests that the answer should be closer to the Behrend bound. The bounds for triangle removal are unlikely to impinge on these upper bounds for some time, if at all.

[7]To give some indication, we note that $W(2) = 4$, $W(3) = 65536$ and $W(4)$ is a tower of 2s of height 65536.

A discussion of this theorem will form the subject of Section 3.2. The key observation here is that the strong regularity lemma is used to prove an intermediate statement (Lemma 3.2 below) which then implies the induced removal lemma. This intermediate statement may be proved without recourse to the full strength of the strong regularity lemma. There are also some strong parallels with the proof of Theorem 1.4 which we will draw attention to in due course.

In Section 3.3, we present the proof of Alon and Shapira's infinite removal lemma. In another paper, Alon and Shapira [11] showed that the dependence in the infinite removal lemma can depend heavily on the family $\mathcal{H}$. They proved that for every function $\delta : (0,1) \rightarrow (0,1)$, there exists a family $\mathcal{H}$ of graphs such that any $\delta_{\mathcal{H}} : (0,1) \rightarrow (0,1)$ which satisfies the infinite removal lemma for $\mathcal{H}$ satisfies $\delta_{\mathcal{H}} = o(\delta)$. However, such examples are rather unusual and the proof presented in Section 3.3 of the infinite removal lemma implies that for many commonly studied families $\mathcal{H}$ of graphs the bound on $\delta_{\mathcal{H}}^{-1}$ is only tower-type, improving the wowzer-type bound from the original proof.

Our discussions of the graph removal lemma and the induced removal lemma will occupy the bulk of this survey but we will also talk about some further recent developments in the study of removal lemmas. These include arithmetic removal lemmas (Section 4) and the recently developed sparse removal lemmas which hold for subgraphs of sparse random and pseudorandom graphs (Section 5). We will conclude with some further comments on related topics.

2   The graph removal lemma

In this section we will discuss the two proofs of the removal lemma, Theorem 1.1, at length. In Section 2.1, we will go through the regularity lemma and the usual proof of the removal lemma. Then, in Section 2.2, we will consider a simplified variant of the second author's recent proof [38], showing how it connects to the weak regularity lemma of Frieze and Kannan [44, 45].

2.1   The standard proof

We begin with the proof of the regularity lemma and then deduce the removal lemma. For vertex subsets S, T of a graph G, we let $e_G(S,T)$ denote the number of pairs in $S \times T$ that are edges of G and $d_G(S,T) = \frac{e_G(S,T)}{|S||T|}$ denote the fraction of pairs in $S \times T$ that are edges of G. For simplicity of notation, we drop the subscript if the graph G is clear from context. Although non-standard, it will be convenient to define the *edge density* of a graph $G = (V,E)$ to be $d(G) = d(V,V) = \frac{2e(G)}{|V|^2}$, which is the fraction of all ordered pairs of (not necessarily distinct) vertices which are edges. A pair (S, T) of subsets is ε-*regular* if, for all subsets $S' \subset S$ and $T' \subset T$ with $|S'| \geq \varepsilon|S|$ and $|T'| \geq \varepsilon|T|$, we have $|d(S', T') - d(S, T)| \leq \varepsilon$. Informally, a pair of subsets is ε-regular with a small ε if the edges between S and T are uniformly distributed among large subsets.

Let $G = (V, E)$ be a graph and $P : V = V_1 \cup \ldots \cup V_k$ be a vertex partition of G. The partition of P is *equitable* if each pair of parts differ in size by at most 1. The partition P is ε-*regular* if all but at most εk^2 pairs of parts (V_i, V_j) are ε-regular. Note that we are considering all k^2 ordered pairs (V_i, V_j), including those with $i = j$. We next state Szemerédi's regularity lemma [111].

Lemma 2.1 *For every $\varepsilon > 0$, there is $K = K(\varepsilon)$ such that every graph $G = (V, E)$ has an equitable, ε-regular vertex partition into at most K parts. Moreover, we may take K to be a tower of height $O(\varepsilon^{-5})$.*

Let $q : [0, 1] \to \mathbb{R}$ be a convex function. For vertex subsets $S, T \subset V$ of a graph G, let $q(S, T) = q(d(S, T))|S||T|/|V|^2$. For partitions $\mathcal{S} : S = S_1 \cup \ldots \cup S_a$ and $\mathcal{T} : T = T_1 \cup \ldots \cup T_b$, let $q(\mathcal{S}, \mathcal{T}) = \sum_{1 \leq i \leq a, 1 \leq j \leq b} q(S_i, T_j)$. For a vertex partition $P : V = V_1 \cup \ldots \cup V_k$ of G, define the mean-q density to be

$$q(P) = q(P, P) = \sum_{1 \leq i, j \leq k} q(V_i, V_j).$$

We next state some simple properties which follow from Jensen's inequality using the convexity of q. A *refinement* of a partition P of a vertex set V is another partition Q of V such that every part of Q is a subset of a part of P.

Proposition 2.2   *1. For partitions $\mathcal{S}$ and $\mathcal{T}$ of vertex subsets S and T, we have $q(\mathcal{S}, \mathcal{T}) \geq q(S, T)$.*

  2. If Q is a refinement of P, then $q(Q) \geq q(P)$.

  3. If $d = d(G) = d(V, V)$ is the edge density of G, then, for any vertex partition P,
$$q(d) \leq q(P) \leq dq(1) + (1 - d)q(0).$$

The first and second part of Proposition 2.2 show that by refining a vertex partition the mean-q density cannot decrease, while the last part gives the range of possible values for $q(P)$ if we only know the edge density d of G.

The convex function $q(x) = x^2$ for $x \in [0, 1]$ is chosen in the standard proof of the graph regularity lemma and we will do the same for the rest

of this subsection. The following lemma is the key claim for the proof of
the regularity lemma. The set-up is that we have a partition P which is
not ε-regular. For each pair (V_i, V_j) of parts of P which is not ε-regular,
there are a pair of witness subsets V_{ij}, V_{ji} to the fact that the pair of parts
is not ε-regular. We consider the coarsest refinement Q of P so that each
witness subset is the union of parts of Q. The lemma concludes that the
number of parts of Q is at most exponential in the number of parts of P
and, using a Cauchy-Schwarz defect inequality, that the mean-q density
of the partition Q is substantially larger than the mean-q density of P.
Because it simplifies our calculations a little, we will assume, when we say
a partition is equitable, that it is exactly equitable, that is, that all parts
have precisely the same size. This does not affect our results substantially
but simplifies the presentation.

Lemma 2.3 *If an equitable partition $P : V = V_1 \cup \ldots \cup V_k$ is not ε-regular then there is a refinement Q of P into at most $k2^k$ parts for which $q(Q) \geq q(P) + \varepsilon^5$.*

Proof For each pair (V_i, V_j) which is not ε-regular, there are subsets
$V_{ij} \subset V_i$ and $V_{ji} \subset V_j$ with $|V_{ij}| \geq \varepsilon|V_i|$ and $|V_{ji}| \geq \varepsilon|V_j|$ such that
$|d(V_{ij}, V_{ji}) - d(V_i, V_j)| \geq \varepsilon$. For each part V_j such that (V_i, V_j) is not
ε-regular, we have a partiton P_{ij} of V_i into two parts V_{ij} and $V_i \setminus V_{ij}$. Let
P_i be the partition of V_i which is the common refinement of these at most
$k-1$ partitions of V_i, so P_i has at most 2^{k-1} parts. We let Q be the
partition of V which is the union of the k partitions of the form P_i, so Q
has at most $k2^{k-1}$ parts. We have

$$
\begin{aligned}
q(Q) - q(P) &= \sum_{i,j} (q(P_i, P_j) - q(V_i, V_j)) \\
&\geq \sum_{(V_i,V_j) \text{ irregular}} (q(P_i, P_j) - q(V_i, V_j)) \\
&\geq \sum_{(V_i,V_j) \text{ irregular}} (q(P_{ij}, P_{ji}) - q(V_i, V_j)) \\
&= \sum_{(V_i,V_j) \text{ irregular}} \sum_{U \in P_{ij}, W \in P_{ji}} \frac{|U||W|}{|V|^2} (d(U, W) - d(V_i, V_j))^2 \\
&\geq \sum_{(V_i,V_j) \text{ irregular}} \frac{|V_{ij}||V_{ji}|}{|V|^2} (d(V_{ij}, V_{ji}) - d(V_i, V_j))^2 \\
&\geq \varepsilon k^2 \left(\frac{\varepsilon}{k}\right)^2 \varepsilon^2 \\
&= \varepsilon^5,
\end{aligned}
$$

where the first and third inequalities are by noting that the summands are nonnegative and the second inequality follows from the first part of Proposition 2.2, which shows that the mean-q density cannot decrease when taking a refinement. In the fourth inequality, we used that $|V_{ij}| \geq \varepsilon|V_i| \geq \frac{\varepsilon}{k}|V|$ and similarly for $|V_{ji}|$. Finally, the equality in the fourth line follows from the identity

$$\sum_{U \in P_{ij}, W \in P_{ji}} |U||W|d(V_i, V_j) = \sum_{U \in P_{ij}, W \in P_{ji}} |U||W|d(U, W),$$

which counts $e(V_i, V_j)$ in two different ways. This completes the proof.   □

The next lemma, which is rather standard, shows that for any vertex partition Q, there is a vertex equipartition P' with a similar number of parts to Q and mean-square density not much smaller than the mean-square density of Q. It is useful in density increment arguments where at each stage one would like to work with an equipartition. It is proved by first arbitrarily partitioning each part of Q into parts of order $|V|/t$, except possibly one additional remaining smaller part, and then arbitrarily partitioning the union of the smaller remaining parts into parts of order $|V|/t$.

Lemma 2.4 *Let $G = (V, E)$ be a graph and $Q : V = V_1 \cup \ldots \cup V_\ell$ be a vertex partition into ℓ parts. Then, for $q(x) = x^2$, there is an equitable partition P' of V into t parts such that $q(P') \geq q(Q) - 2\frac{\ell}{t}$.*

Combining Lemmas 2.3 and 2.4 with $t = 4\varepsilon^{-5}|Q| \leq \varepsilon^{-5}k2^{k+2}$, we obtain the following corollary.

Corollary 2.5 *If an equitable partition $P : V = V_1 \cup \ldots \cup V_k$ is not ε-regular then there is an equitable refinement P' of P into at most $\varepsilon^{-5}k2^{k+2}$ parts for which $q(P') \geq q(P) + \varepsilon^5/2$.*

We next show how Szemerédi's regularity lemma, Lemma 2.1, can be quickly deduced from this result.

Proof To prove the regularity lemma, we start with the trivial partition P_0 into one part, and iterate the above corollary to obtain a sequence $P_0, P_1, \ldots, P_s$ of equitable partitions with $q(P_{i+1}) \geq q(P_i) + \varepsilon^5/2$ until we arrive at an equitable ε-regular partition P_s. As the mean-square density of each partition has to lie between 0 and 1, after at most $2\varepsilon^{-5}$ iterations we arrive at the equitable ε-regular partition P_s with $s \leq 2\varepsilon^{-5}$. The number of parts increases by one exponential in each iteration, giving the desired number of parts in the regularity partition. This completes the proof of Szemerédi's regularity lemma.   □

The constructions of Gowers [53] and the authors [24] show that the tower-type bound on the number of parts in Szemerédi's regularity lemma is indeed necessary. In particular, the construction in [24] shows that $K(\varepsilon)$ in Lemma 2.1 is at least a tower of twos of height $\Omega(\varepsilon^{-1})$. The constructions are formed by reverse engineering the upper bound proof. We construct a sequence $P_0, \ldots, P_s$ of partitions with $s = \Omega(\varepsilon^{-1})$. As in the upper bound proof, each partition in the sequence uses exponentially more parts than the previous partition in the sequence. We may choose the edges using these partitions and some randomness so as to guarantee that none of these partitions (except the last) are ε-regular. Furthermore, we can guarantee that any partition that is ε-regular must be close to being a refinement of the last partition in this sequence. This implies that the number of parts must be at least roughly $|P_s|$.

We next prove the graph removal lemma, Theorem 1.1, from the regularity lemma.

Proof Let m denote the number of edges of H, so $m \leq \binom{h}{2}$. Let $\gamma = \frac{\varepsilon^h}{4h}$ and $\delta = (2h)^{-2h}\varepsilon^m K^{-h}$, where $K = K(\gamma)$ is as in the regularity lemma. We apply the regularity lemma to G and obtain an equitable, γ-regular partition into $k \leq K$ parts. If the number n of vertices of G satisfies $n < \delta^{-1/h}$, then the number of copies of H in G is at most $\delta n^h < 1$ and G is H-free, in which case there is nothing to prove. So we may assume $n \geq \delta^{-1/h}$. We obtain a subgraph G' of G by removing edges of G between all pairs of parts which are not γ-regular or which have edge density at most ε. As there are at most γk^2 ordered pairs of parts which are not γ-regular and each part has order at most $2n/k$, at most $(\gamma k^2/2)(2n/k)^2 = 2\gamma n^2$ edges are deleted between pairs of parts which are not γ-regular. The number of edges between parts which have edge density at most ε is at most $\varepsilon n^2/2$. Hence, the number of edges of G deleted to obtain G' is at most $2\gamma n^2 + \varepsilon n^2/2 < \varepsilon n^2$. If G' is H-free, then we are done.

Assume for contradiction that G' is not H-free. A copy of H in G' must have its edges going between pairs of parts which are both γ-regular and have density at least ε. Hence, there is a mapping from $V(H)$ to the partition of $V(G)$ so that each edge of H maps to a pair of parts which is both γ-regular and have edge density at least ε. But the following standard counting lemma (see, e.g., Lemma 3.2 in Alon, Fischer, Krivelevich and Szegedy [6] for a minor variant) shows that the number of labeled copies of H in G' (and hence in G) is at least $2^{-h}\varepsilon^m(n/2k)^h > h!\delta n^h$.

This contradicts that G has at most δn^h copies of H, completing the proof.   □

Lemma 2.6 *If H is a graph with vertices $1, \ldots, h$ and m edges and G is a graph with not necessarily disjoint vertex subsets $W_1, \ldots, W_h$ such that $|W_i| \geq \gamma^{-1}$ for $1 \leq i \leq h$ and, for every edge (i, j) of H, the pair (W_i, W_j) is γ-regular with density $d(W_i, W_j) > \varepsilon$ and $\gamma \leq \frac{\varepsilon^h}{4h}$, then G contains at least $2^{-h} \varepsilon^m |W_1| \times \cdots \times |W_h|$ labeled copies of H with the copy of vertex i in W_i.*

The standard proof of this counting lemma uses a greedy embedding strategy. One considers embedding the vertices one at a time, using the regularity condition to maintain the property that at each step where vertex i of H is not yet embedded the set of vertices of G which could potentially be used to embed vertex i is large.

2.2   An improved bound

A partition $P : V = V_1 \cup \ldots \cup V_k$ of the vertex set of a graph $G = (V, E)$ is *weak ε-regular* if, for all subsets $S, T \subset V$, we have

$$\left| e(S, T) - \sum_{1 \leq i, j \leq k} |S \cap V_i||T \cap V_j|d(V_i, V_j) \right| \leq \varepsilon |V|^2.$$

That is, the density between two sets may be approximated by taking a weighted average over the densities between the sets which they intersect.

The Frieze-Kannan weak regularity lemma [44, 45] states that any graph has such a weak regular partition.

Lemma 2.7 *Let $R(\varepsilon) = 2^{c\varepsilon^{-2}}$, where c is an absolute constant. For every graph $G = (V, E)$ and every equitable partition P of G into k parts, there is an equitable partition P' which is a refinement of P into at most $kR(\varepsilon)$ parts which is weak ε-regular.*

Unlike the usual regularity lemma, the bounds in Lemma 2.7 are quite reasonable.[8] It is therefore natural to try to apply it to prove the removal lemma. However, it seems unlikely that this lemma is itself sufficient to prove the removal lemma, since it only gives control over edge densities of a global nature.

---

[8]They are also sharp, that is, there are graphs for which the minimum number of parts in any weak ε-regular partition is $2^{\Omega(\varepsilon^{-2})}$. This was proved in [24] (see also [5]).

However, as noted by Tao [112] (see also [88]), one can prove a stronger theorem by simply iterating the Frieze-Kannan weak regularity lemma.[9] Tao developed this lemma to give an alternative proof of the regularity lemma[10] which extended more easily to hypergraphs [113]. Here we use it to improve the bounds for the removal lemma.

Lemma 2.8 *Let $q : [0,1] \to \mathbb{R}$ be a convex function, G be a graph with $d = d(G)$, $f : \mathbb{N} \to [0,1]$ be a decreasing function, $\gamma > 0$, and $r = (dq(1) + (1-d)q(0) - q(d))/\gamma$. Then there are equitable partitions P and Q with Q a refinement of P satisfying $q(Q) \leq q(P) + \gamma$, Q is weak $f(|P|)$-regular and $|Q| \leq t_r$, where $t_0 = 1$, $t_i = t_{i-1}R(f(t_{i-1}))$ for $1 \leq i \leq r$ and $R(x) = 2^{cx^{-2}}$ as in the Frieze-Kannan weak regularity lemma.*

The proof of Lemma 2.8 is quite similar to the proof of Szemerédi's regularity lemma discussed in the previous subsection. One starts with the trivial partition P_0 of V into one part. We then apply Lemma 2.7 repeatedly to construct a sequence of partitions $P_0, P_1, \ldots$ so that P_{i+1} is weak $f(|P_i|)$-regular. If $q(P_{i+1}) > q(P_i) + \gamma$, then we continue with this process. Otherwise, $q(P_{i+1}) \leq q(P_i) + \gamma$, so we set $Q = P_{i+1}$ and $P = P_i$ and stop the process. This process must stop within r iterations as the third part of Proposition 2.2 shows that the mean-q density lies in an interval of length $r\gamma$.

Rather than using the usual $q(x) = x^2$, we will use the convex function q on $[0,1]$ defined by $q(0) = 0$ and $q(x) = x \log x$ for $x \in (0,1]$. This entropy function is central to the proof since it captures the extra structural information coming from Lemma 2.10 below in a concise fashion. Note that the last part of Proposition 2.2 implies that $d \log d \leq q(P) \leq 0$ for every partition P.

The next lemma is a counting lemma that complements the Frieze-Kannan weak regularity lemma. As one might expect, this lemma gives a global count for the number of copies of H, whereas the counting lemma associated with the usual regularity lemma gives a means of counting copies of H between any $v(H)$ parts of the partition which are pairwise regular. Its proof, which we omit, is by a simple telescoping sum argument.

---

[9]We will say more about this sort of iteration in Section 3.1 below.

[10]More recently, Conlon and Fox [24] showed that it is also closely related to the regular approximation lemma. This lemma, which arose in the study of graph limits by Lovász and Szegedy [75] and also in work on the hypergraph generalization of the regularity lemma by Rödl and Schacht [86], says that by adding and/or deleting a small number of edges in a graph G, we may find another graph G' which admits very fine regular partitions. We refer the reader to [24] and [88] for further details.

Lemma 2.9 *([18], Theorem 2.7 on page 1809) Let H be a graph on $\{1, \ldots, h\}$ with m edges. Let $G = (V, E)$ be a graph on n vertices and $Q : V = V_1 \cup \ldots \cup V_t$ be a vertex partition which is weak ε-regular. The number of homomorphisms from H to G is within $\varepsilon m n^h$ of*

$$\sum_{1 \leq i_1, \ldots, i_h \leq t} \prod_{(r,s) \in E(H)} d(V_{i_r}, V_{i_s}) \prod_{a=1}^{h} |V_{i_a}|.$$

Let P and Q be vertex partitions of a graph G with Q a refinement of P. A pair (V_i, V_j) of parts of P is (α, c)-*shattered* by Q if at least a c-fraction of the pairs $(u, v) \in V_i \times V_j$ go between pairs of parts of Q with edge density between them less than α.

One of the key components of the proof is the following lemma, which says that if P and Q are vertex partitions like those given by Lemma 2.8, then there are many pairs of vertex sets in P which are shattered by Q.

Lemma 2.10 *Let H be a graph on $\{1, \ldots, h\}$ with m edges and let $\alpha > 0$. Suppose G is a graph on n vertices for which there are less than δn^h homomorphisms of H into G, where $\delta = \frac{1}{4}\alpha^m(2k)^{-h}$. Suppose P and Q are equitable vertex partitions of G with $|P| = k \leq n$ and Q is a refinement of P which is weak $f(k)$-regular, where $f(k) = \frac{1}{4m}\alpha^m(2k)^{-h}$. For every h-tuple $V_1, \ldots, V_h$ of parts of P, there is an edge (i, j) of H for which the pair (V_i, V_j) is $(\alpha, \frac{1}{2m})$-shattered by Q.*

Proof As $|P| = k \leq n$, we have $|V_i| \geq \frac{n}{2k}$ for each i. Let Q_i denote the partition of V_i which consists of the parts of Q which are subsets of V_i. Consider an h-tuple $(v_1, \ldots, v_h) \in V_1 \times \cdots \times V_h$ picked uniformly at random. Also consider the event E that, for each edge (i, j) of H, the pair (v_i, v_j) goes between parts of Q_i and Q_j with density at least α. If E occurs with probability at least $1/2$, as Q is weak $f(k)$-regular, Lemma 2.9 implies that the number of homomorphisms of H into G where the copy of vertex i is in V_i for $1 \leq i \leq h$ is at least

$$\frac{1}{2}\alpha^m \prod_{i=1}^{h} |V_i| - mf(k)n^h \geq \left(\frac{1}{2}\alpha^m(2k)^{-h} - mf(k) \right) n^h = \delta n^h,$$

contradicting that there are less than δn^h homomorphisms of H into G. So E occurs with probability less than $1/2$. Hence, for at least $1/2$ of the h-tuples $(v_1, \ldots, v_h) \in V_1 \times \cdots \times V_h$, there is an edge (i, j) of H such that the pair (v_i, v_j) goes between parts of Q_i and Q_j with density less than α. This implies that for at least one edge (i, j) of H, the pair (V_i, V_j) is $(\alpha, \frac{1}{2m})$-shattered by Q. $\square$

We will need the following lemma from [38] which tells us that if a pair of parts from P is shattered by Q then there is an increment in the mean-entropy density. Its proof is by a simple application of Jensen's inequality.

Lemma 2.11 *([38], Lemma 7 on page 570) Let $q : [0,1] \rightarrow \mathbb{R}$ be the convex function given by $q(0) = 0$ and $q(x) = x \log x$ for $x > 0$. Let $\varepsilon_1, \ldots, \varepsilon_r$ and $d_1, \ldots, d_r$ be nonnegative real numbers with $\sum_{i=1}^{r} \varepsilon_i = 1$ and $d = \sum_{i=1}^{r} \varepsilon_i d_i$. Suppose $\beta < 1$ and $I \subset [r]$ is such that $d_i \leq \beta d$ for $i \in I$ and let $s = \sum_{i \in I} \varepsilon_i$. Then*

$$\sum_{i=1}^{r} \varepsilon_i q(d_i) \geq q(d) + (1 - \beta + q(\beta))sd.$$

We are now ready to prove Theorem 1.4 in the following precise form.

Theorem 2.12 *Let H be a graph on $\{1, \ldots, h\}$ with m edges. Let $\varepsilon > 0$ and δ^{-1} be a tower of twos of height $8h^4 \log \varepsilon^{-1}$. If G is a graph on n vertices in which at least εn^2 edges need to be removed to make it H-free, then G contains at least δn^h copies of H.*

Proof Suppose for contradiction that there is a graph G on n vertices in which at least εn^2 edges need to be removed from G to delete all copies of H, but G contains fewer than δn^h copies of H. If $n \leq \delta^{-1/h}$, then the number of copies of H in G is less than $\delta n^h \leq 1$, so G is H-free, contradicting that at least εn^2 edges need to be removed to make the graph H-free. Hence, $n > \delta^{-1/h}$. Note that the number of mappings from $V(H)$ to $V(G)$ which are not one-to-one is $n^h - h!\binom{n}{h} \leq h^2 n^{h-1} < h^2 \delta^{1/h} n^h$. Let $\delta' = 4h! \delta^{1/h}$, so the number of homomorphisms from H to G is at most $\delta' n^h$.

The graph G contains at least $\varepsilon n^2 / m$ edge-disjoint copies of H. Let G' be the graph on the same vertex set which consists entirely of the at least $\varepsilon n^2 / m$ edge-disjoint copies of H. Then $d(G') \geq m \cdot \varepsilon/m = \varepsilon$ and G' consists of $\frac{d(G')}{m} n^2$ edge-disjoint copies of H. We will show that there are at least $\delta' n^h$ homomorphisms from H to G' (and hence to G as well). For the rest of the argument, we will assume the underlying graph is G'.

Let $\alpha = \frac{\varepsilon}{8m}$. Apply Lemma 2.8 to G' with $f(k) = \frac{1}{4m} \alpha^m (2k)^{-h}$ and $\gamma = \frac{d(G')}{2h^4}$. Note that r as in Lemma 2.8 is

$$r = d(G') \log(1/d(G'))/\gamma = 2h^4 \log(1/d(G')) \leq 2h^4 \log \varepsilon^{-1}.$$

Hence, we get a pair of equitable vertex partitions P and Q, with Q a refinement of P, $q(Q) \leq q(P) + \gamma$, Q is weak $f(|P|)$-regular and $|Q|$ is at

most a tower of twos of height $3r \leq 6h^4 \log \varepsilon^{-1}$. Let $V_1, \ldots, V_k$ denote the
parts of P and Q_i denote the partition of V_i consisting of the parts of Q
which are subsets of V_i.

Suppose that (V_a, V_b) is a pair of parts of P with edge density $d = d(V_a, V_b) \geq \varepsilon/m$ which is $(\alpha, \frac{1}{2m})$-shattered by Q. Note that $\alpha \leq d/8$.
Arbitrarily order the pairs $U_i \times W_j \in Q_a \times Q_b$, letting $d_{ij} = d(U_i, W_j)$,
$\varepsilon_{ij} = \frac{|U_i||W_j|}{|V_a||V_b|}$, and I be the set of pairs (i, j) such that $d_{ij} < \alpha$, so that the
conditions of Lemma 2.11 with $\beta = 1/8$ are satisfied. Applying Lemma
2.11, we get, since $q(\beta) = -\frac{1}{8} \log 8 = -\frac{3}{8}$, that

$$q(Q_a, Q_b) - q(V_a, V_b) \geq (1 - \beta + q(\beta)) \frac{1}{2m} d(V_a, V_b) \frac{|V_a||V_b|}{n^2} \geq \frac{1}{4m} \frac{e(V_a, V_b)}{n^2}.$$

The first inequality uses $q(Q_a, Q_b) = p_{ab} \sum_{i,j} \varepsilon_{ij} q(d_{ij})$ and $q(V_a, V_b) = p_{ab} q\left(\sum_{i,j} \varepsilon_{ij} d_{ij}\right)$, where $p_{ab} = |V_a||V_b|/n^2$. It also uses $s := \sum_{ij \in I} \varepsilon_{ij} \geq \frac{1}{2m}$, which follows from the fact that the pair (V_a, V_b) is $(\alpha, \frac{1}{2m})$-shattered.
Note that

$$q(Q) - q(P) = \sum_{1 \leq a, b \leq k} (q(Q_a, Q_b) - q(V_a, V_b)),$$

which shows that $q(Q) - q(P)$ is the sum of nonnegative summands.

There are at most $\frac{\varepsilon}{m} n^2/2$ edges of G' going between pairs of parts of
P with density at most $\frac{\varepsilon}{m}$. Hence, at least $1/2$ of the edge-disjoint copies
of H making up G' have all its edges going between pairs of parts of P of
density at least $\frac{\varepsilon}{m}$. By Lemma 2.10, for each copy of H, at least one of its
edges goes between a pair of parts of P which is $(\alpha, \frac{1}{2m})$-shattered by Q.
Thus,

$$q(Q) - q(P) \geq \sum \frac{1}{4m} e(V_a, V_b)/n^2 \geq \frac{1}{4m} \cdot \frac{d(G')}{2m} = \frac{d(G')}{8m^2} > \gamma,$$

where the sum is over all ordered pairs (V_a, V_b) of parts of P which are
$(\alpha, \frac{1}{2m})$-shattered by Q and with $d(V_a, V_b) \geq \frac{\varepsilon}{m}$. This contradicts $q(Q) \leq q(P) + \gamma$ and completes the proof. $\square$

3   The induced removal lemma

As in the last section, we will again discuss two different proofs of the
induced removal lemma, Theorem 1.3. In Section 3.1, we will discuss the
proof of Alon, Fischer, Krivelevich and Szegedy [6], which uses their strong
regularity lemma and gives a wowzer-type bound. In Section 3.2, we will
examine the authors' recent proof [24] of a tower-type bound. We will
discuss Alon and Shapira's generalization of the induced removal lemma,
which applies to infinite families of graphs, in Section 3.3.

3.1 The usual proof

For an equitable partition $P = V_1 \cup \ldots \cup V_k$ of $V(G)$ and an equitable refinement $Q = \bigcup_{i=1}^{k} \bigcup_{j=1}^{\ell} V_{i,j}$ of P with $V_i = \bigcup_{j=1}^{\ell} V_{i,j}$, we say that Q is ε-*close* to P if the following is satisfied. All $1 \le i \le i' \le k$ but at most εk^2 of them are such that, for all $1 \le j, j' \le \ell$ but at most $\varepsilon \ell^2$ of them, $|d(V_i, V_{i'}) - d(V_{i,j}, V_{i',j'})| < \varepsilon$ holds. This notion roughly says that Q is an approximation of P. The strong regularity lemma of Alon, Fischer, Krivelevich and Szegedy [6] is now as follows.

Lemma 3.1 (Strong regularity lemma) *For every function $f : \mathbb{N} \to (0,1)$ there exists a number $S = S(f)$ with the following property. For every graph $G = (V, E)$, there is an equitable partition P of the vertex set V and an equitable refinement Q of P with $|Q| \le S$ such that the partition P is $f(1)$-regular, the partition Q is $f(|P|)$-regular and Q is $f(1)$-close to P.*

That is, there is a regular partition P and a refinement Q such that Q is very regular and yet the densities between parts $(V_{i,j}, V_{i',j'})$ of Q are usually close to the densities between the parts $(V_i, V_{i'})$ of P containing them.

Let $f(1) = \varepsilon$. Here, and throughout this section, we let $q(x) = x^2$ be the square function as in the proof of Szemerédi's regularity lemma in the previous section. The condition that Q be ε-close to P is equivalent, up to a polynomial change in ε, to $q(Q) \le q(P) + \varepsilon$. Indeed, if Q is ε-close to P, then $q(Q) \le q(P) + O(\varepsilon)$, while if $q(Q) \le q(P) + \varepsilon$, then Q is $O(\varepsilon^{1/4})$-close to P. A version of this statement is present in Lemma 3.7 of [6]. As it is sufficient and more convenient to work with mean-square density instead of ε-closeness, we do so from now on. That is, we replace the third condition in the strong regularity lemma with the condition that $q(Q) \le q(P) + \varepsilon$.

With this observation, the proof of the strong removal lemma becomes quite straightforward. Note that we may assume that f is a decreasing function by replacing it, if necessary, with the function given by $f'(i) = \min_{1 \le j \le i} f(j)$. We consider a series of partitions $P_1, P_2, \ldots$, where P_1 is an $f(1)$-regular partition and P_{i+1} is an $f(|P_i|)$-regular refinement of the partition P_i. Since P_{i+1} is a refinement of P_i we know that the mean-square density must have increased, that is, $q(P_{i+1}) \ge q(P_i)$. If also $q(P_{i+1}) \le q(P_i) + \varepsilon$ then, since f is decreasing and P_{i+1} is $f(|P_i|)$-regular, we see that all three conditions of the theorem are satisfied with $P = P_i$ and $Q = P_{i+1}$ as the required partitions. Otherwise, we have $q(P_{i+1}) > q(P_i) + \varepsilon$. However, since the mean-square density is bounded above by 1, this can happen at most ε^{-1} times, concluding the proof.

It is not hard to see why this proof results in wowzer-type bounds. At each step, we are applying the regularity lemma to find a partition P_{i+1} which is regular in the number of parts in the previous partition P_i. The bounds coming from the regularity lemma then imply that $|P_{i+1}| = T(f(|P_i|)^{-O(1)})$. But this iterated tower-type bound is essentially how we define the wowzer function.

That this is the correct behaviour for the bounds in the strong regularity lemma was proved independently by Conlon and Fox [24] and by Kalyanasundaram and Shapira [62], though both proofs use slightly different ideas and result in slightly different bounds. With the function $f : \mathbb{N} \to (0,1)$ taken to be $f(n) = \varepsilon/n$, the proof given in [24] shows that the number of parts in the smaller partition P may need to be as large as wowzer in a power of ε^{-1}, while that given in [62] proves that it must be at least wowzer in $\sqrt{\log \varepsilon^{-1}}$.

The following easy corollary of the strong regularity lemma [6] is the key to proving the induced graph removal lemma.

Lemma 3.2 *For each $0 < \varepsilon < 1/3$ and decreasing function $f : \mathbb{N} \to (0, 1/3)$, there is $\delta' = \delta'(\varepsilon, f)$ such that every graph $G = (V, E)$ with $|V| \geq \delta'^{-1}$ has an equitable partition $V = V_1 \cup \ldots \cup V_k$ and vertex subsets $W_i \subset V_i$ such that $|W_i| \geq \delta'|V|$, each pair (W_i, W_j) with $1 \leq i \leq j \leq k$ is $f(k)$-regular and all but at most εk^2 pairs $1 \leq i \leq j \leq k$ satisfy $|d(V_i, V_j) - d(W_i, W_j)| \leq \varepsilon$.*

In fact, Lemma 3.2 is a little bit stronger than the original version in [6] in that each set W_i is $f(k)$-regular with itself.[11] The original version follows from the strong regularity lemma, applied with

$$f'(k) = \min\left(f(k), \frac{\varepsilon}{4}, \frac{1}{2}\binom{k+2}{2}^{-1}\right),$$

by taking the partition $V = V_1 \cup \ldots \cup V_k$ to be the partition P in the strong regularity lemma and the subset W_i to be a random part $V_{i,p} \subset V_i$ of the refinement Q of P in the strong regularity lemma. Since $f'(k) \leq \frac{1}{2}\binom{k+2}{2}^{-1}$, it is straightforward to check that all pairs (W_i, W_j) are $f(k)$-regular with probability greater than $\frac{1}{2}$. Moreover, the expected number of pairs with $1 \leq i < j \leq k$ for which $|d(V_i, V_j) - d(W_i, W_j)| > \varepsilon$ is at most

$$\frac{\varepsilon}{4}k^2 + \frac{\varepsilon}{4}k^2 = \frac{\varepsilon}{2}k^2.$$

---

[11] This stronger version may be derived from an extra application of the regularity lemma within each of the pieces W_i, together with a suitable application of Ramsey's theorem. This is essentially the process carried out in [6], though they do not state their final result in the same form as Lemma 3.2.

Here, the two $\frac{\varepsilon}{4}$ factors come from the definition of $f'(k)$-closeness. The first factor comes from the fact that at most an $\frac{\varepsilon}{4}$-fraction of the pairs (V_i, V_j) do not have good approximations while the second factor comes from the fact that for all other pairs there are at most an $\frac{\varepsilon}{4}$ fraction of pairs (W_i, W_j) which do not satisfy $|d(V_i, V_j) - d(W_i, W_j)| \leq \varepsilon$. Therefore, by Markov's inequality, the probability that the number of bad pairs is greater than εk^2 is less than $\frac{1}{2}$. We therefore see that with positive probability there is a choice of W_i satisfying the required weaker version of Lemma 3.2.

If we assume the full strength of Lemma 3.2 as stated, that is, that each W_i is also $f(k)$-regular with itself, it is easy to deduce the induced removal lemma. Let $h = |V(H)|$ and take $f(k) = \frac{\varepsilon^h}{4h}$. If there is a mapping $\phi : V(H) \to \{1, \ldots, k\}$ such that for all adjacent vertices v, w of H, the edge density between $W_{\phi(v)}$ and $W_{\phi(w)}$ is at least ε and for all distinct nonadjacent vertices v, w of H, the edge density between $W_{\phi(v)}$ and $W_{\phi(w)}$ is at most $1 - \varepsilon$, then the following standard counting lemma (see, e.g., Lemma 3.2 in Alon, Fischer, Krivelevich and Szegedy [6] for a minor variant) shows that G contains at least δn^h induced copies of H, where $\delta = \frac{1}{h!}(\varepsilon/4)^{\binom{h}{2}}\delta'^h$. As with Lemma 2.6, the standard proof of this counting lemma uses a greedy embedding strategy.

Lemma 3.3 *If H is a graph with vertices $1, \ldots, h$ and G is a graph with not necessarily disjoint vertex subsets $W_1, \ldots, W_h$ such that every pair (W_i, W_j) with $1 \leq i < j \leq h$ is γ-regular with $\gamma \leq \frac{\eta^h}{4h}$, $|W_i| \geq \gamma^{-1}$ for $1 \leq i \leq h$ and, for $1 \leq i < j \leq k$, $d(W_i, W_j) > \eta$ if (i, j) is an edge of H and $d(W_i, W_j) < 1 - \eta$ otherwise, then G contains at least $\left(\frac{\eta}{4}\right)^{\binom{h}{2}} |W_1| \times \cdots \times |W_h|$ induced copies of H with the copy of vertex i in W_i.*

Hence, we may assume that there is no such mapping ϕ. We then delete the edges between V_i and V_j if the edge density between W_i and W_j is less than ε and add the edges between V_i and V_j if the density between W_i and W_j is more than $1 - \varepsilon$. The total number of edges added or removed is at most $5\varepsilon n^2$ and no induced copy of H remains. Replacing ε by $\varepsilon/8$ in the above argument gives the induced removal lemma.

3.2 An improved bound

The main goal of this section is to prove Theorem 1.5, which gives a bound on δ^{-1} which is a tower in h of height polynomial in ε^{-1}. We in fact prove the key corollary of the strong regularity lemma, Lemma 3.2, with a tower-type bound. This is sufficient to prove the desired tower-type bound for the induced graph removal lemma.

As in Section 2.2, the key idea will be to take a weak variant of Sze-merédi's regularity lemma and iterate it. The particular variant we will use, due to Duke, Lefmann and Rödl [28], was originally used by them to derive a fast approximation algorithm for the number of copies of a fixed graph in a large graph.

A *k-cylinder* (or cylinder for short) in a graph G is a product of k vertex subsets. Given a k-partite graph $G = (V, E)$ with k-partition $V = V_1 \cup \ldots \cup V_k$, we will consider a partition $\mathcal{K}$ of the cylinder $V_1 \times \cdots \times V_k$ into cylinders $K = W_1 \times \cdots \times W_k$, $W_i \subset V_i$ for $i = 1, \ldots, k$ and we let $V_i(K) = W_i$. We say that a cylinder is *ε-regular* if all $\binom{k}{2}$ pairs of subsets (W_i, W_j), $1 \leq i < j \leq k$, are ε-regular. The partition $\mathcal{K}$ is *ε-regular* if all but an ε-fraction of the k-tuples $(v_1, \ldots, v_k) \in V_1 \times \cdots \times V_k$ are in ε-regular cylinders in the partition $\mathcal{K}$.

The weak regularity lemma of Duke, Lefmann and Rödl [28] is now as follows. Note that, like the Frieze-Kannan weak regularity lemma, it has only a single-exponential bound on the number of parts. We will sometimes refer to this lemma as the cylinder regularity lemma.

Lemma 3.4 *Let $0 < \varepsilon < 1/2$ and $\beta = \beta(\varepsilon) = \varepsilon^{k^2 \varepsilon^{-5}}$. Suppose $G = (V, E)$ is a k-partite graph with k-partition $V = V_1 \cup \ldots \cup V_k$. Then there exists an ε-regular partition $\mathcal{K}$ of $V_1 \times \cdots \times V_k$ into at most β^{-1} parts such that, for each $K \in \mathcal{K}$ and $1 \leq i \leq k$, $|V_i(K)| \geq \beta|V_i|$.*

We would now like to iterate this lemma to get a stronger version, the strong cylinder regularity lemma. Like Lemmas 2.8 and 3.1, this will yield two closely related cylinder partitions P and Q with P regular and Q regular in a function of $|P|$. To state the lemma, we first strengthen the definition of regular cylinders so that pieces are also regular with themselves.

A k-cylinder $W_1 \times \cdots \times W_k$ is *strongly ε-regular* if all pairs (W_i, W_j) with $1 \leq i, j \leq k$ are ε-regular. A partition $\mathcal{K}$ of $V_1 \times \cdots \times V_k$ into cylinders is *strongly ε-regular* if all but $\varepsilon|V_1| \times \cdots \times |V_k|$ of the k-tuples $(v_1, \ldots, v_k) \in V_1 \times \cdots \times V_k$ are contained in strongly ε-regular cylinders $K \in \mathcal{K}$.

We now state the strong cylinder regularity lemma. Here $t_i(x)$ is a variant of the tower function defined by $t_0(x) = x$ and $t_{i+1}(x) = 2^{t_i(x)}$. Also, given a cylinder partition $\mathcal{K}$, $Q(\mathcal{K})$ is the coarsest vertex partition such that every set $V_i(K)$ with $i \in [k]$ and $K \in \mathcal{K}$ is the union of parts of $Q(\mathcal{K})$.

Lemma 3.5 *For $0 < \varepsilon < 1/3$, positive integer s, and decreasing function $f : \mathbb{N} \to (0, \varepsilon]$, there is $S = S(\varepsilon, s, f)$ such that the following holds. For every graph G, there is an integer $s \leq k \leq S$, an equitable partition $P : V = V_1 \cup \ldots \cup V_k$ and a strongly $f(k)$-regular partition $\mathcal{K}$ of the cylinder $V_1 \times \cdots \times V_k$ into cylinders satisfying that the partition $Q = Q(\mathcal{K})$ of V has at most S parts and $q(Q) \leq q(P) + \varepsilon$. Furthermore, there is an absolute constant c such that letting $s_1 = s$ and $s_{i+1} = t_4 \left((s_i / f(s_i))^c \right)$, we may take $S = s_\ell$ with $\ell = 2\varepsilon^{-1} + 1$.*

To prove this lemma, we need to find a way to guarantee that the parts of the cylinder partition are regular with themselves as required in the definition of strong cylinder regularity. For a graph $G = (V, E)$, a vertex subset $U \subset V$ is *ε-regular* if the pair (U, U) is ε-regular. The following lemma, which demonstrates that any graph contains a large vertex subset which is ε-regular, is the first step.

Lemma 3.6 *For each $0 < \varepsilon < 1/2$, let $\delta = \delta(\varepsilon) = 2^{-\varepsilon^{-(10/\varepsilon)^4}}$. Every graph $G = (V, E)$ contains an ε-regular vertex subset U with $|U| \geq \delta |V|$.*

One way to prove this lemma is to first find a large collection C of disjoint subsets of equal order which are pairwise α-regular with $\alpha = (\varepsilon/3)^2$. This can be done by an application of Szemerédi's regularity lemma and Turán's theorem, but then the bounds are quite weak. Instead, one can easily deduce this from Lemma 3.4. A further application of Ramsey's theorem allows one to get a subcollection C' of size $s \geq 2\alpha^{-1}$ such that the edge density between each pair of distinct subsets in C' lies in an interval of length at most α. The union of the sets in C' is then an ε-regular subset of the desired order.

It is crucial in this lemma that δ^{-1} be of bounded tower height in ε^{-1}. While our bound gives a double exponential dependence, we suspect that the truth is more likely to be a single exponential. We leave this as an open problem.

Repeated applications of Lemma 3.6 allow us to pull out large, regular subsets until a small fraction of vertices remain. By distributing the remaining vertices amongst these subsets, we only slightly weaken their regularity, while giving a partition of any graph into large parts each of which is ε-regular with itself. This will be sufficient for our purposes.

Lemma 3.7 *For each $0 < \varepsilon < 1/2$, let $\delta = \delta(\varepsilon) = 2^{-\varepsilon^{-(20/\varepsilon)^4}}$. Every graph $G = (V, E)$ has a vertex partition $V = V_1 \cup \ldots \cup V_k$ such that for each i, $1 \leq i \leq k$, $|V_i| \geq \delta |V|$ and V_i is an ε-regular set.*

We are now ready to prove the strong cylinder regularity lemma.

Proof of Lemma 3.5: We may assume $|V| \geq S$, as otherwise we can let P and Q be the trivial partitions into singletons, and it is easy to see the lemma holds. We will define a sequence of partitions $P_1, P_2, \ldots$ of equitable partitions, with P_{j+1} a refinement of P_j and $q(P_{j+1}) > q(P_j) + \varepsilon/2$. Let P_1 be an arbitrary equitable partition of V consisting of $s_1 = s$ parts. Suppose we have already found an equitable partition $P_j : V = V_1 \cup \ldots \cup V_k$ with $k \leq s_j$.

Let $\beta(x, \ell) = x^{\ell^2 x^{-5}}$ as in Lemma 3.4 and $\delta(x) = 2^{-x^{-(20/x)^4}}$ as in Lemma 3.7. We apply Lemma 3.7 to each part V_i of the partition P_j to get a partition of each part $V_i = V_{i1} \cup \ldots \cup V_{ih_i}$ of P_j into parts each of cardinality at least $\delta|V_i|$, where $\delta = \delta(\gamma)$ and $\gamma = f(k) \cdot \beta$ with $\beta = \beta(f(k), k)$, such that each part V_{ih} is γ-regular. Note that δ^{-1} is at most triple-exponential in a polynomial in $k/f(k)$. For each k-tuple $\ell = (\ell_1, \ldots, \ell_k) \in [h_1] \times \cdots \times [h_k]$, by Lemma 3.4 there is an $f(k)$-regular partition $\mathcal{K}_\ell$ of the cylinder $V_{1\ell_1} \times \cdots \times V_{k\ell_k}$ into at most β^{-1} cylinders such that, for each $K \in \mathcal{K}_\ell$, $|V_{i\ell_i}(K)| \geq \beta|V_{i\ell_i}|$. The union of the $\mathcal{K}_\ell$ forms a partition $\mathcal{K}$ of $V_1 \times \cdots \times V_k$ which is strongly $f(k)$-regular.

Recall that $Q = Q(\mathcal{K})$ is the partition of V which is the common refinement of all parts $V_i(K)$ with $i \in [k]$ and $K \in \mathcal{K}$. The number of parts of $\mathcal{K}$ is at most $\delta^{-k}\beta^{-1}$ and hence the number of parts of Q is at most $k2^{1/(\delta^k \beta)}$. Thus, the number of parts of Q is at most quadruple-exponential in a polynomial in $k/f(k)$. Let P_{j+1} be an equitable partition into $4\varepsilon^{-1}|Q|$ parts with $q(P_{j+1}) \geq q(Q) - \frac{\varepsilon}{2}$, which exists by Lemma 2.4. Hence, there is an absolute constant c such that

$$|P_{j+1}| \leq t_4 \left((k/f(k))^c \right) \leq s_{j+1}.$$

If $q(Q) \leq q(P_j) + \varepsilon$, then we may take $P = P_j$ and $Q = Q(\mathcal{K})$, and these partitions satisfy the desired properties. Otherwise, $q(P_{j+1}) \geq q(Q) - \frac{\varepsilon}{2} > q(P_j) + \frac{\varepsilon}{2}$, and we continue the sequence of partitions. Since $q(P_1) \geq 0$ and the mean-square density goes up by more than $\varepsilon/2$ at each step and is always at most 1, this process must stop within $2/\varepsilon$ steps, and we obtain the desired partitions. $\qquad\square$

Let $G = (V, E)$, $P : V = V_1 \cup \ldots \cup V_k$ be an equipartition and $\mathcal{K}$ be a partition of the cylinder $V_1 \times \cdots \times V_k$ into cylinders. For $K = W_1 \times \cdots \times W_k \in \mathcal{K}$, define the density $d(K) = \frac{|W_1| \times \cdots \times |W_k|}{|V_1| \times \cdots \times |V_k|}$. The cylinder K is ε-close to P if $|d(W_i, W_j) - d(V_i, V_j)| \leq \varepsilon$ for all but at most εk^2 pairs $1 \leq i \neq j \leq k$. The cylinder partition $\mathcal{K}$ is ε-close to P if $\sum d(K) \leq \varepsilon$, where the sum is over all $K \in \mathcal{K}$ that are not ε-close to P. As with the definition of closeness used in the strong regularity lemma, this definition is closely related to the condition that $q(Q) \leq q(P) + \varepsilon$, where here $Q = Q(\mathcal{K})$.

The connection we shall need to prove Lemma 3.2 is contained in the following statement.

Lemma 3.8 *Let $G = (V, E)$ and $P : V = V_1 \cup \ldots \cup V_k$ be an equipartition with $k \geq 2\varepsilon^{-1}$ and $|V| \geq 4k\varepsilon^{-1}$. Let $\mathcal{K}$ be a partition of the cylinder $V_1 \times \cdots \times V_k$ into cylinders. If $Q = Q(\mathcal{K})$ satisfies $q(Q) \leq q(P) + \varepsilon$, then $\mathcal{K}$ is $(2\varepsilon)^{1/4}$-close to P.*

Proof It will be helpful to assume that all parts of the equipartition P have equal size – this affects the calculations only slightly. It will also be helpful to introduce a slight variant of the mean-square density as follows. Let $q'(P) = \sum_{i<j} d^2(V_i, V_j)p_{ij}$, where $p_{ij} = |V_i||V_j| / \sum_{a<b}|V_a||V_b|$. Thus, $q'(P)$ is the mean of the square densities between the pairs of distinct parts. It is easy to check that $q'(P)$ is close to $q(P)$. Indeed, we have $q'(P) - q(P) = \frac{1}{k}(q'(P) - \bar{q})$, where $\bar{q} = \sum_{i=1}^{k} d^2(V_i)/k$ is the average of the square densities inside the parts. Hence, $|q'(P) - q(P)| \leq \frac{1}{k}$. We similarly have $|q'(Q) - q(Q)| \leq \frac{1}{k}$. Let

$$q(\mathcal{K}) = \binom{k}{2}^{-1} \sum_{i<j} \sum_{K \in \mathcal{K}} d^2(V_i(K), V_j(K))d(K).$$

We have the following equalities

$$
\begin{aligned}
q(\mathcal{K}) - q'(P) &= \binom{k}{2}^{-1} \sum_{i<j} \sum_{K \in \mathcal{K}} \left(d^2(V_i(K), V_j(K)) - d^2(V_i, V_j)\right) d(K) \\
&= \binom{k}{2}^{-1} \sum_{i<j} \sum_{K \in \mathcal{K}} \left(d(V_i(K), V_j(K)) - d(V_i, V_j)\right)^2 d(K),
\end{aligned}
$$

where the last equality uses $d(V_i, V_j) = \sum_{K \in \mathcal{K}} d(V_i(K), V_j(K))d(K)$. This equality shows that $q(\mathcal{K}) \geq q'(P)$ as it expresses their difference as a sum of nonnegative terms. Furthermore, it shows that if $\mathcal{K}$ is not β-close to P, then $q(\mathcal{K}) \geq q'(P) + \binom{k}{2}^{-1} \cdot \frac{\beta k^2}{2} \cdot \beta^2 \cdot \beta \geq q'(P) + \beta^4$. In particular, if $q(\mathcal{K}) \leq q'(P) + 2\varepsilon$, then $\mathcal{K}$ is $(2\varepsilon)^{1/4}$-close to P. So assume for contradiction that $q(\mathcal{K}) > q'(P) + 2\varepsilon$.

A similar equality implies $q'(Q) \geq q(\mathcal{K})$. We therefore have

$$
\begin{aligned}
q(Q) - q(P) &= (q(Q) - q'(Q)) + (q'(Q) - q(\mathcal{K})) \\
&\quad + (q(\mathcal{K}) - q'(P)) + (q'(P) - q(P)) \\
&\geq -\frac{1}{k} + 0 + (q(\mathcal{K}) - q'(P)) - \frac{1}{k} \\
&> \varepsilon,
\end{aligned}
$$

contradicting the assumption of Lemma 3.8 and completing the proof.   □

With this in hand, we can readily deduce a tower-type bound for Lemma 3.2.

Lemma 3.9 *For each $0 < \varepsilon < 1/3$ and decreasing function $f : \mathbb{N} \to (0, \varepsilon]$, there is $\delta' = \delta'(\varepsilon, f)$ such that every graph $G = (V, E)$ with $|V| \geq \delta'^{-1}$ has an equitable partition $V = V_1 \cup \ldots \cup V_k$ and vertex subsets $W_i \subset V_i$ such that $|W_i| \geq \delta'|V|$, each pair (W_i, W_j) with $1 \leq i \leq j \leq k$ is $f(k)$-regular and all but at most εk^2 pairs $1 \leq i \leq j \leq k$ satisfy $|d(V_i, V_j) - d(W_i, W_j)| \leq \varepsilon$. Furthermore, we may take $\delta' = \frac{1}{8S^2}$, where $S = S(\frac{\varepsilon^4}{2}, s, f)$ is defined as in Lemma 3.5 and $s = 2\varepsilon^{-1}$.*

Proof Let $\alpha = \frac{\varepsilon^4}{2}$, $s = 2\varepsilon^{-1}$, and $\delta' = \frac{1}{8S^2}$, where $S = S(\alpha, s, f)$ is as in Lemma 3.5. We apply Lemma 3.5 with α in place of ε. We get an equipartition $P : V = V_1 \cup \ldots \cup V_k$ with $s \leq k \leq S$ and a strongly $f(k)$-regular partition $\mathcal{K}$ of $V_1 \times \cdots \times V_k$ into cylinders such that the refinement $Q = Q(\mathcal{K})$ of P has at most $S = S(\alpha, s, f)$ parts and satisfies $q(Q) \leq q(P) + \alpha$. Since $|V| \geq \delta'^{-1} = 8S^2$, and P is an equipartition into $k \leq S$ parts, the cardinality of each part $V_i \in P$ satisfies $|V_i| \geq \frac{|V|}{2S}$. By Lemma 3.8, as $(2\alpha)^{1/4} = \varepsilon$, the cylinder partition $\mathcal{K}$ is ε-close to P. Hence, at most an ε-fraction of the k-tuples $(v_1, \ldots, v_k) \in V_1 \times \cdots \times V_k$ belong to parts $K = W_1 \times \cdots \times W_k$ of $\mathcal{K}$ that are not ε-close to P. Since $Q(\mathcal{K})$ has at most S parts, the fraction of k-tuples $(v_1, \ldots, v_k) \in V_1 \times \cdots \times V_k$ that belong to parts $K = W_1 \times \cdots \times W_k$ of $\mathcal{K}$ with $|W_i| < \frac{1}{4S}|V_i|$ for at least one $i \in [k]$ is at most $\frac{1}{4S} \cdot S = \frac{1}{4}$. Therefore, at least a fraction $1 - f(k) - \varepsilon - \frac{1}{4} > 0$ of the k-tuples $(v_1, \ldots, v_k) \in V_1 \times \cdots \times V_k$ belong to parts $K = W_1 \times \cdots \times W_k$ of $\mathcal{K}$ satisfying K is strongly $f(k)$-regular, $|W_i| \geq \frac{1}{4S}|V_i| \geq \delta'|V|$ for $i \in [k]$ and K is ε-close to P. Since a positive fraction of the k-tuples belong to such K, there is at least one such K. This K has the desired properties. Indeed, the number of pairs $1 \leq i \neq j \leq k$ for which $|d(W_i, W_j) - d(V_i, V_j)| > \varepsilon$ is at most εk^2 and hence the number of pairs $1 \leq i \leq j \leq k$ for which $|d(W_i, W_j) - d(V_i, V_j)| > \varepsilon$ is at most $\varepsilon k^2/2 + k \leq \varepsilon k^2$. This completes the proof.   □

By using the induced counting lemma, Lemma 3.3, we may now conclude the proof as in Section 3.1 to obtain the following quantitative version of Theorem 1.3.

Theorem 3.10 *There exists a constant c such that, for any graph H on h vertices and $0 < \varepsilon < 1/2$, if $\delta^{-1} = t_j(h)$, where $j = c\varepsilon^{-4}$, then any* *graph G on n vertices with at most δn^h induced copies of H may be made induced H-free by adding and/or deleting at most εn^2 edges.*

3.3   Infinite removal lemma

In order to characterize the natural graph properties which are testable, the induced removal lemma was extended by Alon and Shapira [12] to the following infinite version. For a family $\mathcal{H}$ of graphs, a graph G is induced $\mathcal{H}$-free if G does not contain any graph H in $\mathcal{H}$.

Theorem 3.11 *For every (possibly infinite) family of graphs $\mathcal{H}$ and $\varepsilon > 0$, there are n_0, h_0, and δ such that the following holds. If a graph $G = (V, E)$ on $n \geq n_0$ vertices has at most δn^h induced copies of each graph $H \in \mathcal{H}$ on $h \leq h_0$ vertices, then G can be made induced $\mathcal{H}$-free by adding and/or deleting at most εn^2 edges.*

Proof The proof is a natural extension of the proof of the induced removal lemma and similarly uses the key corollary, Lemma 3.2, of the strong regularity lemma. The main new idea is to pick an appropriate function f to apply Lemma 3.2. The choice of the function f will depend heavily on the family $\mathcal{H}$.

For a graph H and an edge-coloring c of the edges of the complete graph with loops R on $[k]$ with colors white, black and grey, we write $H \to_c R$ if there is a mapping $\phi : V(H) \to [k]$ such that for each edge (u, v) of H we have that $c(\phi(u), \phi(v))$ is black or grey and for each pair (u, v) of distinct vertices of H which do not form an edge we have that $c(\phi(u), \phi(v))$ is white or grey. We write $H \not\to_c R$ if $H \to_c R$ does not hold.

Let $P : V = V_1 \cup \ldots \cup V_k$ be a vertex partition of G. A key observation is that if we *round* G by the partition P and the coloring c to obtain a graph G' on the same vertex set as G by adding edges to make (V_i, V_j) complete if (i, j) is black, deleting edges to make (V_i, V_j) empty if (i, j) is white and we have that $H \not\to_c R$, then G' does not contain H as an induced subgraph.

For any (possibly infinite) family of graphs $\mathcal{H}$ and any integer r, let $\mathcal{H}_r$ be the following set of colored complete graphs with loops: a colored complete graph with loops R belongs to $\mathcal{H}_r$ if and only if it has at most r vertices and there is at least one $H \in \mathcal{H}$ such that $H \to_c R$. For any family $\mathcal{H}$ of graphs and integer r for which $\mathcal{H}_r \neq \emptyset$, let

$$\Psi_{\mathcal{H}}(r) = \max_{R \in \mathcal{H}_r} \min_{H \in \mathcal{H}: H \to_c R} |V(H)|.$$

If $\mathcal{H}_r = \emptyset$, define $\Psi_{\mathcal{H}}(r) = 1$. Note that $\Psi_{\mathcal{H}}(r)$ is a monotonically increasing function of r. Let

$$f(r) = \frac{\varepsilon^{\Psi_{\mathcal{H}}(r)}}{4\Psi_{\mathcal{H}}(r)}.$$

Note that the function f only depends on ε and $\mathcal{H}$.

Let $\delta' = \delta'(\varepsilon, f)$ be as in Lemma 3.2, which only depends on ε and $\mathcal{H}$. Also let $k_0 = 2\delta'^{-1}$, $h_0 = \Psi_{\mathcal{H}}(k_0)$, $n_0 = 1/(\delta' f(k_0))$ and $\delta = \frac{1}{h_0!}(\varepsilon/4)^{h_0^2}\delta'^{h_0}$. We have that k_0, h_0, n_0 and $\delta > 0$ only depend on ε and $\mathcal{H}$. By assumption, G has $n \geq n_0$ vertices.

We apply Lemma 3.2 to G. We get an equitable vertex partition P: $V = V_1 \cup \ldots \cup V_k$ of G and subsets $W_i \subset V_i$ with $|W_i| \geq \delta'|V|$ such that, for $1 \leq i \leq j \leq k$, the pair (W_i, W_j) is $f(k)$-regular and all but at most εk^2 pairs $1 \leq i \leq j \leq k$ satisfy $|d(V_i, V_j) - d(W_i, W_j)| \leq \varepsilon$. As $\delta'|V| \leq |W_i| \leq |V_i| \leq 2n/k$, we have $k \leq 2\delta'^{-1} \leq k_0$.

Consider the coloring c of the complete graph with loops R on $[k]$ where a pair (i, j) of vertices is black if $d(W_i, W_j) \geq 1 - \varepsilon$, white if $d(W_i, W_j) \leq \varepsilon$ and grey if $\varepsilon < d(W_i, W_j) < 1 - \varepsilon$. Suppose, for the sake of contradiction, that there is a graph H with $H \to_c R$. From the definition of Ψ, there is a graph H on $h \leq \Psi_{\mathcal{H}}(k)$ vertices with $H \to_c R$. As $k \leq k_0$, the number of vertices of H satisfies $h \leq h_0$. As each pair (W_i, W_j) is $f(k)$-regular and $|W_i| \geq \delta'|V| \geq f(k)^{-1}$, applying the induced counting lemma, Lemma 3.3, with $\gamma = f(k)$, we get at least

$$\frac{1}{h!}\left(\frac{\varepsilon}{4}\right)^{\binom{h}{2}}(\delta'|V|)^h \geq \delta n^h$$

induced copies of H in G, contradicting the supposition of the theorem. Thus, there is no graph H with $H \to_c R$.

We round the graph G by the partition P and the coloring c as described earlier in the proof to obtain a graph G'. By the key observation, for each graph H with $H \not\to_c R$, the graph G' does not contain H as an induced subgraph. Hence, G' is induced $\mathcal{H}$-free.

Moreover, not many edges were changed from G to obtain G'. Indeed, as there are at most εk^2 pairs $1 \leq i \leq j \leq k$ which satisfy $|d(V_i, V_j) - d(W_i, W_j)| > \varepsilon$, the number of edge modifications made between such pairs is at most $\varepsilon k^2 \cdot (2n/k)^2 = 4\varepsilon n^2$. Between the other pairs we have made at most $2\varepsilon\binom{n}{2} \leq \varepsilon n^2$ edge modifications. In total, at most $5\varepsilon n^2$ edge modifications were made to obtain G' from G. Replacing ε by $\varepsilon/5$ in the above argument completes the proof. $\square$

4 Arithmetic removal

The notion of arithmetic removal was introduced by Green [58]. By establishing an appropriate variant of the regularity lemma in the context of abelian groups, he proved the following result.

Theorem 4.1 *For any natural number $k \geq 3$ and any $\varepsilon > 0$, there exists $\delta > 0$ such that if G is an abelian group of order n and $A_1, \ldots, A_k$ are subsets of G such that there are at most δn^{k-1} solutions to the equation $a_1 + a_2 + \cdots + a_k = 0$ with $a_i \in A_i$ for all i then it is possible to remove at most εn elements from each set A_i to form sets A_i' so that there are no solutions to the equation $a_1' + a_2' + \cdots + a_k' = 0$ with $a_i' \in A_i'$ for all i.*

It is an exercise to show that Green's result implies Roth's theorem. While Green's proof of this result relied on Fourier analytic techniques, an alternative proof was found by Král', Serra and Vena [71], who showed that the following more general result follows from an elegant reduction to the removal lemma in directed graphs.

Theorem 4.2 *For any natural number $k \geq 3$ and any $\varepsilon > 0$, there exists $\delta > 0$ such that if G is a group of order n, $g \in G$ and $A_1, \ldots, A_k$ are subsets of G such that there are at most δn^{k-1} solutions to the equation $a_1 a_2 \cdots a_k = g$ with $a_i \in A_i$ for all i then it is possible to remove at most εn elements from each set A_i to form sets A_i' so that there are no solutions to the equation $a_1' a_2' \cdots a_k' = g$ with $a_i' \in A_i'$ for all i.*

This is stronger than Theorem 4.1 in two ways. Firstly, it applies to all groups and not just to abelian groups. Secondly, it applies to non-homogeneous equations, that is, $a_1 a_2 \cdots a_k = g$ for a general g, whereas Green only treats the homogeneous case where $g = 1$. However, it is easy to see that the general version follows from the homogeneous case by substituting $A_k g$ for A_k. To give some idea of their proof, we will need the following definition.

A directed graph is a graph where each edge has been given a direction. Formally, the edge set may be thought of as a collection of ordered pairs. We will always assume that the directed graph has no loops and does not contain parallel directed edges, though we do allow anti-parallel edges, that is, both the edge $\vec{uv}$ and the edge $\vec{vu}$. The following analogue of the graph removal lemma for directed graphs was proved by Alon and Shapira [9] as part of their study of property testing in directed graphs.

Theorem 4.3 *For any directed graph H and any $\varepsilon > 0$, there exists $\delta > 0$ such that any directed graph on n vertices which contains at most $\delta n^{v(H)}$ copies of H may be made H-free by removing at most εn^2 edges.*

We will show how to prove Theorem 4.2 with $g = 1$ using Theorem 4.3. Suppose that G is a group of order n and $A_1, \ldots, A_k$ are subsets of G such that there are at most δn^{k-1} solutions to the equation $a_1 a_2 \cdots a_k = 1$ with $a_i \in A_i$ for all i. Consider the auxiliary directed graph Γ whose vertex set is $G \times \{1, 2, \ldots, k\}$. We place an edge from (x, i) to $(y, i + 1)$, where addition is taken modulo k, if there exists $a_i \in A_i$ such that $x a_i = y$. It is easy to see that any directed cycle in Γ corresponds to a solution of the equation $a_1 a_2 \cdots a_k = 1$. Moreover, every such solution will result in n different directed cycles in Γ, namely, those with vertices $(x, 1), (x a_1, 2), (x a_1 a_2, 3), \ldots, (x a_1 \cdots a_{k-1}, k)$.

Since G has at most δn^{k-1} solutions to $a_1 a_2 \cdots a_k = 1$, this implies that there are at most δn^k directed cycles in Γ. By Theorem 4.3, for an appropriately chosen δ, we may therefore remove at most $\frac{\varepsilon}{k} n^2$ edges to make it free of directed cycles of length k. In A_i, we now remove the element a_i if at least $\frac{n}{k}$ edges of the form $(x, i)(x a_i, i + 1)$ have been removed. Note that this results in us removing at most εn elements from each A_i. Suppose now that the remaining sets A_i' are such that there is a solution $a_1' a_2' \ldots a_k' = 1$ with $a_i' \in A_i'$ for all i. Then, as above, there are at least n cycles $(x, 1), (x a_1', 2), \ldots, (x a_1' \cdots a_{k-1}', k)$ corresponding to this solution. Since we must have removed one edge from each of these cycles, we must have removed at least $\frac{n}{k}$ edges of the form $(y, i)(y a_i', i + 1)$ for some i. But this implies that $a_i' \notin A_i'$, yielding the required contradiction.

It was observed by Fox [38] that δ^{-1} in Theorem 4.3 may, like the graph removal lemma, be taken to be at most a tower of twos of height logarithmic in ε^{-1}. This may in turn be used to give a similar bound for δ^{-1} in Theorem 4.2.

In [71], Král', Serra and Vena also showed how to prove a removal lemma for systems of equations which are graph representable, in the sense that they can be put in a natural correspondence with a directed graph. An example of such a system is

$$x_1 x_2 x_4^{-1} x_3^{-1} = 1$$
$$x_1 x_2 x_5^{-1} = 1.$$

This idea of associating a system of linear equations with a directed graph representation was extended to hypergraphs independently by Král', Serra and Vena [72] and by Shapira [101, 102] in order to prove the following theorem (some partial results had been obtained earlier by Král', Serra and Vena [70], Szegedy [109] and Candela [20]).

Theorem 4.4 *For any natural numbers k and ℓ and any $\varepsilon > 0$, there exists $\delta > 0$ such that if F is the field of size n, M is an $\ell \times k$ matrix with* *coefficients in* F, $b \in F^\ell$ *and* $A_1, \dots, A_k$ *are subsets of* F *such that there are at most* $\delta n^{k-\ell}$ *solutions* $a = (a_1, \dots, a_k)$ *of the system* $Ma = b$ *then it is possible to remove at most* εn *elements from each set* A_i *to form sets* A_i' *so that there are no solutions* $a' = (a_1', \dots, a_k')$ *to the equation* $Ma' = b$ *with* $a_i' \in A_i'$ *for all* i.

An easy application of this result shows that a removal lemma for systems of linear equations holds in the set $[n]$, confirming a conjecture of Green [58]. We remark that this result easily implies Szemerédi's theorem. Both proofs use a colored variant of the hypergraph removal lemma due to Austin and Tao [13], though the representations which they use to transfer the problem to hypergraphs are different.

It would be interesting to know whether an analogous statement holds for all groups. A partial extension of these results to abelian groups is proved in [73] (see also [109]) but already in this case there are technical difficulties which do not arise for finite fields.

5 Sparse removal

Given graphs Γ and H, let $N_H(\Gamma)$ be the number of copies of H in Γ. A possible generalization of the graph removal lemma, which corresponds to the case $\Gamma = K_n$, could state that if G is a subgraph of Γ with $N_H(G) \leq \delta N_H(\Gamma)$ then G may be made H-free by deleting at most $\varepsilon e(\Gamma)$ edges. Unfortunately, this is too much to hope in general. However, if the graph Γ is sufficiently well-behaved, such an extension does hold. We will discuss two such results here.

5.1 Removal in random graphs

The binomial random graph $G_{n,p}$ is formed by taking n vertices and considering each pair of vertices in turn, choosing each connecting edge to be in the graph independently with probability p. These graphs were introduced by Erdős and Rényi [33, 34] in the late fifties[12] and their study has grown enormously since then (see, for example, the monographs [17, 61]).

Usually, one is interested in finding a threshold function $p^* := p^*(n)$ where the probability that the random graph $G_{n,p}$ has a particular property $\mathcal{P}$ changes from $o(1)$ to $1 - o(1)$ as we pass from random graphs chosen

---

[12]The notion was also introduced independently by several other authors at about the same time but, quoting Bollobás [17], "Erdős and Rényi introduced the methods which underlie the probabilistic treatment of random graphs. The other authors were all concerned with enumeration problems and their techniques were essentially deterministic."

with probability $p \ll p^*$ to those chosen with probability $p \gg p^*$. For example, a threshold for the random graph to be connected is at $p^*(n) = \frac{\ln n}{n}$.

One theme that has received a lot of attention in recent years is the question of determining thresholds for the appearance of certain combinatorial properties. One well-studied example is the Ramsey property. Given a graph H and a natural number $r \geq 2$, we say that a graph G is (H, r)-*Ramsey* if in any r-coloring of the edges of G there is guaranteed to be a monochromatic copy of H. Ramsey's theorem [83] is itself the statement that K_n is (H, r)-Ramsey for n sufficiently large. The following celebrated result of Rödl and Ruciński [84, 85] from 1995 (see also [61], Chapter 8) determines the threshold for the appearance of the Ramsey property in random graphs. For a graph H, we let $v(H)$ and $e(H)$ denote the number of vertices and edges, respectively, of H.

Theorem 5.1 *For any graph H that is not a forest consisting of stars and paths of length 3 and every positive integer $r \geq 2$, there exist constants $c, C > 0$ such that*

$$\lim_{n \to \infty} \mathbb{P}\big(G_{n,p} \text{ is } (H, r)\text{-Ramsey}\big) = \begin{cases} 0, & \text{if } p < cn^{-1/m_2(H)}, \\ 1, & \text{if } p > Cn^{-1/m_2(H)}, \end{cases}$$

where

$$m_2(H) = \max\left\{ \frac{e(H') - 1}{v(H') - 2} : H' \subseteq H \text{ and } v(H') \geq 3 \right\}.$$

The threshold occurs at the largest value of p^* such that there is some subgraph H' of H for which the number of copies of H' is approximately the same as the number of edges. For p significantly smaller than p^*, the number of copies of H' will also be significantly smaller than the number of edges. This property allows us (by a rather long and difficult argument [84]) to show that the edges of the graph may be colored in such a way as to avoid any monochromatic copies of H'. For p significantly larger than p^*, every edge of the random graph is contained in many copies of every subgraph of H. The intuition, which takes substantial effort to make rigorous [85], is that these overlaps are enough to force the graph to be Ramsey.

Many related questions were studied in the late nineties. In particular, people were interested in determining the threshold for the following Turán property. Given a graph H and a real number $\varepsilon > 0$, we say that a graph G is (H, ε)-*Turán* if every subgraph of G with at least

$$\left(1 - \frac{1}{\chi(H) - 1} + \varepsilon\right) e(G)$$

edges contains a copy of H. The classical Erdős-Stone-Simonovits theorem [35, 36, 116] states that the graph K_n is (H, ε)-Turán for n sufficiently large. Resolving a conjecture of Haxell, Kohayakawa, Łuczak and Rödl [59, 66], Conlon and Gowers [25] and, independently, Schacht [99] proved the following theorem. It is worth noting that the result of Conlon and Gowers applies in the strictly balanced case, that is, when $m_2(H') < m_2(H)$ for all $H' \subset H$, while Schacht's result applies to all graphs. However, the class of strictly balanced graphs includes most of the graphs one would naturally consider, such as cliques or cycles.

Theorem 5.2 *For any graph H[13] and any $\varepsilon > 0$, there exist positive constants c and C such that*

$$\lim_{n \to \infty} \mathbb{P}\big(G_{n,p} \text{ is } (H, \varepsilon)\text{-}Tur\acute{a}n\big) = \begin{cases} 0, & \text{if } p < cn^{-1/m_2(H)}, \\ 1, & \text{if } p > Cn^{-1/m_2(H)}. \end{cases}$$

The results of [25] and [99] (see also [43]) allow one to prove thresholds for the appearance of many different combinatorial properties. For example, the results extend without difficulty to prove analogues of Theorems 5.1 and 5.2 for hypergraphs. The results also apply to give thresholds in different contexts – one example is an extension of Szemerédi's theorem to random subsets of the integers.

Perhaps surprisingly, the methods used in [25] and [99] are very different and have different strengths and weaknesses. We have already mentioned that Schacht's results applied to all graphs while the results of Conlon and Gowers only applied to strictly balanced graphs. On the other hand, the results of [25] also allowed one to transfer structural statements to the sparse setting, including the stability version of the Erdős-Stone-Simonovits theorem [104] and the graph removal lemma. More recently, Samotij [95] modified Schacht's method to extend this sparse stability theorem to all graphs. The result is the following theorem.

Theorem 5.3 *For any graph H and any $\varepsilon > 0$, there exist positive constants δ and C such that if $p \geq Cn^{-1/m_2(H)}$ then the following holds a.a.s. in $G_{n,p}$. Every H-free subgraph of $G_{n,p}$ with at least $\left(1 - \frac{1}{\chi(H)-1} - \delta\right) p\binom{n}{2}$ edges may be made $(\chi(H) - 1)$-partite by deleting at most εpn^2 edges.*

Recently, a third method was developed by Balogh, Morris and Samotij [15] and, simultaneously and independently, by Saxton and Thomason

---
[13] Note that if $H = K_2$, we take $m_2(H) = \frac{1}{2}$.

[98] for proving sparse random analogues of combinatorial theorems. One of the results of their research is a proof of the KLR conjecture of Kohayakawa, Łuczak and Rödl [66] (the conjecture had already been known in several special cases – see [51] and its references). This is a technical statement which allows one to prove an embedding lemma complementing the sparse regularity lemma of Kohayakawa [64] and Rödl. A variant of this conjecture has also been proved by Conlon, Gowers, Samotij and Schacht [26] using the methods of [25, 99]. One of the applications of this latter result is the following sparse random analogue of the graph removal lemma (this was already proved for triangles in [65] and for strictly balanced graphs in [25]).

Theorem 5.4 *For any graph H and any $\varepsilon > 0$, there exist positive constants δ and C such that if $p \geq Cn^{-1/m_2(H)}$ then the following holds a.a.s. in $G_{n,p}$. Every subgraph of $G_{n,p}$ containing at most $\delta p^{e(H)} n^{v(H)}$ copies of H may be made H-free by removing at most $\varepsilon p n^2$ edges.*

Note that for any ε there exists a positive constant c such that if $p \leq cn^{-1/m_2(H)}$, the removal lemma is trivial. This is because, for c sufficiently small, the number of copies of the densest subgraph H' of H will a.a.s. be smaller than $\varepsilon p n^2$. Theorem 5.4 shows that it also holds for $p \geq Cn^{-1/m_2(H)}$. This leaves a small intermediate range of p where it might also be expected that a sparse removal lemma a.a.s. holds. That this is so was conjectured by Łuczak [77].

For balanced graphs H, we may close the gap by letting δ be sufficiently small depending on C, ε and H. Indeed, as $p \leq Cn^{-1/m_2(H)}$, the number of copies of H is a.a.s. on the order of $p^{e(H)} n^{v(H)} \leq C^{e(H)} p n^2$. Therefore, taking $\delta < \varepsilon C^{-e(H)}$, we see that the number of copies of H is a.a.s. less than $\varepsilon p n^2$. Deleting one edge from each copy of H in the graph then makes it H-free.

A sparse random analogue of the hypergraph removal lemma was shown in [25] when $\mathcal{H} = K_{k+1}^{(k)}$. This result also extends to cover all strictly balanced hypergraphs.[14] It would be interesting to extend this result to all hypergraphs.

It is worth noting that the sparse random version of the triangle removal lemma does not imply a sparse random version of Roth's theorem. This is because the reduction which allows us to pass from a subset of the integers with no arithmetic progressions of length 3 to a graph containing

---

[14]We note that for k-uniform hypergraphs the relevant function is $m_k(\mathcal{H}) = \max\left\{ \frac{e(\mathcal{H}')-1}{v(\mathcal{H}')-k} \right\}$, where the maximum is taken over all subgraphs $\mathcal{H}'$ of $\mathcal{H}$ with at least $k+1$ vertices.

few triangles gives us a graph with dependencies between its edges. This issue does not occur with pseudorandom graphs, which we discuss in the next section.

5.2 Removal in pseudorandom graphs

Though there have long been explicit examples of graphs which behave like the random graph $G_{n,p}$, the first systematic study of what it means for a given graph to be like a random graph was initiated by Thomason [114, 115]. Following him,[15] we say that a graph on vertex set V is (p, β)-*jumbled* if, for all vertex subsets $X, Y \subseteq V$,

$$|e(X,Y) - p|X||Y|| \leq \beta\sqrt{|X||Y|}.$$

The random graph $G_{n,p}$ is, with high probability, (p, β)-jumbled with $\beta = O(\sqrt{pn})$. This is also optimal in that a graph on n vertices with $p \leq 1/2$ cannot be (p, β)-jumbled with $\beta = o(\sqrt{pn})$. The Paley graph is an example of an explicit graph which is optimally jumbled. This graph has vertex set $\mathbb{Z}_p$, where $p \equiv 1 \pmod 4$ is prime, and edge set given by connecting x and y if their difference is a quadratic residue. It is (p, β)-jumbled with $p = \frac{1}{2}$ and $\beta = O(\sqrt{n})$. Many more examples are given in the excellent survey [74].

A fundamental result of Chung, Graham and Wilson [22] states that for graphs of density p, where p is a fixed positive constant, the property of being $(p, o(n))$-jumbled is equivalent to a number of other properties that one would typically expect in a random graph. For example, if the number of cycles of length 4 is as one would expect in a binomial random graph then, surprisingly, this is enough to imply that the edges are very well-spread.

For sparser graphs, the equivalences are less clear cut, but the notion of jumbledness defined above is a natural property to study. Given a graph property $\mathcal{P}$ that one would expect of a random graph, one can ask for the range of p and β for which a (p, β)-jumbled graph satisfies $\mathcal{P}$.

To give an example, it is known that there is a constant c such that if $\beta \leq cp^2 n$ then any (p, β)-jumbled graph contains a triangle. It is also known that this is sharp, since an example of Alon [2] gives a triangle-free graph with $p = \Omega(n^{-1/3})$ which is optimally jumbled, so that $\beta = O(\sqrt{pn}) = O(p^2 n)$.

As in the previous section, one can ask for conditions on p and β which guarantee that a (p, β)-jumbled graph satisfies certain combinatorial

---

[15]Strictly speaking, Thomason considered a slightly different notion, namely, that $|e(X) - p\binom{|X|}{2}| \leq \beta|X|$ for all $X \subseteq V$, but the two are closely related.

properties. For the property of being (K_3, ε)-Turán, this question was addressed by Sudakov, Szabó and Vu [108] (see also [21]), who showed that it was enough that $\beta \leq cp^2 n$ for an appropriate c. This is clearly sharp, since for larger values of β we cannot even guarantee that the graph contains a triangle. More generally, they proved the following theorem.[16]

Theorem 5.5 *For any natural number $t \geq 3$ and any $\varepsilon > 0$, there exists $c > 0$ such that if $\beta \leq cp^{t-1} n$ then any (p, β)-jumbled graph is (K_t, ε)-Turán.*

Except in the case of triangles, there are no known constructions which demonstrate that this theorem is tight. However, it is conjectured [108] that the bound on β in Theorem 5.5 is the correct condition for finding copies of K_t in a (p, β)-jumbled graph. This would in turn imply that Theorem 5.5 is tight.

For the triangle removal lemma, the following pseudorandom analogue was recently proved by Kohayakawa, Rödl, Schacht and Skokan [69].

Theorem 5.6 *For any $\varepsilon > 0$, there exist positive constants δ and c such that if $\beta \leq cp^3 n$ then any (p, β)-jumbled graph G on n vertices has the following property. Any subgraph of G containing at most $\delta p^3 n^3$ triangles may be made triangle-free by removing at most $\varepsilon p n^2$ edges.*

The condition on β in this theorem is stronger than that employed for triangles in Theorem 5.5. As a result, Alon's construction does not apply and it is an open problem to determine whether the condition $\beta \leq cp^3 n$ is optimal or if it can be improved to $\beta \leq cp^2 n$. Kohayakawa, Rödl, Schacht and Skokan conjecture the latter, though we feel that the former is a genuine possibility.

In a recent paper, Conlon, Fox and Zhao [27] found a way to prove a counting lemma for embedding any fixed small graph into a regular subgraph of a sufficiently pseudorandom host graph. Like the KŁR conjecture for random graphs, this serves to complement the sparse regularity lemma of Kohayakawa [64] and Rödl in the pseudorandom context. As corollaries, they extended Theorems 5.5 and 5.6 to all graphs and proved sparse pseudorandom extensions of several other theorems, including Ramsey's theorem and the Erdős-Simonovits stability theorem.

---

[16]Their results were only stated for the special class of (p, β)-jumbled graphs known as (n, d, λ)-graphs. These are graphs on n vertices which are d-regular and such that all eigenvalues of the adjacency matrix, save the largest, have absolute value at most λ. The expander mixing lemma implies that these graphs are (p, β)-jumbled with $p = \frac{d}{n}$ and $\beta = \lambda$. However, it is not hard to verify that their method applies in the more general case.

To state these theorems, we define the *degeneracy* $d(H)$ of a graph H to be the smallest nonnegative integer d for which there exists an ordering of the vertices of H such that each vertex has at most d neighbors which appear earlier in the ordering. Equivalently, it may be defined as $d(H) = \max\{\delta(H') : H' \subseteq H\}$, where $\delta(H)$ is the minimum degree of H.[17]

The pseudorandom analogue of the graph removal lemma proved in [27] is now as follows.[18]

Theorem 5.7 *For any graph H and any $\varepsilon > 0$, there exist positive constants δ and c such that if $\beta \leq cp^{d(H)+\frac{5}{2}}n$ then any (p,β)-jumbled graph G on n vertices has the following property. Any subgraph of G containing at most $\delta p^{e(H)}n^{v(H)}$ copies of H may be made H-free by removing at most εpn^2 edges.*

It is not hard to show, by using the random graph, that there are (p,β)-jumbled graphs with $\beta = O(p^{(d(H)+2)/4}n)$ which contain no copies of H. We therefore see that the exponent of p is sharp up to a multiplicative constant. However, in many cases, we expect it to be sharp up to an additive constant.

For certain classes of graph, Theorem 5.7 can be improved. For example, if we know that the degeneracy of the graph is the same as the maximum degree, such as what happens for the complete graph K_t, it is sufficient that $\beta \leq cp^{d(H)+1}n$. In particular, for K_3, we reprove Theorem 5.6. For cycles, the improvement is even more pronounced, since $\beta \leq cp^{t_\ell}n$, where $t_3 = 3$, $t_4 = 2$, $t_\ell = 1 + \frac{1}{\ell-3}$ if $\ell \geq 5$ is odd and $t_\ell = 1 + \frac{1}{\ell-4}$ if $\ell \geq 6$ is even, is sufficient for removing the cycle C_ℓ.

By following the proof of Král', Serra and Vena [71], these bounds on the cycle removal lemma in pseudorandom graphs[19] allow us to prove an analogue of Theorem 4.2 for pseudorandom subsets of any group G. The *Cayley graph* $G(S)$ of a subset S of a group G has vertex set G and (x,y) is an edge of G if $x^{-1}y \in S$. We say that a subset S of a group G is (p,β)-jumbled if the Cayley graph $G(S)$ is (p,β)-jumbled. When G is abelian, if $\left|\sum_{x\in S}\chi(x)\right| \leq \beta$ for all nontrivial characters $\chi: G \to \mathbb{C}$, then S is $(\frac{|S|}{|G|},\beta)$-jumbled (see [69, Lemma 16]).

---

[17]In [27], a slightly different parameter, the 2-degeneracy $d_2(H)$, is used. Though there are many cases in which this parameter is more appropriate, the degeneracy will be sufficient for the purposes of our discussion here.

[18]For other properties, such as that of being (H,r)-Ramsey or that of being (H,ε)-Turán, an exactly analogous theorem holds with the same condition $\beta \leq cp^{d(H)+\frac{5}{2}}n$. Any of the improvements subsequently discussed for specific graphs H also apply for these properties.

[19]Rather, a colored or directed version of this theorem.

Theorem 5.8 *For any natural number $k \geq 3$ and any $\varepsilon > 0$, there exist positive constants δ and c such that the following holds. Suppose $B_1, \ldots, B_k$ are subsets of a group G of order n such that each B_i is (p, β)-jumbled with $\beta \leq cp^{t_k}n$. If subsets $A_i \subseteq B_i$ for $i = 1, \ldots, k$ are such that there are at most $\delta|B_1| \cdots |B_k|/n$ solutions to the equation $x_1 x_2 \cdots x_k = 1$ with $x_i \in A_i$ for all i, then it is possible to remove at most $\varepsilon|B_i|$ elements from each set A_i so as to obtain sets A_i' for which there are no solutions to $x_1 x_2 \cdots x_k = 1$ with $x_i \in A_i'$ for all i.*

This result easily implies a Roth-type theorem in quite sparse pseudorandom subsets of a group. We say that a subset B of a group G is (ε, k)-*Roth* if, for all integers $a_1, \ldots, a_k$ which satisfy $a_1 + \cdots + a_k = 0$ and $\gcd(a_i, |G|) = 1$ for $1 \leq i \leq k$, every subset $A \subseteq B$ which has no nontrivial solution to $x_1^{a_1} x_2^{a_2} \cdots x_k^{a_k} = 1$ has $|A| \leq \varepsilon|B|$.

Corollary 5.9 *For any natural number $k \geq 3$ and any $\varepsilon > 0$, there exists $c > 0$ such that the following holds. If G is a group of order n and B is a (p, β)-jumbled subset of G with $\beta \leq cp^{t_k}n$, then B is (ε, k)-Roth.*

Note that Roth's theorem on 3-term arithmetic progressions in dense sets of integers follows from the special case of this result with $B = G = \mathbb{Z}_n$, $k = 3$ and $a_1 = a_2 = 1$, $a_3 = -2$. The rather weak pseudorandomness condition in Corollary 5.9 shows that even quite sparse pseudorandom subsets of a group have the Roth property.

6 Further topics

6.1 The Erdős-Rothschild problem

A problem of Erdős and Rothschild [30] asks one to estimate the maximum number $h(n, c)$ such that every n-vertex graph with at least cn^2 edges, each of which is contained in at least one triangle, must contain an edge that is in at least $h(n, c)$ edges. Here, and throughout this subsection, we assume $c > 0$ is a fixed absolute constant. The fact that $h(n, c)$ tends to infinity already follows from the triangle removal lemma.[20]

To see this, suppose that G is an n-vertex graph with cn^2 edges such that every edge is in at least one and at most $h := h(n, c)$ triangles. The total number of triangles in G is at most $hcn^2/3$. Therefore, if h does not tend to infinity, the triangle removal lemma tells us that there is a collection E of $o(n^2)$ edges such that every triangle contains at least one of them. Since each edge in G is in at least one triangle, we know that there

---

[20]Even the statement that $h(n, c) > 1$ is already enough to imply Roth's theorem.

are at least $cn^2/3$ triangles. It follows that some edge in E is contained in at least $\omega(1)$ edges.

Using Fox's bound [38] for the triangle removal lemma, this implies that $h(n,c) \geq e^{a \log^* n}$, where $\log^* n$ is the iterated logarithm. This is defined by $\log^* x = 0$ if $x \leq 1$ and $\log^* x = \log^*(\log x) + 1$ otherwise. This improves on the bound $h(n,c) \geq (\log^* n)^a$ which follows from Ruzsa and Szemerédi's original proof of the triangle removal lemma.

On the other hand, Alon and Trotter (see [31]) showed that for any positive $c < \frac{1}{4}$ there is $c' > 0$ such that $h(n,c) < c'\sqrt{n}$. The condition $c < \frac{1}{4}$ is easily seen to be best possible since any n-vertex graph with more than $n^2/4$ edges contains an edge in at least $n/6$ triangles [29, 63]. Erdős conjectured that perhaps this behaviour is correct. That is, that for any positive $c < \frac{1}{4}$ there exists $\varepsilon > 0$ such that $h(n,c) > n^\varepsilon$ for all sufficiently large n. This was recently disproved by Fox and Loh [39] as follows.

Theorem 6.1 *For n sufficiently large, there is an n-vertex graph with $\frac{n^2}{4}(1 - e^{-(\log n)^{1/6}})$ edges such that every edge is in a triangle and no edge is in more than $n^{14/\log\log n}$ triangles.*

To give some idea of the construction, consider a tripartite graph between sets A, B and C, each of which is a copy of a lattice cube with appropriate sidelength r and dimension d. We join points in A and B if their distance is close to the expected distance between random points in A and B. By concentration, this implies that the density of edges between A and B is close to 1. We join points in C to points in A or B if their distance is close to half the expected distance. It is not hard to see that every edge between A and B is then contained in few triangles. At the same time, every edge will be in at least one triangle, as can be seen by considering the midpoint of any two connected points a and b. This yields a construction with roughly $\frac{n^2}{9}$ edges but the result of Fox and Loh may be obtained by shrinking the vertex set C (or blowing up A and B) in an appropriate fashion.

6.2 Induced matchings

Call a graph $G = (V,E)$ an (r,t)-*Ruzsa-Szemerédi graph* $((r,t)$-RS graph for short) if its edge set can be partitioned into t induced matchings in G, each of size r. The total number of edges of such a graph is rt. The most interesting problem concerns the existence of such graphs when r and t are both relatively large as a function of the number of vertices. The construction of Ruzsa and Szemerédi [94] using Behrend's construction demonstrates that such a graph on n vertices exists with $r = e^{-c\sqrt{\log n}}n$

and $t = n/3$. The Ruzsa-Szemerédi result on the $(6,3)$-problem is equivalent to showing that no (r,t)-RS graph on n vertices exists with r and t linear in n.

For r linear in the number n of vertices, it is still an open problem if there exists an (r,t)-RS graph with $t = n^\varepsilon$. The best known construction in this case, due to Fischer et al. [37], is an example with $r = n/3$ and $t = n^{c/\log\log n}$. However, for $r = n^{1-o(1)}$, substantial progress was made recently by Alon, Moitra and Sudakov [8] by extending ideas used in the construction of Fox and Loh [39] discussed in the previous subsection. They give a construction of n-vertex graphs with $rt = (1-o(1))\binom{n}{2}$ and $r = n^{1-o(1)}$. That is, there are nearly complete graphs, with edge density $1 - o(1)$, such that its edge set can be partitioned into large induced matchings, each of order $n^{1-o(1)}$. They give several applications of this construction to combinatorics, complexity theory and information theory.

6.3   Testing small graphs

A *property* of graphs is a family of graphs closed under isomorphism. A graph G on n vertices is ε-*far* from satisfying a property P if no graph which can be constructed from G by adding and/or removing at most εn^2 edges satisfies P. An ε-*tester* for P is a randomized algorithm which, given the quantity n and the ability to make queries whether a desired pair of vertices spans an edge in G, distinguishes with probability at least $2/3$ between the case that G satisfies P and the case that G is ε-far from satisfying P. Such an ε-tester is a *one-sided ε-tester* if when G satisfies P the ε-tester outputs that this is the case. The property P is called *testable* if, for every fixed $\varepsilon > 0$, there exists a one-sided ε-tester for P whose total number of queries is bounded only by a function of ε which is independent of the size of the input graph. This means that the running time of the algorithm is also bounded by a function of ε only and is independent of the input size. We measure query-complexity by the number of vertices sampled, assuming we always examine all edges spanned by them. The infinite removal lemma, Theorem 3.11, of Alon and Shapira [12] shows that every hereditary graph property, that is, a graph property closed under taking induced subgraphs, is testable. Many of the best studied graph properties are hereditary.

If the query complexity of an ε-tester is polynomial in ε^{-1}, we say that the property is *easily testable*. It is an interesting open problem to characterize the easily testable hereditary properties. Alon [3] considered the case where $P = P_H$ is the property that the graph does not contain H as a subgraph. He showed that P_H is easily testable if and only if H is bipartite. Alon and Shapira [10] considered the case where $P = P_H^*$ is the property that the graph does not contain H as an *induced* subgraph. They showed that for any graph H except for the path with at most four vertices, the cycle of length four and their complements, the property P_H^* is not easily testable. The problem of determining whether the property P_H^* is easily testable for the path with four vertices or the cycle of length four (or equivalently its complement) was left open. The case where H is a path with four vertices was recently shown to be easily testable by Alon and Fox [7]. The case where H is a cycle of length four is still open. Alon and Fox also showed that if P is the family of perfect graphs, then P is not easily testable and, in a certain sense, testing for P is at least as hard as testing triangle-freeness.

6.4 Local repairability

The standard proof of the regularity lemma contains a procedure for turning a graph which is almost triangle-free into a graph which is triangle-free. We simply delete the edges between all vertex sets of low density and between all vertex sets which do not form a regular pair. This procedure can be made more explicit still by using an algorithmic version of the regularity lemma [4].

A surprising observation of Austin and Tao [13] is that this repair procedure can be determined in a local fashion. They show that for any graph H and any $\varepsilon > 0$ there exists $\delta > 0$ and a natural number m such that if G is a graph containing at most $\delta n^{v(H)}$ copies of H then there exists a set A of size at most m such that G may be made H-free by removing at most εn^2 edges and the decision of whether to delete a given edge uv may be determined solely by considering the restriction of G to the set $A \cup \{u, v\}$.[21]

The key point, first observed by Ishigami [60], is that the regular partition can be determined in a local fashion by randomly selecting vertex neighborhoods to create the partition. Since a finite set of points determine the partition, this may in turn be used to create a local modification rule which results in an H-free graph.

Similar ideas may also be applied to show that any hereditary graph property, including the property of being induced H-free, is locally repairable in the same sense. This again follows from the observation that random neighborhoods can be used to construct the partitions arising in the strong regularity lemma.

Surprisingly, Austin and Tao show that, even though all hereditary hypergraph properties are testable, there are hereditary properties which

---

[21]Strictly speaking, Austin and Tao [13] consider two forms of local repairability. Here we are considering only the weak version.

are not locally repairable. On the other hand, they show that many natural hypergraph properties, including the property of being $\mathcal{H}$-free, are locally repairable.

6.5   Linear hypergraphs

A *linear hypergraph* is a hypergraph where any pair of edges overlap in at most one vertex. For this special class of hypergraphs, it is not necessary to apply the full strength of hypergraph regularity to prove a corresponding removal lemma [67]. Instead, a straightforward analogue of the usual regularity lemma is sufficient. This results in bounds for δ^{-1} in the linear hypergraph removal lemma which are of tower-type in a power of ε^{-1}.

While this is already a substantial improvement on general hypergraphs, where the best known bounds are Ackermannian,[22] it can be improved further by using the ideas of [38]. This results in a bound of the form $T(a_{\mathcal{H}} \log \varepsilon^{-1})$.

A similar reduction does not exist for induced removal of linear hypergraphs. Because we need to consider all edges, whether present or not, between the vertices of the hypergraph, we must apply the full strength of the strong hypergraph regularity lemma. This results in Ackermannian bounds.

It is plausible that an extension of the methods of Section 2.2 could be used to give a primitive recursive, or even tower-type, bound for hypergraph removal. We believe that such an improvement would be of great interest, not least because it would give the first primitive recursive bound for the multidimensional extension of Szemerédi's theorem. Such an improvement would also be likely to lead to an analogous improvement of the bounds for induced hypergraph removal.

Acknowledgements

The authors would like to thank Noga Alon, Zoltan Füredi, Vojta Rödl and Terry Tao for helpful comments regarding the history of the removal lemma. They would also like to thank Asaf Shapira and the anonymous referee for helpful comments.

---

[22]We have already seen two levels of the Ackermann function, the tower function and the wowzer function. Generally, the kth level is defined by $A_k(1) = 2$ and $A_k(i + 1) = A_{k-1}(A_k(i))$. Taking $A_1(i) = 2^i$, we see that $A_2(i) = T(i)$ and $A_3(i) = W(i)$. The upper bound on δ^{-1} in the k-uniform hypergraph removal lemma given by the hypergraph regularity proofs are of the form $A_k(\varepsilon^{-O(1)})$ or worse.

References

[1] M. Ajtai and E. Szemerédi, *Sets of lattice points that form no squares,* Stud. Sci. Math. Hungar. **9** (1974), 9–11.

[2] N. Alon, *Explicit Ramsey graphs and orthonormal labellings,* Electron. J. Combin. **1** (1994), R12, 8pp.

[3] N. Alon, *Testing subgraphs in large graphs,* Random Structures Algorithms **21** (2002), 359–370.

[4] N. Alon, R. A. Duke, H. Lefmann, V. Rödl and R. Yuster, *The algorithmic aspects of the regularity lemma,* J. Algorithms **16** (1994), 80–109.

[5] N. Alon, W. Fernandez de la Vega, R. Kannan and M. Karpinski, *Random sampling and approximation of MAX-CSPs,* J. Comput. System Sci. **67** (2003), 212–243.

[6] N. Alon, E. Fischer, M. Krivelevich and M. Szegedy, *Efficient testing of large graphs,* Combinatorica **20** (2000), 451–476.

[7] N. Alon and J. Fox, *Testing perfectness is hard,* submitted.

[8] N. Alon, A. Moitra and B. Sudakov, *Nearly complete graphs decomposable into large induced matchings and their applications,* in Proc. of STOC 2012, 1079–1090, to appear in *J. European Math. Soc.*

[9] N. Alon and A. Shapira, *Testing subgraphs in directed graphs,* J. Comput. System Sci. **69** (2004), 353–382.

[10] N. Alon and A. Shapira, *A characterization of easily testable induced subgraphs,* Combin. Probab. Comput. **15** (2006), 791–805.

[11] N. Alon and A. Shapira, *Every monotone graph property is testable,* in Proc. of STOC 2005, 128–137, SIAM J. Comput. (Special Issue on STOC '05) **38** (2008), 505–522.

[12] N. Alon and A. Shapira, *A characterization of the (natural) graph properties testable with one-sided error,* in Proc. of FOCS 2005, 429–438, SIAM J. Comput. (Special Issue on FOCS '05) **37** (2008), 1703–1727.

[13] T. Austin and T. Tao, *Testability and repair of hereditary hypergraph properties,* Random Structures Algorithms **36** (2010), 373–463.

[14] C. Avart, V. Rödl and M. Schacht, *Every monotone 3-graph property is testable*, SIAM J. Discrete Math. **21** (2007), 73–92.

[15] J. Balogh, R. Morris and W. Samotij, *Independent sets in hypergraphs*, submitted.

[16] F. Behrend, *On sets of integers which contain no three terms in arithmetic progression*, Proc. Nat. Acad. Sci. **32** (1946), 331–332.

[17] B. Bollobás, *Random graphs*, second edition, Cambridge Studies in Advanced Mathematics 73, Cambridge University Press, Cambridge, 2001.

[18] C. Borgs, J. T. Chayes, L. Lovász, V. T. Sós and K. Vesztergombi, *Convergent sequences of dense graphs I: subgraph frequencies, metric properties and testing*, Adv. Math. **219** (2008), 1801–1851.

[19] W. G. Brown, P. Erdős and V. T. Sós, *On the existence of triangulated spheres in 3-graphs, and related problems*, Period. Math. Hungar. **3** (1973), 221–228.

[20] P. Candela, *Developments at the interface between combinatorics and Fourier analysis*, PhD thesis, University of Cambridge, 2009.

[21] F. R. K. Chung, *A spectral Turán theorem*, Combin. Probab. Comput. **14** (2005), 755–767.

[22] F. R. K. Chung, R. L. Graham and R. M. Wilson, *Quasi-random graphs*, Combinatorica **9** (1989), 345–362.

[23] L. H. Clark, R. C. Entringer, J. E. McCanna and L. A. Székely, *Extremal problems for local properties of graphs*, in Combinatorial mathematics and combinatorial computing (Palmerston North, 1990), Australas. J. Combin. **4** (1991), 25–31.

[24] D. Conlon and J. Fox, *Bounds for graph regularity and removal lemmas*, Geom. Funct. Anal. **22** (2012), 1192–1256.

[25] D. Conlon and W. T. Gowers, *Combinatorial theorems in sparse random sets*, submitted.

[26] D. Conlon, W. T. Gowers, W. Samotij and M. Schacht, *On the KLR conjecture in random graphs*, submitted.

[27] D. Conlon, J. Fox and Y. Zhao, *Extremal results in sparse pseudorandom graphs*, submitted.

[28] R. A. Duke, H. Lefmann and V. Rödl, *A fast approximation algorithm for computing the frequencies of subgraphs in a given graph,* SIAM J. Comput. **24** (1995), 598–620.

[29] C. S. Edwards, *A lower bound for the largest number of triangles with a common edge,* 1977, unpublished manuscript.

[30] P. Erdős, *Some problems on finite and infinite graphs,* in Logic and combinatorics (Arcata, Calif., 1985), 223–228, Contemp. Math. **65**, Amer. Math. Soc., Providence, RI, 1987.

[31] P. Erdős, *Some of my favourite problems in various branches of combinatorics,* in Combinatorics 92 (Catania, 1992), Matematiche (Catania) **47** (1992), 231–240.

[32] P. Erdős, P. Frankl and V. Rödl, *The asymptotic number of graphs not containing a fixed subgraph and a problem for hypergraphs having no exponent,* Graphs Combin. **2** (1986), 113–121.

[33] P. Erdős and A. Rényi, *On random graphs I,* Publ. Math. Debrecen **6** (1959), 290–297.

[34] P. Erdős and A. Rényi, *On the evolution of random graphs,* Magyar Tud. Akad. Mat. Kutató Int. Közl. **5** (1960), 17–61.

[35] P. Erdős and M. Simonovits, *A limit theorem in graph theory,* Studia Sci. Math. Hungar. **1** (1966), 51–57.

[36] P. Erdős and A. H. Stone, *On the structure of linear graphs,* Bull. Amer. Math. Soc. **52** (1946), 1087–1091.

[37] E. Fischer, I. Newman, S. Raskhodnikova, R. Rubinfeld and A. Samorodnitsky, *Monotonicity testing over general poset domains,* in Proceedings of the 2002 ACM Symposium on Theory of Computing, 474–483, ACM, New York, 2002.

[38] J. Fox, *A new proof of the graph removal lemma,* Ann. of Math. **174** (2011), 561–579.

[39] J. Fox and P. Loh, *On a problem of Erdős and Rothschild on edges in triangles,* to appear in Combinatorica.

[40] P. Frankl and Z. Füredi, *Exact solution of some Turán-type problems,* J. Combin. Theory Ser. A **45** (1987), 226–262.

[41] P. Frankl, R. L. Graham and V. Rödl, *On subsets of abelian groups with no 3-term arithmetic progression,* J. Combin. Theory Ser. A **45** (1987), 157–161.

[42] P. Frankl and V. Rödl, *Extremal problems on set systems,* Random Structures Algorithms **20** (2002), 131–164.

[43] E. Friedgut, V. Rödl and M. Schacht, *Ramsey properties of discrete random structures,* Random Structures Algorithms, **37** (2010), 407–436.

[44] A. Frieze and R. Kannan, *The regularity lemma and approximation schemes for dense problems,* in Proceedings of the 37th IEEE FOCS (1996), 12–20.

[45] A. Frieze and R. Kannan, *Quick approximation to matrices and applications,* Combinatorica **19** (1999), 175–220.

[46] Z. Füredi, *The maximum number of edges in a minimal graph of diameter 2,* J. Graph Theory **16** (1992), 81–98.

[47] Z. Füredi, *Extremal hypergraphs and combinatorial geometry,* in Proceedings of the International Congress of Mathematicians, Vol. 1, 2 (Zürich, 1994), 1343–1352, Birkhäuser, Basel, 1995.

[48] H. Furstenberg, *Ergodic behavior of diagonal measures and a theorem of Szemerédi on arithmetic progressions,* J. Analyse Math. **31** (1977), 204–256.

[49] H. Furstenberg and Y. Katznelson, *An ergodic Szemerédi theorem for commuting transformations,* J. Analyse Math. **34** (1978), 275–291.

[50] H. Furstenberg, Y. Katznelson and D. Ornstein, *The ergodic theoretical proof of Szemerédi's theorem,* Bull. Amer. Math. Soc. **7** (1982), 527–552.

[51] S. Gerke and A. Steger, *The sparse regularity lemma and its applications,* in Surveys in Combinatorics 2005, 227–258, London Math. Soc. Lecture Note Ser. 327, Cambridge Univ. Press, Cambridge, 2005.

[52] O. Goldreich, S. Goldwasser and D. Ron, *Property testing and its applications to learning and approximation,* J. ACM **45** (1998), 653–750.

[53] W. T. Gowers, *Lower bounds of tower type for Szemerédi's uniformity lemma,* Geom. Funct. Anal. **7** (1997), 322–337.

[54] W. T. Gowers, *A new proof of Szemerédi's theorem for arithmetic progressions of length four,* Geom. Funct. Anal. **8** (1998), 529–551.

[55] W. T. Gowers, *A new proof of Szemerédi's theorem,* Geom. Funct. Anal. **11** (2001), 465–588.

[56] W. T. Gowers, *Quasirandomness, counting and regularity for 3-uniform hypergraphs,* Combin. Probab. Comput. **15** (2006), 143–184.

[57] W. T. Gowers, *Hypergraph regularity and the multidimensional Szemerédi theorem,* Ann. of Math. **166** (2007), 897–946.

[58] B. Green, *A Szemerédi-type regularity lemma in abelian groups, with applications,* Geom. Funct. Anal. **15** (2005), 340–376.

[59] P. E. Haxell, Y. Kohayakawa and T. Łuczak, *Turán's extremal problem in random graphs: forbidding odd cycles,* Combinatorica **16** (1996), 107–122.

[60] Y. Ishigami, *A simple regularization of hypergraphs,* submitted.

[61] S. Janson, T. Łuczak and A. Ruciński, *Random graphs,* Wiley-Interscience Series in Discrete Mathematics and Optimization, Wiley-Interscience, New York, 2000.

[62] S. Kalyanasundaram and A. Shapira, *A wowzer-type lower bound for the strong regularity lemma,* Proc. London Math. Soc., to appear.

[63] N. Khadžiivanov and V. Nikiforov, *Solution of a problem of P. Erdős about the maximum number of triangles with a common edge in a graph,* C. R. Acad. Bulgare Sci. **32** (1979), 1315–1318.

[64] Y. Kohayakawa, *Szemerédi's regularity lemma for sparse graphs,* in Foundations of computational mathematics (Rio de Janeiro, 1997), Springer, Berlin, 1997, 216–230.

[65] Y. Kohayakawa, T. Łuczak and V. Rödl, *Arithmetic progressions of length three in subsets of a random set,* Acta Arith. **75** (1996), 133–163.

[66] Y. Kohayakawa, T. Łuczak and V. Rödl, *On K^4-free subgraphs of random graphs,* Combinatorica **17** (1997), 173–213.

[67] Y. Kohayakawa, B. Nagle, V. Rödl and M. Schacht, *Weak regularity and linear hypergraphs,* J. Combin. Theory Ser. B **100** (2010), 151–160.

[68] Y. Kohayakawa, B. Nagle, V. Rödl, J. Skokan and M. Schacht, *The hypergraph regularity method and its applications,* Proc. Natl. Acad. Sci. USA **102** (2005), 8109–8113.

[69] Y. Kohayakawa, V. Rödl, M. Schacht and J. Skokan, *On the triangle removal lemma for subgraphs of sparse pseudorandom graphs,* in An Irregular Mind (Szemerédi is 70), Bolyai Society Math. Studies 21, Springer, 2010, 359–404.

[70] D. Král', O. Serra and L. Vena, *A removal lemma for linear systems over finite fields,* in Sixth conference on discrete mathematics and computer science (Spanish), 417–423, Univ. Lleida, Lleida, 2008.

[71] D. Král', O. Serra and L. Vena, *A combinatorial proof of the removal lemma for groups,* J. Combin. Theory Ser. A **116** (2009), 971–978.

[72] D. Král', O. Serra and L. Vena, *A removal lemma for systems of linear equations over finite fields,* Israel J. Math. **187** (2012), 193–207.

[73] D. Král', O. Serra and L. Vena, *On the removal lemma for linear systems over abelian groups,* European J. Combin. **34** (2013), 248–259.

[74] M. Krivelevich and B. Sudakov, *Pseudo-random graphs,* in More sets, graphs and numbers, Bolyai Soc. Math. Stud. 15, Springer, Berlin, 2006, 199–262.

[75] L. Lovász and B. Szegedy, *Szemerédi's lemma for the analyst,* Geom. Funct. Anal. **17** (2007), 252–270.

[76] L. Lovász and B. Szegedy, *Testing properties of graphs and functions,* Israel J. Math. **178** (2010), 113–156.

[77] T. Łuczak, *Randomness and regularity,* in International Congress of Mathematicians, Vol. III, 899–909, Eur. Math. Soc., Zürich, 2006.

[78] B. Nagle and V. Rödl, *Regularity properties for triple systems,* Random Structures Algorithms **23** (2003), 264–332.

[79] B. Nagle, V. Rödl and M. Schacht, *The counting lemma for regular k-uniform hypergraphs,* Random Structures Algorithms **28** (2006), 113–179.

[80] Y. Peng, V. Rödl, and A. Ruciński, *Holes in graphs,* Electron. J. Combin. **9** (2002), R1, 18pp.

[81] Y. Peng, V. Rödl and J. Skokan, *Counting small cliques in 3-uniform hypergraphs,* Combin. Probab. Comput. **14** (2005), 371–413.

[82] D. H. J. Polymath, *A new proof of the density Hales-Jewett theorem,* Ann. of Math. **175** (2012), 1283–1327.

[83] F. P. Ramsey, *On a problem of formal logic,* Proc. London Math. Soc. **30** (1930), 264–286.

[84] V. Rödl and A. Ruciński, *Lower bounds on probability thresholds for Ramsey properties,* in Combinatorics, Paul Erdős is eighty, Vol. 1, 317–346, Bolyai Soc. Math. Stud., János Bolyai Math. Soc., Budapest, 1993.

[85] V. Rödl and A. Ruciński, *Threshold functions for Ramsey properties,* J. Amer. Math. Soc. **8** (1995), 917–942.

[86] V. Rödl and M. Schacht, *Regular partitions of hypergraphs: regularity lemmas,* Combin. Probab. Comput. **16** (2007), 833–885.

[87] V. Rödl and M. Schacht, *Generalizations of the removal lemma,* Combinatorica **29** (2009), 467–501.

[88] V. Rödl and M. Schacht, *Regularity lemmas for graphs,* in Fete of Combinatorics and Computer Science, Bolyai Soc. Math. Stud. **20**, Springer, 2010, 287–325.

[89] V. Rödl and J. Skokan, *Regularity lemma for uniform hypergraphs,* Random Structures Algorithms **25** (2004), 1–42.

[90] V. Rödl and J. Skokan, *Counting subgraphs in quasi-random 4-uniform hypergraphs,* Random Structures Algorithms **26** (2005), 160–203.

[91] V. Rödl and J. Skokan, *Applications of the regularity lemma for uniform hypergraphs,* Random Structures Algorithms **28** (2006), 180–194.

[92] K. F. Roth, *On certain sets of integers,* J. London Math. Soc. **28** (1953), 104–109.

[93] R. Rubinfield and M. Sudan, *Robust characterization of polynomials with applications to program testing,* SIAM J. Comput. **25** (1996), 252–271.

[94] I. Z. Ruzsa and E. Szemerédi, *Triple systems with no six points carrying three triangles,* in Combinatorics (Keszthely, 1976), Coll. Math. Soc. J. Bolyai 18, Volume II, 939–945.

[95] W. Samotij, *Stability results for random discrete structures,* to appear in Random Structures Algorithms.

[96] T. Sanders, *On Roth's theorem on progressions,* Ann. of Math. **174** (2011), 619–636.

[97] T. Sanders, *On the Bogolyubov-Ruzsa lemma,* Anal. PDE **5** (2012), 627–655.

[98] D. Saxton and A. Thomason, *Hypergraph containers,* submitted.

[99] M. Schacht, *Extremal results for random discrete structures,* submitted.

[100] T. Schoen and I. Shkredov, *Roth's theorem in many variables,* submitted.

[101] A. Shapira, *Green's conjecture and testing linear-invariant properties,* in Proceedings of the 2009 ACM International Symposium on Theory of Computing, 159–166, ACM, New York, 2009.

[102] A. Shapira, *A proof of Green's conjecture regarding the removal properties of sets of linear equations,* J. London Math. Soc. **81** (2010), 355–373.

[103] I. Shkredov, *On a generalization of Szemerédi's theorem,* Proc. London Math. Soc. **93** (2006), 723–760.

[104] M. Simonovits, *A method for solving extremal problems in graph theory, stability problems,* in Theory of graphs (Proc. Colloq. Tihany,1966), Academic Press, New York, 1968, 279–319.

[105] J. Solymosi, *Note on a generalization of Roth's theorem,* in Discrete and computational geometry, Algorithms Combin. Vol. 25, Springer, 2003, 825–827.

[106] J. Solymosi, *A note on a question of Erdős and Graham,* Combin. Probab. Comput. **13** (2004), 263–267.

[107] J. Solymosi, *Regularity, uniformity, and quasirandomness,* Proc. Natl. Acad. Sci. USA **102** (2005), 8075–8076.

[108] B. Sudakov, T. Szabó and V. H. Vu, *A generalization of Turán's theorem*, J. Graph Theory **49** (2005), 187–195.

[109] B. Szegedy, *The symmetry preserving removal lemma*, Proc. Amer. Math. Soc. **138** (2010), 405–408.

[110] E. Szemerédi, *Integer sets containing no k elements in arithmetic progression*, Acta Arith. **27** (1975), 299–345.

[111] E. Szemerédi, *Regular partitions of graphs*, in Colloques Internationaux CNRS 260 – Problèmes Combinatoires et Théorie des Graphes, Orsay, (1976), 399–401.

[112] T. Tao, *Szemerédi's regularity lemma revisited*, Contrib. Discrete Math. **1** (2006), 8–28.

[113] T. Tao, *A variant of the hypergraph removal lemma*, J. Combin. Theory Ser. A **113** (2006), 1257–1280.

[114] A. Thomason, *Pseudorandom graphs*, in Random graphs '85 (Poznań, 1985), 307–331, North-Holland Math. Stud. 144, North-Holland, Amsterdam, 1987.

[115] A. Thomason, *Random graphs, strongly regular graphs and pseudorandom graphs*, Surveys in combinatorics 1987 (New Cross, 1987), 173–195, London Math. Soc. Lecture Note Ser. 123, Cambridge Univ. Press, Cambridge, 1987.

[116] P. Turán, *Eine Extremalaufgabe aus der Graphentheorie*, Mat. Fiz. Lapok **48** (1941), 436–452.

Mathematical Institute
University of Oxford
Oxford OX1 3LB, UK
david.conlon@maths.ox.ac.uk

Department of Mathematics
MIT
Cambridge, MA 02139-4307, USA
fox@math.mit.edu

The geometry of covering codes: small complete caps and saturating sets in Galois spaces

Massimo Giulietti

Abstract

Complete caps and saturating sets in projective Galois spaces are the geometrical counterpart of linear codes with covering radius 2. The smaller the cap/saturating set, the better the covering properties of the code. In this paper we survey the state of the art of the research on these geometrical objects, with particular emphasis on the recent developments and on the connections with algebraic curves over finite fields.

1 Introduction

Galois spaces, that is affine and projective spaces of dimension $N \geq 2$ defined over a finite (Galois) field $\mathbb{F}_q$, are well known to be rich in nice geometric, combinatorial and group-theoretic properties that have also found wide and relevant applications in several branches of combinatorics, especially to design theory and graph theory, as well as in more practical areas, notably coding theory and cryptography.

The systematic study of Galois spaces was initiated in the late 1950's by the pioneering work of B. Segre [77]. The trilogy [53, 55, 58] covers the general theory of Galois spaces including the study of objects which are linked to linear codes. Typical such objects are plane arcs and their generalizations, especially caps, saturating sets and arcs in higher dimensions, whose code-theoretic counterparts are distinguished types of error-correcting and covering linear codes, such as MDS codes. Their investigation has received a great stimulus from coding theory, especially in the last decades; see the survey papers [56, 57].

An important issue in this context is to ask for explicit constructions of small complete caps and saturating sets. A *cap* in a Galois space is a set of points no three of which are collinear. A *saturating set* is a set of points whose secants (lines through at least two points of the set) cover the whole space. A cap is *complete* if it is also a saturating set. From these geometric objects there arise linear codes which turn out to have very good covering properties, provided that the size of the set of points is small with respect to the dimension N and the order q of the ambient space; see Section 2.

The aim of this paper is to provide a survey on the state of the art of the research on small complete caps and saturating sets, with particular emphasis on recent developments. In the last decade a number of new results have appeared, and new notions have emerged as powerful tools in dealing with the covering problem, including bicovering arcs, translation caps, and (m)-saturating sets. Also, although caps and saturating sets are rather combinatorial objects, constructions and proofs sometimes rely heavily on concepts and results from Algebraic Geometry in positive characteristic.

The paper is organized as follows. In Section 2 we explain the close relationship between linear codes with covering radius 2 and caps and saturating sets in Galois spaces. We also present in more detail the notion of a multiple covering of the farthest-off points, and relate it to that of an (m)-saturating set.

In Section 3 we summarize without proofs some of the material on algebraic curves and function fields that will be relevant to our proofs.

In Section 4 we deal with caps and saturating sets in the plane. A cap in a Galois plane is often called a *plane arc*. The theory of plane arcs is well developed and quite rich in constructions (see [54, 56, 57, 86, 87] and the references therein, as well as the monograph [55]); however, we decided to focus on plane arcs arising from cubic plane curves, which will be relevant for the recursive constructions of small complete caps presented in the subsequent sections. For arcs contained in elliptic curves, a new description relying on the properties of the Tate-Lichtenbaum pairing is given. Also, a construction by Szőnyi of complete arcs contained in cuspidal cubics is generalized.

Section 5 is about three-dimensional spaces. The even order case was substantially settled by Segre himself in 1959 [75], whereas the problem of constructing complete caps of size close to the trivial lower bound is still wide open for odd q's. Here we describe how a construction by Pellegrino [71] can give rise to very small complete caps in the odd order case.

In Section 6 we deal with inductive methods. In particular, we discuss under what conditions the product construction and the blowing-up construction preserve the completeness of a cap. Then the notions of a translation cap in the even order case, and that of a bicovering arc in the odd order case, come into play as powerful tools to construct small complete caps in higher dimensions. Davydov's recursive construction of saturating sets is also described.

In the last section we discuss small (m)-saturating sets in Galois planes.

2   Covering codes with radius 2

Let $\mathbb{F}_q$ be the finite field with q elements. A q-ary linear code $\mathbf{C}$ of length n and dimension k is a k-dimensional linear subspace of $\mathbb{F}_q^n$. The number of non-zero positions in a vector $\mathbf{v} \in \mathbb{F}_q^n$ is called the Hamming weight $w(\mathbf{v})$ of $\mathbf{v}$. For $\mathbf{v}_1, \mathbf{v}_2 \in \mathbb{F}_q^n$ the Hamming distance $d(\mathbf{v}_1, \mathbf{v}_2)$ is the weight $w(\mathbf{v}_1 - \mathbf{v}_2)$. The Hamming sphere centered in $\mathbf{x} \in \mathbb{F}_q^n$ and with radius ρ, that is, the set of vectors $\mathbf{v} \in \mathbb{F}_q^n$ with $d(\mathbf{x}, \mathbf{v}) \leq \rho$, is denoted by $S(\mathbf{x}, \rho)$.

The minimum distance of the code $\mathbf{C}$ is

$$d(\mathbf{C}) := \min\{w(\mathbf{x}) \mid \mathbf{x} \in \mathbf{C}, \, \mathbf{x} \neq 0\},$$

and a q-ary linear code of length n, dimension k and minimum distance d is called an $[n, k, d]_q$-code. The distance $d(\mathbf{v}, \mathbf{C})$ of a vector $\mathbf{v} \in \mathbb{F}_q^n$ from $\mathbf{C}$ is the minimum distance $d(\mathbf{v}, \mathbf{x})$ as $\mathbf{x}$ ranges over the words in $\mathbf{C}$. An $[n, k, d]_q$-code is said to be t-error correcting, where t is the integer part of $(d-1)/2$.

The covering radius of $\mathbf{C}$ is the minimum integer $R(\mathbf{C})$ such that for any vector $\mathbf{v} \in \mathbb{F}_q^n$ there exists $\mathbf{x} \in \mathbf{C}$ with $d(\mathbf{v}, \mathbf{x}) \leq R(\mathbf{C})$. An $[n, k, d]_q$-code with covering radius R is sometimes called an $[n, k, d]_q R$-code. Clearly, $R(\mathbf{C}) \geq t$ holds, and when equality is attained the code $\mathbf{C}$ is said to be perfect.

One of the parameters which characterizes the covering quality of an $[n, k, d]_q R$-code $\mathbf{C}$ is its *covering density* $\mu(\mathbf{C})$, first introduced in [19] as the average number of codewords at distance less than or equal to R from a vector in $\mathbb{F}_q^n$:

$$\mu(\mathbf{C}) = \frac{V_q(n, R(\mathbf{C}))}{q^{n-k}},$$

where

$$V_q(n, j) = \sum_{i=0}^{j} \binom{n}{i}(q-1)^i$$

is the size of a sphere of radius j in $\mathbb{F}_q^n$.

The covering density $\mu(\mathbf{C})$ is always greater than or equal to 1, and equality holds precisely when $\mathbf{C}$ is perfect. For fixed codimension $n-k$ and covering radius R, the shorter the code the better its covering density. This is why one of the central problems concerning covering codes is that of determining the minimal length n for which there is an $[n, n-r, d]_q R$-code with given r, q, d, and R; see the seminal paper [15], as well as [21, 22, 25, 26, 30, 31, 32, 33, 34, 35, 36, 37, 39, 44]. Here, such minimal length is denoted as $l(r, R; q)_d$, whereas $l(r, R; q)$ stands for the the minimum

$l(r, R; q)_d$ as d ranges over the set of possible integers for which an $[n, n - r, d]_q R$-code exists.

In this work we will restrict our attention to codes with covering radius $R = 2$. For these codes, $d \leq 5$ holds. When $d > 3$, the columns of a parity check matrix of an $[n, n - r, d]_q 2$-code can be considered as points of a complete cap of size n in $\mathrm{PG}(r - 1, q)$, the Galois projective space of dimension $r - 1$ over $\mathbb{F}_q$; see [56]. The only caps associated to a code with $d = 5$ are the complete 5-cap in $\mathrm{PG}(3, 2)$ giving rise to a binary $[5, 1, 5]$-code, and the complete 11-cap in $\mathrm{PG}(4, 3)$ corresponding to the Golay $[11, 6, 5]$-code over $\mathbb{F}_3$; see e.g. [44]. The case $d = 3$ corresponds to saturating sets of size n in $\mathrm{PG}(r - 1, q)$ containing 3 collinear points. Clearly,

$$l(r, 2; q)_3 \leq l(r, 2; q)_4 + 1$$

holds. Finally, if $\mathbf{C}$ is an $[n, n - r, d]_q 2$-code with $d \leq 2$, then $\mathbf{C}$ can be shortened by removing proportional columns from its parity check matrix. This increases the minimum distance without affecting the covering radius. Thus,

$$l(r, 2; q)_1 = l(r, 2; q)_2 = l(r, 2; q)_3 + 1.$$

It is easily seen that a lower bound on $l(r, 2; q)_3$ is $\sqrt{2}q^{r/2}$. In geometrical terms, any saturating set in an N-dimensional (affine or projective) Galois space over $\mathbb{F}_q$ has at least

$$\sqrt{2}q^{(N-1)/2} \tag{2.1}$$

points. Throughout the paper, (2.1) will be referred to as the *trivial lower bound* on the size of a saturating set. Clearly, this lower bound holds for complete caps as well.

If $\mathrm{AG}(r - 1, q)$, the Galois affine space of dimension $r - 1$ over $\mathbb{F}_q$, is embedded in $\mathrm{PG}(r - 1, q)$, then a complete cap K in $\mathrm{AG}(r - 1, q)$ can be viewed as a cap in $\mathrm{PG}(r - 1, q)$. The corresponding $[n, n - r, 4]$-code has covering radius $R = 2$ if and only if K is complete in $\mathrm{PG}(r - 1, q)$ as well. However, it is worth pointing out that if this does not happen then the code has still good covering properties, as the number of words at distance greater than 2 from the code is less then $\frac{1}{q}$ of the total number of words in $\mathbb{F}_q^n$; see [46].

2.1   Multiple coverings of the farthest-off points

For a code $\mathbf{C}$ with covering radius R it is sometimes useful that for every word $\mathbf{v}$ at distance R from $\mathbf{C}$ there is more than one codeword in the Hamming sphere $S(\mathbf{v}, R)$. An $[n, k, d]_q R$-code $\mathbf{C}$ is said to be a linear (n, q^k, R, m) *multiple covering of the farthest-off points* (MCF for short) if for each $\mathbf{v} \in \mathbb{F}_q^n$ with $d(\mathbf{v}, \mathbf{C}) = R$ the size of $S(\mathbf{v}, R) \cap \mathbf{C}$ is at least m. One motivation for studying MCF codes arises from the generalized football pool problem; see e.g [50, 52, 68] and the references therein. Results on MCF codes, mostly concerning the binary and the ternary cases, can be found in [49, 51, 59, 60, 72, 73].

For a q-ary linear (n, q^k, R, m) MCF code $\mathbf{C}$, a natural parameter to consider is the average number η of codewords belonging to $S(\mathbf{v}, R) \cap \mathbf{C}$, where $\mathbf{v}$ is a fixed element in $\mathbb{F}_q^n$ with distance R from $\mathbf{C}$. Clearly, $\eta \geq m$ holds and equality is attained precisely when each $\mathbf{v} \in \mathbb{F}_q^n$ with $d(\mathbf{x}, \mathbf{C}) = R$ belongs to exactly m spheres centered on codewords. The m-density $\mu(\mathbf{C}, m)$ of $\mathbf{C}$ is, by definition, the ratio η/m. Clearly, $\mu(\mathbf{C}, m) \geq 1$ holds. If the minimum distance d of $\mathbf{C}$ is at least $2R - 1$, then it is easily seen that

$$\mu(\mathbf{C}, m) \leq \frac{\binom{n}{R(\mathbf{C})} \cdot (q-1)^{R(\mathbf{C})}}{m(q^{n-k} - V_q(n, R(\mathbf{C}) - 1))}.$$

Assume that $\mathbf{C}$ is a q-ary linear (n, q^k, R, m) MCF code with $R = 2$ and $d \geq 3$. Then a precise formula for $\mu(\mathbf{C}, m)$ is

$$\mu(\mathbf{C}, m) = \frac{\binom{n}{2} \cdot (q-1)^2 - 3A_3}{m(q^{n-k} - 1 - n(q-1))},$$

where A_3 is the number of codewords in $\mathbf{C}$ with weight 3. Interestingly, $\mathbf{C}$ has a nice geometrical counterpart.

Definition 2.1 A proper subset S of $\mathrm{PG}(N, q)$ is said to be (m)-saturating if

- S generates $\mathrm{PG}(N, q)$, and

- for every point Q in $\mathrm{PG}(N, q) \setminus S$, the number of secants of S through Q is at least m, counted with multiplicity.

Here, the multiplicity of a secant $\mathcal{L}$ through Q is computed as $\binom{\#(\mathcal{L} \cap S)}{2}$.

For a fixed positive integer N, it is easy to see that (m)-saturating sets in $\mathrm{PG}(N, q)$ of size n and q-ary linear $(n, q^{n-N-1}, 2, m)$ MCF codes with $d \geq 3$ are equivalent objects. For the code $\mathbf{C}$ associated to an (m)-saturating set of size n we have that

$$\mu(\mathbf{C}, m) = \frac{\frac{n-1}{2}(q-1) - \frac{3B_3(S)}{n}}{m \cdot \left(\frac{\#\mathrm{PG}(N,q)}{n} - 1\right)},$$

where

$$B_3(S) = \#\{\text{triples of collinear points in } S\}.$$

Equivalently,

$$\mu(\mathbf{C}, m) = \frac{1}{m \cdot \#(\mathrm{PG}(N,q) \setminus S)} \cdot$$

$$\sum_{Q \in \mathrm{PG}(N,q) \setminus S} (\text{number of secants of } S \text{ through } Q),$$

where the number of secants of S through Q is counted with multiplicity.

The code $\mathbf{C}$ associated to an (m)-saturating set S in $\mathrm{PG}(N,q)$ then has optimal m-density $\mu(\mathbf{C}, m) = 1$ if and only if

$$\binom{\#S}{2}(q-1) - 3B_3(S) = m \cdot \#(\mathrm{PG}(N,q) \setminus S),$$

or, equivalently, if for every point Q in $\mathrm{PG}(N,q) \setminus S$ the number of secants of S through Q is precisely m (counted with multiplicity).

Examples of q-ary linear $(n, q^k, 2, m)$ MCF codes with optimal m-density are 1-error correcting codes whose dual is a 2-weight code; see [88, Definition 8.1] and [17, Corollary 4.3]. The strong connection linking 2-weight codes to both finite geometry and graph theory is described in [17].

The following generalization of the length function $l(r, 2; q)$ is considered:

$$l(r, 2, m; q) := \min\{n \mid \text{there exists a linear } (n, q^{n-r}, 2, m) \text{ MCF code}\}.$$

Equivalently, $l(r, 2, m; q)$ equals the minimum size of an (m)-saturating set in the projective space $\mathrm{PG}(r-1, q)$.

3   Irreducible algebraic curves and function fields

Throughout this section, by a curve we will mean a projective absolutely irreducible algebraic curve defined over an algebraically closed field $\mathbb{K}$. A function field over a field L is an extension F of L such that F is a finite algebraic extension of $L(\alpha)$, with α transcendental over L. For basic definitions on function fields we refer to [80].

To a curve $\mathcal{C}$ one can associate a function field over $\mathbb{K}$, namely the field of rational functions of $\mathcal{C}$, which will be denoted by $\mathbb{K}(\mathcal{C})$. If $\mathcal{C}$ is a plane curve with affine equation $f(X, Y) = 0$, then $\mathbb{K}(\mathcal{C})$ is the field $\mathbb{K}(\bar{x}, \bar{y})$ generated by $\mathbb{K}$ and by the rational functions $\bar{x}$ and $\bar{y}$ corresponding to the affine coordinates of the points in $\mathcal{C}$; $\bar{x}$ and $\bar{y}$ are algebraically dependent as $f(\bar{x}, \bar{y}) = 0$ holds.

A curve C is defined over $\mathbb{F}_q$ if the ideal of C is generated by polynomials with coefficients in $\mathbb{F}_q$. In this case, $\mathbb{F}_q(C)$ denotes the subfield of $\mathbb{K}(C)$ consisting of the rational functions defined over $\mathbb{F}_q$. This subfield is a function field over $\mathbb{F}_q$, and if C is a plane curve with affine equation $f(X, Y) = 0$ then $\mathbb{F}_q(C) = \mathbb{F}_q(\bar{x}, \bar{y})$.

The set of $\mathbb{F}_q$-rational points of C, that is the set of points whose coordinates lie in the base field $\mathbb{F}_q$, is denoted as $C(\mathbb{F}_q)$. A place γ of $\mathbb{K}(C)$ determines a unique point of C, called the center of γ. If C is non-singular, this correspondence is a bijection. When C is defined over $\mathbb{F}_q$, the q-Frobenius map Φ_q can be defined in $\mathbb{K}(C)$, and a place of $\mathbb{K}(C)$ left invariant by Φ_q is said to be $\mathbb{F}_q$-rational. The center of an $\mathbb{F}_q$-rational place lies in $C(\mathbb{F}_q)$.

Given a place γ of $\mathbb{K}(C)$ and a rational function $\alpha \in \mathbb{K}(C)$ such that γ is not a pole of α, the image of α by the residue class map with respect to γ is $\alpha(\gamma) \in \mathbb{K}$. When C is defined over $\mathbb{F}_q$, if $\alpha \in \mathbb{F}_q(C)$ and γ is $\mathbb{F}_q$-rational, then $\alpha(\gamma)$ belongs to $\mathbb{F}_q$.

To a function field F over $\mathbb{K}$ one can associate a curve C defined over $\mathbb{K}$ such that $\mathbb{K}(C)$ is $\mathbb{K}$-isomorphic to F. The genus of F as a function field coincides with the genus of C as a curve.

If F is the function field of a curve defined over $\mathbb{F}_q$, then by the Hasse-Weil bound, the number N of $\mathbb{F}_q$-rational places of F satisfies

$$q + 1 - 2g\sqrt{q} \leq N \leq q + 1 + 2g\sqrt{q}, \tag{3.1}$$

where g is the genus of F.

If F' is a finite extension of a function field F, then a place γ' of F' is said to be lying over a place γ of F if $\gamma \subset \gamma'$. This holds precisely when $\gamma = \gamma' \cap F$.

4   The plane case

4.1   Small complete arcs from cubic curves

Since the seminal work by B. Segre, complete arcs in Galois planes have played a prominent role in Finite Geometry. As mentioned in the introduction, a number of thorough surveys have been written on this topic, and it is not our purpose to touch all the aspects of the theory. Instead, we will mainly focus on arcs arising from irreducible cubic curves. There are a few reasons behind our choice. Firstly, complete arcs contained in cubic curves are the smallest complete arcs that have been explicitly constructed so far for arbitrarily large q. Secondly, arcs in cubics are the base for some recent inductive constructions of complete caps in higher dimensions that will be described in Section 6.2. Finally, although the theory of complete arcs contained in cubic curves was already well established in the late 80's, some significant refinements of the known results have been obtained very recently.

Throughout this section we assume that the characteristic of $\mathbb{F}_q$ is $p > 3$. Let $\mathcal{X}$ be an irreducible plane cubic curve defined over $\mathbb{F}_q$ and let P_0 be an $\mathbb{F}_q$-rational non-singular point of $\mathcal{X}$. It is a classical result from algebraic geometry that it is possible to define a binary operation $\oplus$ on the set G of the non-singular $\mathbb{F}_q$-rational points of $\mathcal{X}$, in such a way that $(G, \oplus)$ is an abelian group with neutral element P_0. The key property here is that three distinct points in G are collinear if and only if their sum is the neutral element in G. This allows to obtain arcs contained in G in a rather easy way.

Proposition 4.1 *Let H be a subgroup of G of index m with $(3, m) = 1$, and let P be a point in $G \setminus H$. Then the coset $K = H \oplus P$ is an arc.*

Proof If three distinct points P_1, P_2, P_3 in K are collinear, then $3K = 0$ holds in the factor group G/H. Then $3P \in H$. Taking into account that $(3, m) = 1$, this implies $P \in H$, a contradiction.   □

We will consider three types of irreducible plane cubics: (I) singular with an $\mathbb{F}_q$-rational cusp; (II) singular with an $\mathbb{F}_q$-rational node and at least one $\mathbb{F}_q$-rational inflection; (III) non-singular with at least one $\mathbb{F}_q$-rational inflection (or *elliptic*). A complete discussion and classification of cubic curves over a finite field can be found in [55]; see also [16].

If $\mathcal{X}$ is singular with a cusp, we can assume (up to projective equivalence) that the affine equation of $\mathcal{X}$ is $Y = X^3$. If the neutral element of $(G, \oplus)$ is chosen to be the affine point $(0, 0)$, then $(G, \oplus)$ is isomorphic to $(\mathbb{F}_q, +)$ via the map $v \mapsto (v, v^3)$.

If $\mathcal{X}$ is singular with a node and at least one $\mathbb{F}_q$-rational inflection, then a canonical equation for $\mathcal{X}$ is $XY = (X - 1)^3$. If the neutral element of $(G, \oplus)$ is chosen to be the affine point $(1, 0)$, then $(G, \oplus)$ is isomorphic to $(\mathbb{F}_q^*, \cdot)$ via the map $v \mapsto (v, (v - 1)^3/v)$.

Finally, if $\mathcal{X}$ is elliptic, we can assume that the affine equation of $\mathcal{X}$ is $Y^2 = X^3 + AX + B$ with $A, B \in \mathbb{F}_q$ such that $4A^3 + 27B^2 \neq 0$. The standard choice for the neutral element of $(G, \oplus)$ is the ideal point of the Y-axis, which will be denoted as O. As the genus of $\mathcal{X}$ is equal to 1, by the Hasse-Weil bound (3.1) the order of G lies in the interval $[q + 1 - 2\sqrt{q}, q + 1 + 2\sqrt{q}]$.

Theorem 4.2 ([93]) *Let $q = p^r$, and let n be an integer in $[q + 1 - 2\sqrt{q}, q + 1 + 2\sqrt{q}]$. There exists an elliptic cubic curve $\mathcal{X}$ over $\mathbb{F}_q$ with $\#\mathcal{X}(\mathbb{F}_q) = n$ in the following cases:*

(1) $n \not\equiv q + 1 \pmod{p}$;

(2) $n = q + 1$, r odd;

(3) $n = q + 1$, $p \not\equiv 1 \pmod{4}$, r even;

(4) $n = q + 1 \pm \sqrt{q}$, $p \not\equiv 1 \pmod{3}$, r even;

(5) $n = q + 1 \pm 2\sqrt{q}$, r even;

(6) $n = q + 1 \pm \sqrt{2q}$, $p = 2$, r odd;

(7) $n = q + 1 \pm \sqrt{3q}$, $p = 3$, r odd.

The possible structures of G in each of the above cases were determined independently by Rück [74] and Voloch [89]. In cases (4), (6), and (7) the group G is cyclic. If (5) holds, then G is the direct product of two cyclic groups of the same order. If either (2) or (3) holds, then when $q \not\equiv 3 \pmod{4}$, the group G is necessarily cyclic; if $q \equiv 3 \pmod{4}$ it is also possible that G is the direct product of the group of order 2 by a cyclic group of order $|G|/2$. Finally, in case (1),

$$G \cong C_{p^{\nu_p(|G|)}} \times \prod_{\ell \neq p} \left(C_{\ell^{r_\ell}} \times C_{\ell^{s_\ell}} \right)$$

where $r_\ell + s_\ell = \nu_\ell(|G|)$ and $\min\{r_\ell, s_\ell\} \leq \nu_\ell(q - 1)$. Here, C_s denotes the cyclic group of order s, and for an integer M and a prime divisor ℓ, $\nu_\ell(M)$ denotes the highest power of ℓ dividing M.

In order to investigate the covering properties of the arc K of Proposition 4.1 it would be useful to write K in an algebraically parametrized form. This can be easily done when $\mathcal{X}$ is singular, using the above isomorphism of $(G, \oplus)$ and $(\mathbb{F}_q, +)$ or $(\mathbb{F}_q^*, \cdot)$.

If $\mathcal{X} : Y = X^3$, then

$$K = \left\{ \left(L(a) + t, (L(a) + t)^3 \right) \mid a \in \mathbb{F}_q \right\} \tag{4.1}$$

for some separable linearized polynomial L of degree m, and for some $t \in \mathbb{F}_q$ with $t \notin \{L(a) \mid a \in \mathbb{F}_q\}$.

If $\mathcal{X} : XY = (X - 1)^3$, then

$$K = \left\{ \left(ta^m, \frac{(ta^m - 1)^3}{ta^m} \right) \mid a \in \mathbb{F}_q^* \right\} \tag{4.2}$$

for some $t \in \mathbb{F}_q^*$ such that t not an m-th power in $\mathbb{F}_q^*$.

When $\mathcal{X}$ is elliptic no such explicit parametrization exists. However, when m is a prime dividing $q - 1$, and the abelian group $(\mathcal{X}(\mathbb{F}_q), \oplus)$ has a unique subgroup of index m, a useful description of K can be obtained. Assume first that, in addition, m^2 does not divide the size of $\mathcal{X}(\mathbb{F}_q)$; in this case, the Tate-Lichtenbaum pairing can be defined on $\mathcal{X}(\mathbb{F}_q)$ (see e.g. [92, Section 11.3]). We consider the factor groups $\mathcal{X}(\mathbb{F}_q)/H$ and $\mathbb{F}_q^*/(\mathbb{F}_q^*)^m$, where $(\mathbb{F}_q^*)^m$ denotes the subgroup of m-th powers in $\mathbb{F}_q^*$. Write

$$\mathcal{X}(\mathbb{F}_q)/H = \{K_1, \ldots, K_m\}, \qquad \mathbb{F}_q^*/(\mathbb{F}_q^*)^m = \{E_1, \ldots, E_m\}.$$

Note that H coincides with the group $m\mathcal{X}(\mathbb{F}_q)$ of the m-th powers in $\mathcal{X}(\mathbb{F}_q)$. By the definition of the Tate-Lichtenbaum pairing there exists

- a rational function $\alpha(\bar{x}, \bar{y}) \in \mathbb{F}_q(\mathcal{X})$ whose divisor is $mT - mS$ for some $T \in \mathcal{X}(\mathbb{F}_q) \setminus H$, and some $S \in H \setminus \{O\}$, and

- a bijection π of the set of indices $\{1, \ldots, m\}$,

such that $Q \in K_i$ if and only if $\alpha(Q) \in E_{\pi(i)}$, with the only exceptions of T and S.

By rearranging indices if necessary, we can assume that π is the identity, $T \notin K$, and $K = K_1$. Then

$$K = \{(x, y) \mid (x, y) \in \mathcal{X}(\mathbb{F}_q), \alpha(x, y) = ct^m \text{ for some } t \in \mathbb{F}_q\}, \qquad (4.3)$$

where c is a fixed element in E_1.

When m^2 divides the size of $\mathcal{X}(\mathbb{F}_q)$ the Tate-Lichtenbaum pairing cannot be used, and a different definition of α is needed so that (4.3) holds; see [3].

The idea for proving that the secants of K cover a point P off $\mathcal{X}$ goes back to Segre [76] and Lombardo-Radice [66] (see also [87]): construct an algebraic curve $\mathcal{C}_P$ defined over $\mathbb{F}_q$ describing the collinearity of two points of K and P, show that $\mathcal{C}_P$ has an absolutely irreducible component defined over $\mathbb{F}_q$, apply the Hasse-Weil bound (3.1) to guarantee the existence of a suitable $\mathbb{F}_q$-rational point in $\mathcal{C}_P$, and finally deduce that P is collinear with two points in K.

If $\mathcal{X}$ is singular, then $\mathcal{C}_P$ is easily constructed as a (singular) plane curve. Assume that $P = (a, b)$. If $\mathcal{X} : Y = X^3$ and K is as in (4.1), then

$$\mathcal{C}_P : b + (L(X) + t)(L(Y) + t)^2 + (L(X) + t)^2(L(Y) + t)$$
$$- a((L(X) + t)^2 + (L(X) + t)(L(Y) + t) + (L(Y) + t)^2) = 0.$$

The curve $\mathcal{C}_P$ is absolutely irreducible, and the genus g of $\mathcal{C}_P$ is less than or equal to $3m^2 - 3m + 1$; see [83] for the case $L(T) = T^{p^h} - T$ for some integer h, and [2] for the general case.

If $\mathcal{X}$ is the nodal cubic with equation $\mathcal{X} : XY = (X-1)^3$ and K is as in (4.2), then
$$\mathcal{C}_P : f_P(X,Y) = 0,$$
with
$$f_P(X,Y) = a(t^3 X^{2m} Y^m + t^3 X^m Y^{2m} - 3t^2 X^m Y^m + 1) - bt^2 X^m Y^m$$
$$- t^4 X^{2m} Y^{2m} + 3t^2 X^m Y^m - tX^m - tY^m.$$

In [85, 86] it is claimed without proof that $\mathcal{C}_P$ is absolutely irreducible of genus less than some absolute constant times m^2. However, it does not seem that any standard irreducibility criterion can be applied in a straightforward way, including Segre's criterion ([76]; see also [79, Lemma 8]). Actually, for $a^3 = -1$ and $b = 1 - (a-1)^3$, the polynomial $f_P(X,Y)$ is reducible; in fact,

$$f_P(X,Y) = -(a^2 + t^2 X^m Y^m - at Y^m)(a^2 + t^2 X^m Y^m - at X^m).$$

Here we are able provide a proof of the existence of an absolutely irreducible component of $\mathcal{C}_P$ defined over $\mathbb{F}_q$, provided that $a \neq 0$. Assume first that either $a^3 \neq -1$ or $b \neq 1 - (a-1)^3$. Then straightforward computation shows that the plane quartic curve $\mathcal{Q}_P$ with equation

$$a(t^3 X^2 Y + t^3 XY^2 - 3t^2 XY + 1)$$
$$- bt^2 XY - t^4 X^2 Y^2 + 3t^2 XY - tX - tY = 0$$

neither has a linear component, nor splits into two irreducible conics. Thus, $\mathcal{Q}_P$ is absolutely irreducible. Let $\mathbb{F}_q(\mathcal{Q}_P) = \mathbb{F}_q(\bar{x}, \bar{y})$ be the function field over $\mathbb{F}_q$ of $\mathcal{Q}_P$. It is not difficult to show that in $\mathbb{F}_q(\mathcal{Q}_P)$ there exist a place γ_1, centered at the ideal point of the Y-axis, such that the valuation of $\bar{y}$ at γ_1 is equal to -1, and a place γ_2, centered at the ideal point of the X-axis, such that the valuation of $\bar{y}$ at γ_2 is equal to 0 whereas that of $\bar{x}$ is equal to -1. Therefore, by [80, Corollary 3.7.4] the extension $\mathbb{F}_q(\mathcal{Q}_P)(\bar{z})$ of $\mathbb{F}_q(\mathcal{Q}_P)$ defined by the equation $\bar{z}^m = \bar{y}$ is a Kummer extension of $\mathbb{F}_q(\mathcal{Q}_P)$. In particular, the full constant field of $\mathbb{F}_q(\mathcal{Q}_P)(\bar{z})$ is $\mathbb{F}_q$. Now, let γ_2' be a place of $\mathbb{F}_q(\mathcal{Q}_P)(\bar{z})$ lying over γ_2. By [80, Proposition 3.7.3] the ramification index $e(\gamma_2' \mid \gamma_2)$ is equal to 1, whence the valuation of $\bar{x}$ at γ_2' equals -1. We can now apply the previous argument again, in order to deduce that the extension $\mathbb{F}_q(\mathcal{Q}_P)(\bar{z})(\bar{u})$ of $\mathbb{F}_q(\mathcal{Q}_P)(\bar{z})$ defined by the equation $\bar{u}^m = \bar{x}$ is a Kummer extension of $\mathbb{F}_q(\mathcal{Q}_P)(\bar{z})$, and that its full constant field is $\mathbb{F}_q$. As clearly $\mathbb{F}_q(\mathcal{Q}_P)(\bar{z})(\bar{u}) = \mathbb{F}_q(\bar{u}, \bar{z})$ with $f_P(\bar{u}, \bar{z}) = 0$, it follows that $\mathcal{C}_P : f_P(X,Y) = 0$ is an absolutely irreducible curve. By Riemann's Inequality [80, Corollary 3.11.4], the genus of $\mathcal{C}_P$ is less than

$4m^2$. If both $a^3 = -1$ and $b = 1 - (a-1)^3$ hold, then the component of C_P with equation $a^2 + t^2 X^m Y^m - at X^m = 0$ is a generalized Fermat curve over $\mathbb{F}_q$ (see e.g [43]); therefore, it is absolutely irreducible and its genus is less than m^2.

When $\mathcal{X} : Y^2 = X^3 + AX + B$ is elliptic and K is as in (4.3) the situation is more involved, as C_P can be described as a curve in 6-dimensional space. The equations of C_P are the following:

$$C_P : \begin{cases} Y^2 = X^3 + AX + B \\ \alpha(X,Y) = cW^m \\ V^2 = U^3 + AU + B \\ \alpha(U,V) = cZ^m \\ UY - XV - a(Y-V) + b(X-U) = 0, \end{cases} \qquad (4.4)$$

where (X,Y,U,V,W,Z) are affine coordinates in 6-dimensional space. It was shown in [3] that the curve C_P is absolutely irreducible, provided that the j-invariant $j(\mathcal{X})$ of $\mathcal{X}$ is not zero (or, equivalently, $A \neq 0$). The proof involves some function field theory and Galois theory; again, the genus of C_P is bounded by an absolute constant times m^2; see [3]. It should be noted that in [90] the same result was stated in a more general setting, that is, without assuming that m is a prime dividing $q - 1$; in this case both K and the curve C_P cannot be described as explicitly as in (4.3) and (4.4). However, the possibility that C_P splits into components that are not defined over $\mathbb{F}_q$ was not considered in [90], and actually it might have been overlooked by the author [91].

As a consequence of the absolute irreducibility of the curves C_P above, the following result holds; for more precise statements see [2, 3, 86].

Theorem 4.3 *Let K be as in* (4.1), (4.2) *or* (4.3). *There exists an absolute constant C such that if the size of K exceeds $Cq^{3/4}$, then the secants of K cover all the points of* $\mathrm{AG}(2,q) \setminus \mathcal{X}(\mathbb{F}_q)$, *with the possible exceptions of some points on the line $X = 0$ when K is as in* (4.2).

The secants of K cover all the points on $\mathcal{X}(\mathbb{F}_q)$ only when $m = 2$, whereas when $m > 2$ some extra points on $\mathcal{X}(\mathbb{F}_q)$ are needed in order to obtain a complete arc. We recall the definition of a maximal 3-independent subset of a finite abelian group G, as given in [90]; a different terminology is used in [85, 86]. A subset X of G is said to be *maximal 3-independent* if

(a) $x_1 + x_2 + x_3 \neq 0$ for all $x_1, x_2, x_3 \in X$, and

(b) for each $y \in G \setminus X$ there exist $x_1, x_2 \in X$ with $x_1 + x_2 + y = 0$.

If in (b) $x_1 \neq x_2$ can be assumed, then X is said to be *good*. Now, let X be a maximal 3-independent subset of the factor group $\mathcal{X}(\mathbb{F}_q)/H$. Then the union S of the cosets of H corresponding to X is a good maximal 3-independent subset of $(\mathcal{X}(\mathbb{F}_q), \oplus)$; see [90], Lemma 1, together with Remark 5(5). In geometrical terms, S is an arc containing K whose secants cover all the points in $\mathcal{X}(\mathbb{F}_q)$. Actually, by Theorem 4.3, if K is large enough with respect to q then S is a complete arc in $\mathrm{AG}(2, q)$.

The smallest possible order of magnitude for a maximal 3-independent subset of an abelian group G is $(\#G)^{1/2}$ [85, 86], and for some families of groups this lower bound is actually attained. For instance, if G is elementary abelian of order ℓ^r for some prime ℓ and some integer $r \geq 2$, then there exists a maximal 3-independent subset of G of size $2\sqrt{\ell^r} - 3$ if r is even and $\sqrt{\ell^{r-1}} + \sqrt{\ell^{r+1}} - 3$ if r is odd [83]. In the case where G is the direct product of two groups $G_1 \times G_2$ of order at least 4, neither of which is elementary 3-abelian, there exists a maximal 3-independent subset of G of size less than or equal to $(\#G_1) + (\#G_2)$; see [85]. If $G = C_\ell$ is cyclic of prime order $\ell > 7$, then there exists a maximal 3-independent subset of G of size $s \leq (\ell + 1)/3$ [90]. As a result of a computer assisted computation, it turned out that for primes ℓ with $37 \leq \ell \leq 1187$ there exists maximal 3-independent subsets of significantly smaller size; see [3, Table 1]. We remark that by [85, Corollary 2.2] it is possible to construct a subset X of C_ℓ of size less than $30\sqrt{\ell}$, with the following properties:

(a) $x_1 + x_2 + x_3 \neq 0$ for all $x_1, x_2, x_3 \in X$;

(c) for each $y \in C_\ell \setminus X$, the set $X \cup \{y\}$ does not satisfy (a).

However, it is possible that this subset X is not a maximal 3-independent subset of C_ℓ. In fact, there could exist some $y \in C_\ell \setminus X$ with either $2y \in -X$ or $3y = 0$, but $y + x_1 + x_2 \neq 0$ for all $x_1, x_2 \in X$.

For K as in Theorem 4.3, the smallest possible size of a complete arc S obtained as the union of K and some other cosets of H is about

$$Cq^{3/4} \cdot \sqrt{\#(\mathcal{X}(\mathbb{F}_q)/H)} \sim Cq^{3/4} \cdot q^{1/8} = Cq^{7/8}.$$

In [85, 86], Szőnyi developed the idea of constructing good maximal 3-independent subsets of $\mathcal{X}(\mathbb{F}_q)$ by deleting a point Q from K and extend $K \setminus \{Q\}$ directly in $\mathcal{X}(\mathbb{F}_q)$ to a good maximal 3-indipendent subset of size roughly $\#K + \#(\mathcal{X}(\mathbb{F}_q)/H)$. This produces significantly smaller complete arcs, provided that $\mathcal{X}(\mathbb{F}_q) = H \times H_1$ for some subgroup H_1 of order m. More precisely, there is an absolute constant C such that

- if $(q - 1)$ has a divisor m such that $(m, q - 1) = 1$ and m is slightly less than $\frac{1}{2\sqrt{2}} q^{1/4}$, then there exists a complete arc in $\mathrm{PG}(2, q)$ with size about $Cq^{3/4}$ lying, with few exceptions, in the nodal cubic with equation $XY = (X - 1)^3$;

- If $(q - 1)$ has a prime divisor m slightly less than $\frac{1}{8}q^{1/4}$, then there exists a complete arc in $\mathrm{PG}(2, q)$ with size about $Cq^{3/4}$ lying, with few exceptions, in an elliptic cubic.

Remark Complete arcs of size $Cq^{3/4}$ contained in cubic curves are the smallest complete arcs that have been explicitly constructed so far for arbitrarily large q's. However, significantly smaller complete arcs can be obtained via either probabilistic or computational methods. In [61] it is shown that there are absolute constants C and D such that every plane $\mathrm{PG}(2, q)$ has a complete arc of cardinality at most $D\sqrt{q}\log^C q$; moreover, an algorithm which constructs such an arc in polynomial time is given. In a series of recent papers [8, 9, 11] complete arcs of size smaller than $\sqrt{q}\log q$ have been constructed by computer for every $q \leq 42013$ and for many sporadic q's with $42013 < q < 80000$. A recent account of the computational results on the size of a complete arc in $\mathrm{PG}(2, q)$ can be found in [9].

Remark Some of the results of this section can be extended to the case $p \leq 3$. When $p = 3$, the completeness of arcs arising from elliptic curves can be proved, even though there a few technicalities to deal with [3, 90], partly due to the canonical equation of $\mathcal{X}$ involving one more monomial. The results of [83] about arcs as in (4.1) hold for $p = 2$ as well.

Remark Results on arcs contained in cubics with no rational inflection or cubics with an isolated double point have not appeared in the literature so far. The latter case is currently under investigation by the same authors of [2], also in connection with complete caps in higher dimensions.

4.2 Small saturating sets

A saturating set S in the projective space $\mathrm{PG}(2, q)$ is a point-set whose secants cover $\mathrm{PG}(2, q)$, that is, any point of $\mathrm{PG}(2, q)$ belongs to a line joining two distinct points of S. Saturating sets are sometimes called 1-*saturating* sets, *dense* sets, or *saturated* sets. As well as being a natural geometrical problem, the construction of small saturating sets in $\mathrm{PG}(2, q)$ is relevant in other areas of combinatorics. In fact, saturating sets are related to defining sets of block designs [13], and, as recently pointed out in [62], small saturating sets are connected to the degree/diameter problem in graph theory [67].

For q a square there is a nice example of a saturating set of size $3\sqrt{q}$, namely the union of three non-concurrent lines of a subplane of $PG(2,q)$ of order $\sqrt{q}$. This construction was refined in [21], where saturating sets of size $3\sqrt{q} - 1$ were obtained. It has been recently observed that if q is a fourth power, then a saturating set of size $2(\sqrt{q} + \sqrt[4]{q} + 1)$ can be obtained as the union of two disjoint Baer subplanes in $PG(2, \sqrt{q})$, viewed as a subgeometry of $PG(2,q)$; see [29, Theorem 3.4]. By a similar argument, if $q = p^6$ with p a prime, then it is possible to construct a saturating set of size $2(\sqrt{q} + \sqrt[3]{q} + \sqrt[6]{q} + 1)$ provided that there exist two disjoint $\mathbb{F}_{\sqrt{q}}$-projectively equivalent copies of the blocking set of $PG(2, \sqrt{q})$ described in [55, Lemma 13.8(iii)]:

$$ B = \{(1, x, x^p) \mid x \in \mathbb{F}_{\sqrt{q}}\} \cup \{(0, 1, y) \mid y \in \mathbb{F}_{\sqrt{q}}, \ y^{p^2+p+1} = 1\}. $$

The existence of a projectivity γ defined over $\mathbb{F}_{\sqrt{q}}$ such that B is disjoint from $\gamma(B)$ was checked with a computer for $p \leq 73$, see [29, Corollary 3.6]. It is an open problem to establish a general result.

When q is not a square, the trivial lower bound (2.1) is far away from the size of the known examples. The existence of saturating sets of size $\lfloor 5\sqrt{q \log q} \rfloor$ was shown by means of probabilistic methods; see [13, 64]. Besides complete arcs arising from cubic curves (see Section 4.1), other saturating sets of size approximately $Cq^{\frac{3}{4}}$, with C a constant independent of q, have been explicitly described in a number of papers; see [6, 48, 82]. A construction by Davydov and Östergård [35, Thm. 3] provides saturating sets of size $2q/p + p$, where p is the characteristic of $\mathbb{F}_q$; note that in the special case where $q = p^3$, $p \geq 17$, the size of these sets is less than $q^{\frac{3}{4}}$. Some of the above constructions are summarized in [86, Section 4] and [13, Sections 3, 4].

A general explicit construction of saturating sets in $PG(2,q)$ of size about $3q^{\frac{2}{3}}$ was obtained in 2007; see [45]. For a large non-square q with $q \neq p^3$, these are the smallest explicitly constructed saturating sets, whereas for $q = p^3$ the size is the same as that of the example by Davydov and Östergård. Using the same technique, smaller saturating sets were obtained for specific values of q; in some cases they provide an improvement on the probabilistic bound.

The constructions in [45] are essentially algebraic, and use linearized polynomials over the finite field $\mathbb{F}_q$. For properties of linearized polynomials see [65, Chapter 3]. In the affine line $AG(1,q)$, take a subset E whose points are coordinatized by an additive subgroup H of $\mathbb{F}_q$. Then H consists of the roots of a linearized polynomial $L_H(X)$. Let S_1 be the union of two copies of E, embedded in two parallel lines in $AG(2,q)$, namely the lines with equation $Y = 0$ and $Y = 1$. The condition for a point $P = (a, b)$

in $AG(2, q)$ to belong to some secant of S_1 is that the algebraic curve $\mathcal{C}_p$ with equation

$$\mathcal{C}_P : L_H(X) - bL_H(Y) + a = 0$$

has at least one $\mathbb{F}_q$-rational affine point. This certainly occurs when the algebraic curve with equation

$$L_H(X) - bL_H(Y) = 0, \tag{4.5}$$

has precisely q $\mathbb{F}_q$-rational affine points, which leads to the purely algebraic problem of determining the values of b for which this condition holds. This is the case if and only if $-b$ belongs to the set $\mathbb{F}_q \setminus \mathcal{M}_H$, where

$$\mathcal{M}_H := \left\{ \left(\frac{L_{H_1}(\beta_1)}{L_{H_2}(\beta_2)} \right)^p \right\}. \tag{4.6}$$

Here, H_1 and H_2 range over all subgroups of H of index p, while $\beta_i \in H \setminus H_i$. This shows that the points which are not covered by the secants of S_1 are the points $P = (a, b)$ with $-b \in \mathcal{M}_H$. The final step of the construction consists in adding a possibly small number of points $Q_1, \ldots, Q_t$ to S_1 to obtain a saturating set. For the general case, this is done by just ensuring that the secants $Q_i Q_j$ cover all points uncovered by the secants of S_1.

Theorem 4.4 ([45]) *Let $q = p^r$, and let H be any additive subgroup of $\mathbb{F}_q$ of size p^s, with $2s \leq r$. The set*

$$S = \{(L_H(a) : 1 : 1), (L_H(a) : 0 : 1) \mid a \in \mathbb{F}_q\}$$
$$\cup \{(0 : \mu : 1) \mid \mu \in \mathcal{M}_H\} \cup \{(0 : 1 : 0), (1 : 0 : 0)\}$$

is a saturating set of size at most

$$\frac{2q}{p^s} + \frac{(p^s - 1)^2}{p - 1} + 1.$$

For special cases, the above construction can give better results when more than two copies of E are used.

Theorem 4.5 ([45]) *Let $q = p^r$, with r odd. Let H be any additive subgroup of $\mathbb{F}_q$ of size p^s, with $2s + 1 = r$. For any integer $v \geq 1$ there exists a saturating set S in $PG(2, q)$ such that*

$$\#S \leq (v + 1)p^{s+1} + (\#\mathcal{M}_H)^v (q - 1)^{1-v} + 2.$$

Corollary 4.6 *Let* $q = p^{2s+1}$. *Then there exists a saturating set in* $PG(2, q)$ *of size less than or equal to*

$$\min \left\{ (v+1)p^{s+1} + \frac{(p^s - 1)^{2v}}{(p-1)^v (p^{(2s+1)} - 1)^{(v-1)}} + 2 \right\}$$

where the minimum is taken over all $v \in \{1, \ldots, 2s+1\}$.

The sizes of the smallest known saturating sets in $PG(2, q)$, $q \leq 1217$, are reported in [29, Table 1].

Theorem 4.7 ([29]) *Assume that* $q \leq 1217$. *The minimum size of a saturating set in* $PG(2, q)$ *is less than* $a_q \sqrt{q}$, *with* $a_q = 3$ *if* $q \leq 109$, $a_q = 7/2$ *if* $109 < q \leq 349$, *and* $a_q = 4$ *if* $349 < q \leq 1217$.

5 The three-dimensional case

5.1 Small complete caps

In $PG(3, q)$, with q even, the trivial lower bound (2.1) on the size of a complete cap is substantially sharp. The existence of a complete cap of size $3q + 2$ was shown by Segre [75]. Segre's construction was generalized by Pambianco and Storme [69], who obtained complete caps of size $2q + n$ for each n such that there exists a complete arc with n points in $PG(2, q)$ (see also [27]). When q is odd, the size of the smallest complete cap in $PG(3, q)$ is known for $q \leq 5$; see [57]. Complete caps of size approximately $q^2/2$ [1, 53] and $q^2/3$ [40, 41] which share many points with an elliptic quadric have been obtained by generalizing the idea of Segre [76] and Lombardo Radice [66] for constructing plane arcs.

Segre's construction for q even consists of three conics, plus two points. In 1998, Pellegrino tried to extend this idea to the odd order case; see [71]. He claimed that it is possible to choose s conics with $\binom{s-1}{2} \leq \frac{q+1}{4}$ in such a way that their union, together with few extra points, is a complete cap of size less than or equal to $\frac{\sqrt{q}}{2} q + 2$. A major gap in Pellegrino's completeness proof has recently emerged; see [12]. Nonetheless, his idea turned out to be useful to construct small complete caps by computer; in fact, the problem of establishing whether a Pellegrino's cap is complete can be somehow reduced to a problem in the affine plane $AG(2, q)$, which makes it feasible to investigate rather large q's (see Proposition 5.1 below).

For a fixed primitive element ω in $\mathbb{F}_q$, and for $a, z \in \mathbb{F}_q$ with $a \neq 0$, let $\mathcal{C}_{a,z}$ be the irreducible conic in $AG(3, q)$ with equations

$$\begin{cases} X^2 - \omega Y^2 = a \\ Z = z. \end{cases}$$

A cap in $AG(3, q)$ consisting of the union of some conics $\mathcal{C}_{a,z}$, plus some affine points on the line with equations $X = 0, Y = 0$, will be referred to as a *Pellegrino cap*.

Proposition 5.1 ([12]) *Let $\mathcal{C}_{a_1,z_1}$, $\mathcal{C}_{a_2,z_2}$, $\mathcal{C}_{a_3,z_3}$ be such that z_1, z_2 and z_3 are pairwise distinct, and $a_1 a_2 a_3 \neq 0$. Then there exist three collinear points in*

$$\mathcal{C} = \bigcup_{i=1}^{3} \mathcal{C}_{a_i,z_i}$$

if and only if

$$
a_1^2 (z_2 - z_3)^4 + a_2^2 (z_1 - z_3)^4 + a_3^2 (z_1 - z_2)^4
$$
$$
- 2a_1 a_2 (z_1 - z_3)^2 (z_2 - z_3)^2 - 2a_1 a_3 (z_1 - z_2)^2 (z_3 - z_2)^2
$$
$$
- 2a_2 a_3 (z_3 - z_1)^2 (z_2 - z_1)^2 \tag{5.1}
$$

is either zero or a non-square in $\mathbb{F}_q$.

Three conics $\mathcal{C}_{a_i,z_i}$, $i = 1, 2, 3$, with pairwise distinct z_i's, are said to be *collinear* if (5.1) holds. We identify a conic $\mathcal{C}_{a_i,z_i}$ with the pair (a_i, z_i) in $\mathbb{F}_q^2$. Given a subset

$$M = \{(a_i, z_i) \in \mathbb{F}_q^2 \mid i = 1, \ldots, n, a_i \neq 0 \text{ for any } i, z_i \neq z_j \text{ when } i \neq j\}$$

of $\mathbb{F}_q^2$, a pair $(a, z) \in \mathbb{F}_q^2$ with $a \neq 0$ is said to be *covered* by M if either $z = z_i$ for some $i = 1, \ldots, n$ or (a, z) is collinear with two elements in M. A set M is said to be *complete* if no three pairs in M are collinear, and in addition any pair (a, z) with $a \neq 0$ is covered by M.

It is easily seen that if M is complete, then there exists a complete cap in $AG(3, q)$ of size at most $\#M(q + 1) + 2$, which consists of the conics corresponding to the pairs in M, possibly together with one or two extra points on the line with equations $X = 0, Y = 0$. A computer search for small complete M's has been performed in [12].

Theorem 5.2 ([12]) *Assume that $19 \leq q \leq 30000$, q odd. Then there exists a complete Pellegrino cap in $AG(3, q)$ with size at most $a_q(q+1)+2$, where*

$$
a_q = \begin{cases}
4 & \text{if } 19 \leq q \leq 37, \\
5 & \text{if } 41 \leq q \leq 121, \\
6 & \text{if } 127 \leq q \leq 509, \\
7 & \text{if } 521 \leq q \leq 2347, \\
8 & \text{if } 2351 \leq q \leq 5227, \\
9 & \text{if } 5231 \leq q \leq 29989,
\end{cases}
$$

with the only exceptions of q = 10531, 18493, 18973, 23677, 24077, 24121, 25163, 25639, 26227, 28643, *where* $a_q = 10$.

It is worth noticing that for $q \leq 97$, smaller complete caps were obtained in [27].

5.2 Small saturating sets

The smallest size of a saturating set in $\mathrm{PG}(3,q)$ for $q \leq 5$ is as follows (see [4, 5, 7, 29]).

q	2	3	4	5
$l(4,2;q)$	5	8	9	11

In general, the trivial lower bound (2.1) for the size of a saturating set in $\mathrm{PG}(3,q)$ turns out to be substantially sharp. In fact, for $q \geq 4$, a saturating set S in $\mathrm{PG}(3,q)$ of size $2q+1$ can be obtained as follows (see [15] for q even, and [21, Theorem 5.1] for the general case):

$$S = \mathcal{C}^* \cup \mathcal{L}^* \cup \{Q\},$$

where $\mathcal{C}^*$ is a set of q points consisting of a plane conic $\mathcal{C}$ minus a point P, $\mathcal{L}^*$ is a set of q points collinear with P and disjoint from the plane containing $\mathcal{C}^*$, and $Q \neq P$ is any point on the tangent line of $\mathcal{C}$ at P. For $q \geq 4$, no example of a smaller saturating set seems to be known.

6 Constructions in higher dimensions

6.1 Recursive constructions for complete caps

Since the theory of caps in small dimensional spaces is well developed and quite rich of constructions, to obtain complete caps in higher dimensions a natural idea is to try to use some kind of lifting methods for plane arcs and space caps.

We recall two recursive constructions for caps in affine spaces.

- *Blowing-up.*

  For a positive integer r, fix a basis of $\mathbb{F}_{q^r}$ as a linear space over $\mathbb{F}_q$, and let π_{N_0} denote the corresponding identification of $\mathrm{AG}(N_0, q^r)$ with $\mathrm{AG}(rN_0, q)$. If K is a cap in $\mathrm{AG}(N_0, q^r)$, then $\pi_{N_0}(K)$ is a cap in $\mathrm{AG}(rN_0, q)$.

- *Product construction.*

  For two positive integers N_1, N_2 let π_{N_1,N_2} denote the natural identification of $AG(N_1, q) \times AG(N_2, q)$ with $AG(N_1 + N_2, q)$. If K_1 is a cap in $AG(N_1, q)$ and K_2 is a cap in $AG(N_2, q)$, then

  $$\pi_{N_1,N_2}(K_1 \times K_2)$$

  is a cap in $AG(N_1 + N_2, q)$.

  From now on, $\mathcal{P}(q)$ denotes the complete cap in $AG(2, q)$ consisting of the $\mathbb{F}_q$-rational points of the parabola with equation $Y = X^2$; namely,

  $$\mathcal{P}(q) = \{(a, a^2) \mid a \in \mathbb{F}_q\} \subset AG(2, q).$$

Theorem 6.1 ([70]) *Let q be even. Then the blow-up of $\mathcal{P}(q^r)$ is a complete cap of size q^r in $AG(2r, q)$.*

Theorem 6.2 ([28]) *Let q be even. Then the product of the blow-up of $\mathcal{P}(q^r)$ by any complete cap of size k in $AG(N_2, q)$ is a complete cap of size kq^r in $AG(2r + N_2, q)$.*

By Theorem 6.2 for $N_2 = 1$, if q is even then there exists a complete cap of size $2q^r$ in $AG(2r + 1, q)$; that is, for even q and odd dimension N, the trivial lower bound (2.1) is substantially sharp.

In order to obtain small complete caps in even dimensions as well, the notion of a translation cap is useful.

Definition 6.3 A cap in $AG(N, q)$ is a translation cap if its points form a whole orbit under the action of a translation group in $AG(N, q)$.

Translation caps can only exist in spaces of even order [47, Remark 2.2]. Interestingly, any translation cap is contained in a complete translation cap [47, Proposition 2.4]. Also, the size of a translation cap K in $AG(N, q)$ is at most $q^{N/2}$ [47, Proposition 2.5], and when this upper bound is attained K is said to be a *maximal translation cap*. Any maximal translation cap in $AG(2, q)$ consists of points $\{(a, a^{2^i}) \mid a \in \mathbb{F}_q\}$, for some integer i with $(i, \log_2 q) = 1$ [47, Remark 2.6]. Maximal translation caps in any even dimension $2r$ can be obtained by blowing up maximal translation caps in $AG(2, q^r)$.

Theorem 6.4 ([47]) *Let q be even. Assume that K_1 be a maximal translation cap in $AG(N_1, q)$. Then the product of K_1 by any complete cap of size k in $AG(N_2, q)$ is a complete cap of size $kq^{N_1/2}$ in $AG(N_1 + N_2, q)$.*

When q is a square, it is possible that there exists a maximal translation cap K in $AG(3,q)$. In this case, the product of K and the blow-up of $\mathcal{P}(q^r)$ is a maximal translation cap in $AG(3+2r,q)$; then by Theorem 6.4 there is a complete cap of size $2q^{(N-1)/2}$ in $AG(N,q)$ for any even dimension $N \geq 4$. The existence of a maximal translation cap in $AG(3,q)$ has been verified for $q \leq 64$, q a square; see [10]. It is an open problem to establish whether this holds for any even square q.

Corollary 6.5 *Let q be even.*

(i) *if N is odd, then there exists a complete cap in $AG(N,q)$ of size $2q^{(N-1)/2}$;*

(ii) *if N is even, then there exists a complete cap in $AG(N,q)$ of size $kq^{(N-2)/2}$, where k is the size of a complete arc in $AG(2,q)$;*

(iii) *if q is a square, $q \leq 64$, and N is even, then there exists a complete cap in $AG(N,q)$ of size $2q^{(N-1)/2}$.*

The complete caps in $AG(N,q)$ described in Corollary 6.5(i)-(ii) have been extended to complete caps in $PG(N,q)$ in [28].

Theorem 6.6 ([28]) *Let $t_2(N,q)$ denote the smallest size of a complete cap in $PG(N,q)$. For $q > 8$, q even,*

$$t_2(N,q) \leq \begin{cases} t_2(2,q)q^{\frac{N-2}{2}} + s_{N,q} - N + 1 & \text{if } N \text{ is even} \\ 2q^{\frac{N-1}{2}} + t_2(2,q)q^{\frac{N-3}{2}} + s_{N,q} - N + 2 & \text{if } N \text{ is odd,} \end{cases}$$

where

$$s_{N,q} = 3\left(q^{\lfloor \frac{N-2}{2} \rfloor} + q^{\lfloor \frac{N-2}{2} \rfloor - 1} + \cdots + q\right) + 2.$$

From now on in this section we assume that q is odd.

Theorem 6.7 ([37]) *Let q be odd. Then the blow-up of $\mathcal{P}(q^r)$ is a complete cap in $AG(2r,q)$ if and only if r is odd.*

A natural question to ask in order to obtain small complete caps in spaces of dimension $N \not\equiv 2 \pmod 4$ is whether the product of the blow-up of $\mathcal{P}(q^r)$, r odd, and a complete arc in $AG(2,q)$ is complete. In this context, the concept of a regular point with respect to a complete arc in $AG(2,q)$ is useful. According to Segre [78], given three pairwise distinct points P, P_1, P_2 on a line $\mathcal{L}$ in $AG(2,q)$, P is external or internal to the segment $P_1 P_2$ depending on whether

$$(x - x_1)(x - x_2) \quad \text{is a non-zero square in } \mathbb{F}_q \text{ or not,}$$

where x, x_1 and x_2 are the coordinates of P, P_1 and P_2 with respect to any affine frame of $\mathcal{L}$. Definition 13 in [78] extends as follows.

Definition 6.8 Let A be a complete arc in $\mathrm{AG}(2,q)$. A point $P \in \mathrm{AG}(2,q) \setminus A$ is *regular* with respect to A if P is external to any segment $P_1 P_2$, with $P_1, P_2 \in A$ collinear with P. The point P is said to be *pseudo-regular* with respect to A if it is internal to any segment $P_1 P_2$, with $P_1, P_2 \in A$ collinear with P. If P is neither regular nor pseudo-regular, than P is said to be *bicovered* by A. If every $P \in \mathrm{AG}(2,q) \setminus A$ is bicovered by A, then A is said to be a *bicovering* arc. If there exists precisely one point in $\mathrm{AG}(2,q) \setminus A$ which is not bicovered by A, then A is said to be *almost bicovering*.

For a complete arc A in $\mathrm{AG}(2,q)$, and an odd integer r, let $K_{A,r}$ denote the product of the blow-up of $\mathcal{P}(q^r)$ and A.

Theorem 6.9 ([46]) *Let q be odd, and let A be a bicovering arc of size k in $\mathrm{AG}(2,q)$. Then for every odd integer r, $K_{A,r}$ is a complete cap of size kq^r in $\mathrm{AG}(2r+2,q)$.*

By a slight modification of the product construction it is possible to obtain complete caps from almost bicovering arcs. For an element $c \in \mathbb{F}_q$, let $\mathcal{P}_c(q)$ denote the parabola

$$\mathcal{P}_c(q) = \{(a, a^2 - c) \mid a \in \mathbb{F}_q\} \subset \mathrm{AG}(2,q).$$

Also, for a point $(x_0, y_0) \in \mathrm{AG}(2,q)$, let $K_{(c,x_0,y_0)}$ be the product of the blow-up of $\mathcal{P}_c(q^r)$ and the trivial cap of $\mathrm{AG}(2,q)$ consisting of the single point (x_0, y_0).

Theorem 6.10 ([46]) *Let q be odd, and let A be an almost bicovering arc of size k in $\mathrm{AG}(2,q)$, admitting exactly one regular point (x_0, y_0). Let c be a non-square element in $\mathbb{F}_q$. Then for every odd integer r*

$$K = K_{A,r} \cup K_{(c,x_0,y_0)}$$

is a complete cap of size $(k+1)q^r$ in $\mathrm{AG}(2r+2,q)$.

A similar result holds for A an almost bicovering arc with one pseudo-regular point.

Theorem 6.11 ([46]) *Let q be odd, and let A be an almost bicovering arc of size k in $\mathrm{AG}(2,q)$, admitting exactly one pseudo-regular point (x_0, y_0). Let c be a non-zero square element in $\mathbb{F}_q$. Then for every odd integer r*

$$K = K_{A,r} \cup K_{(c,x_0,y_0)}$$

is a complete cap of size $(k+1)q^r$ in $\mathrm{AG}(2r+2,q)$.

6.2 Small complete caps from bicovering and almost bicovering arcs

By Theorems 6.9–6.11, bicovering and almost bicovering arcs in affine planes are a potential powerful tool to construct small complete caps in $\mathrm{AG}(N, q)$ with q odd, $N \equiv 0 \pmod 4$. To establish whether a complete arc is bicovering can be a difficult task. In the (simplest) case where the arc A consists of the affine points of an irreducible conic, it follows from previous results by B. Segre [78] that A is either bicovering or almost bicovering, provided that $q > 13$. Smaller complete arcs A in $\mathrm{AG}(2, q)$ can be obtained by choosing some points on a conic $\mathcal{C}$ and adding a few extra points (see e.g. [63, 77, 84]); in this case a point $Q \in \mathcal{C} \setminus A$ can hardly be bicovered, as there are few secants of A through Q. No such problem arises when complete arcs contained in cubic curves are considered. In this section, we survey some recent results on the bicovering properties of the arcs described in Section 4.1.

For $q \leq 127$ some computational results were presented in [42].

Theorem 6.12 ([42]) *There exists a bicovering arc in* $\mathrm{AG}(2, q)$ *of size* n_q *contained in an elliptic curve defined over* $\mathbb{F}_q$, *with* (q, n_q) *as follows:*

q	67	73	81	83	89	97	101	103	107	109	113	121	127
n_q	42	45	49	50	54	55	61	60	63	65	66	71	74

Also, there exists an almost bicovering arc in $\mathrm{AG}(2, 53)$ *of size* 34 *contained in an elliptic curve.*

In order to obtain general results one can argue as follows (notation and terminology are as in Section 4.1). Let $\mathcal{X}$ be an irreducible cubic curve, and let $P = (a, b)$ be a point in $\mathrm{AG}(2, q)$ off $\mathcal{X}$. We first consider the case where $\mathcal{X}$ is singular. Let $\mathcal{C}_P : f_P(X, Y) = 0$ be the auxiliary plane curve associated to P. Then to a suitable $\mathbb{F}_q$-rational point $Q = (x, y)$ of the curve $\mathcal{C}_p$ one can associate two $\mathbb{F}_q$-rational points, say $P_1 = P_1(Q)$ and $P_2 = P_2(Q)$ belonging to K and collinear with P. Let $F(x, y)$ be the first coordinate of $P_1(Q)$, and $G(x, y)$ be the first coordinate of $P_2(Q)$. It is easily seen that $F(x, y)$ and $G(x, y)$ are both polynomial functions in x and y. Then P is external to the segment joining P_1 and P_2 if and only if

$$(a - F(x, y))(a - G(x, y)) \quad \text{is a non-zero square in } \mathbb{F}_q.$$

Similarly, P is internal to $P_1 P_2$ precisely when

$$(a - F(x, y))(a - G(x, y)) \quad \text{is a non-square in } \mathbb{F}_q.$$

This means that P is bicovered by K if and only if both the algebraic curves

$$\mathcal{D}_P : \begin{cases} f_P(X, Y) = 0 \\ (a - F(X, Y))(a - G(X, Y)) = Z^2 \end{cases}$$

and

$$\mathcal{D}'_P : \begin{cases} f_P(X,Y) = 0 \\ (a - F(X,Y))(a - G(X,Y)) = cZ^2 \end{cases}$$

for a fixed c non-square in $\mathbb{F}_q$, have a suitable $\mathbb{F}_q$-rational point. If $\mathcal{X}$ is elliptic, then the same argument applies, but as $\mathcal{C}_P$ is a curve in the 6-dimensional space, both curves $\mathcal{D}_P$ and $\mathcal{D}'_P$ are defined in the 7-dimensional space. In order to prove the existence of a suitable $\mathbb{F}_q$-rational point of $\mathcal{D}_P$ and $\mathcal{D}'_P$ one can use the Hasse-Weil bound (3.1). The main difficulty is again the proof of the absolute irreducibility of both curves. Provided that $\mathcal{C}_P$ is absolutely irreducible, this is equivalent to the rational function $(a - F(\bar{x},\bar{y}))(a - G(\bar{x},\bar{y}))$ not being a square in the function field of $\mathcal{C}_P$.

Now we consider points P in $\mathcal{X}(\mathbb{F}_q)$. As pointed out in Section 4.1, it is possible to extend K to an arc contained in $\mathcal{X}(\mathbb{F}_q)$ of size less than $2\#K$, and covering all the points in $\mathcal{X}(\mathbb{F}_q)$. However, in order to ensure that all such points are bicovered, it seems that larger arcs are needed. As it turned out, a suitable choice is the union of the cosets of H corresponding to a maximal 3-independent subset of the factor group $\mathcal{X}(\mathbb{F}_q)/H$.

The case where $\mathcal{X}$ is a nodal cubic and $m = 2$ was investigated in [46], where it was shown that if $q > 76^2$, then

$$A := \left\{ \left(\frac{(a-1)^3}{a^2 - wa}, \frac{a}{a - w} \right) \mid a \text{ a non-square in } \mathbb{F}_q,\ a \neq 0,\ a \neq w \right\},$$

with w is a primitive element of $\mathbb{F}_q$, is a bicovering arc in $\mathrm{AG}(2,q)$ of size $(q-3)/2$.

Cosets of subgroups of the set of non-singular $\mathbb{F}_q$-rational points of a cuspidal cubic are considered in [2], where almost bicovering arcs of size $2pq^{7/8}$ are obtained for $q = p^h$, provided that p is a sufficiently large prime and $h > 8$.

Theorem 6.13 ([2]) *Let $q = p^h$ with $p > 3$ a prime. Let h' be an integer with $1 < h' < h$ and set $m = p^{h'}$. Assume that $q \geq 144m^4$. Let $\mathcal{X} : Y = X^3$, and let H be as in Section 4.1. Then there exists a set S, defined as the union of some cosets of H, which is an almost bicovering arc in $\mathrm{AG}(2,q)$ of size k, with*

$$k = \begin{cases} (2\sqrt{m} - 3)\dfrac{q}{m} & \text{if } h' \text{ is even} \\ (\sqrt{m/p} + \sqrt{mp} - 3)\dfrac{q}{m} & \text{if } h' \text{ is odd.} \end{cases}$$

The point $Q = (0,0)$ is either regular or pseudo-regular with respect to S, according to whether $q \equiv 1 \pmod 4$ or $q \equiv 3 \pmod 4$.

Corollary 6.14 *Let $q = p^h$ with $p > 3$ a prime, $h > 8$. Let $N \equiv 0$ (mod 4), $N \geq 4$. Let v_h be the integer in $\{1, \ldots, 8\}$ such that $v_h \equiv h$ (mod 8), and let t_h be the integer in $\{1, \ldots, 4\}$ such that $t_h \equiv h$ (mod 4). Assume that $p^{t_h} > 144$. Then there exists a complete k-cap in $\mathrm{AG}(N, q)$ with*

$$k < \left(2p^{h - \lfloor (\lceil h/4 \rceil - 1)/2 \rfloor} \right) q^{\frac{N-2}{2}} \leq 2p^{\frac{v_h}{8}} q^{\frac{N}{2} - \frac{1}{8}} \leq 2pq^{\frac{N}{2} - \frac{1}{8}}.$$

In [3] the elliptic case is dealt with.

Theorem 6.15 ([3]) *Let q be odd, and let m be a prime divisor of $q - 1$, with $7 < m < \frac{1}{8} \sqrt[4]{q}$. Assume that the cyclic group of order m admits a maximal 3-independent subset of size s. Then there exists a bicovering k-arc in $\mathrm{AG}(2, q)$ with*

$$s \cdot \left\lfloor \frac{q - 2\sqrt{q} + 1}{m} \right\rfloor \leq k \leq s \cdot \left(\left\lfloor \frac{q - 2\sqrt{q} + 1}{m} \right\rfloor + 31 \right),$$

consisting of the union of s cosets of a subgroup of index m of the abelian group of the $\mathbb{F}_q$-rational points of an elliptic plane curve.

Corollary 6.16 *Let q be odd, and let m be a prime divisor of $q - 1$, with $7 < m < \frac{1}{8} \sqrt[4]{q}$. Assume that the cyclic group of order m admits a maximal 3-independent subset of size s. Then for any positive integer $N \equiv 0$ (mod 4), there exists a complete cap in $\mathrm{AG}(N, q)$ of size k with*

$$s \cdot q^{\frac{N-2}{2}} \cdot \left\lfloor \frac{q - 2\sqrt{q} + 1}{m} \right\rfloor \leq k \leq s \cdot q^{\frac{N-2}{2}} \cdot \left(\left\lfloor \frac{q - 2\sqrt{q} + 1}{m} \right\rfloor + 31 \right).$$

It has been already recalled in Section 4.1 that in the cyclic group of order m there exists a maximal 3-independent subset of size $s \leq (m + 1)/3$. Corollary 6.16 yields the existence of complete caps in $\mathrm{AG}(N, q)$ of size of the same order of magnitude as $Cq^{N/2}$ with $C \leq 1/3$, provided that $q - 1$ has a prime divisor m greater than 7 and smaller than $\sqrt[4]{q}/8$. For specific values of m, the upper bound on C can be improved, as there exist maximal 3-independent subsets of the cyclic group of order m of size significantly less than $m/3$ (see [3, Table 1]).

6.3 Concatenating construction for saturating sets

One of the most efficient constructions for linear codes is Davydov's q^m-*concatenating construction*, first proposed in [20], and then developed in [21, 22, 23, 24, 26, 29, 32, 33, 34, 36, 38] and [81, Supplement]; see also [14], [18, Section 5.4] and the references in these works. For the

sake of simplicity, here we will write *concatenating construction* rather
than q^m-concatenating construction. Starting from a q-ary linear code
C with covering radius R, often called a *seed code*, the concatenating
construction gives an infinite family of q-ary linear codes with arbitrarily
high codimension, covering radius R, and almost the same covering density
as **C**. When $R = 2$, Davydov's construction can be viewed as an inductive
method to construct small saturating sets in higher dimensional spaces.

We illustrate the idea behind the concatenating construction for $R = 2$.
The starting point is a saturating set S of size n_0 in $\mathrm{AG}(N_0, q)$. Then,

- let r be a positive integer such that $n_0 \leq q^r$;

- for each $P \in S$ choose $\beta(P) \in \mathbb{F}_{q^r}$ in such a way that $\beta(P) \neq \beta(P')$
  when $P \neq P'$;

- identify $\mathrm{AG}(N_0, q) \times \mathbb{F}_{q^r} \times \mathbb{F}_{q^r}$ and $\mathrm{AG}(N_0 + 2r, q)$;

- let
$$\bar{S} = \{(P, \epsilon, \beta(P)\epsilon) \mid P \in S, \, \epsilon \in \mathbb{F}_{q^r}\}.$$

Consider a point (Q, γ_1, γ_2) in $\mathrm{AG}(N_0 + 2r, q)$, and assume first that $Q \notin S$.
As S is a saturating set in $\mathrm{AG}(N_0, q)$, there exist $P_1, P_2 \in S$, $t \in \mathbb{F}_q$, with
$Q = P_1 + t(P_2 - P_1)$. Then it is not difficult to check that there exist ϵ_1, ϵ_2
in $\mathbb{F}_{q^r}$ with

$$(Q, \gamma_1, \gamma_2) = (P_1, \epsilon_1, \beta(P_1)\epsilon_1) + t((P_2, \epsilon_2, \beta(P_2)\epsilon_2) - (P_2, \epsilon_2, \beta(P_2)\epsilon_2));$$

that is, (Q, γ_1, γ_2) is covered by $\bar{S}$. In order to cover points (P, γ_1, γ_2)
with $P \in S$, embed $\mathrm{AG}(N_0 + 2r, q)$ in $\mathrm{PG}(N_0 + 2r, q)$ and add to $\bar{S}$ the
$(q^r - 1)/(q - 1)$ points at infinity of the lines joining $(O, 0, 0)$ and $(O, 0, a)$,
where $O = (0, \dots, 0) \in \mathrm{AG}(N_0, q)$ and a ranges over $\mathbb{F}_{q^r}^*$. Finally, to cover
all the points at infinity, it is enough to add the $(q^r - 1)/(q - 1)$ points at
infinity of the lines joining $(O, 0, 0)$ and $(O, a, 0)$, with $a \in \mathbb{F}_{q^r}^*$.

If S is taken as the trivial saturating set of $\mathrm{PG}(1, q)$ consisting of two
points, then the concatenating construction produces saturating sets in
every odd dimension of the same order of magnitude as the trivial lower
bound (2.1).

Theorem 6.17 *Let $N \geq 3$ be odd. Then in $\mathrm{PG}(N, q)$ there exists a satu-*
rating set S_N with

$$\#S_N = 2q^{(N-1)/2} + 2\frac{q^{(N-1)/2} - 1}{q - 1} = 2 \cdot \frac{q^{(N+1)/2} - 1}{q - 1}.$$

As to even dimensions, the seed can be chosen to be a saturating set in the plane.

Theorem 6.18 *Let S be a saturating set in $\mathrm{PG}(2,q)$. Let $N > 2$ be even. Then in $\mathrm{PG}(N,q)$ there exists a saturating set S_N with*

$$\#S_N = q^{(N-2)/2} \cdot \#S + 2 \cdot \frac{q^{(N-2)/2} - 1}{q - 1}.$$

A number of results on $l(N + 1, 2; q)_3$ for N even can be easily deduced from the constructions described in Section 4.2.

The existence of saturating sets in higher dimensions that are substantially smaller that those arising from the concatenating construction is an interesting open problem.

We remark that the numerous different refined versions of the concatenating construction available in the literature (see the references at the beginning of the present section) may result in smaller saturating sets. However, they do not apply to arbitrary seeds; also, the order of magnitude of the size of the resulting saturating set is the same as that arising from the construction presented here.

7 Multiple saturating sets in the plane

In this final section we deal with the problem of constructing small (m)-saturating sets in $\mathrm{PG}(2,q)$. Most of the results are taken from [10].

7.1 Trivial lower bound

Let S be an (m)-saturating set in $\mathrm{PG}(2,q)$ of size n. Then every point in $\mathrm{PG}(2,q) \setminus S$ can be written in m distinct ways as a linear combination of two points in S. The total number of possible combinations is

$$(q - 1) \cdot \binom{n}{2}.$$

Then

$$(q - 1) \cdot \binom{n}{2} \geq m(q^2 + q + 1 - n),$$

which roughly gives $\#S \geq \sqrt{2mq}$.

7.2 Probabilistic upper bound

The existence of saturating sets of size $\lfloor 5\sqrt{q\log q}\rfloor$ was shown by means of probabilistic methods; see Section 4.2. Taking m disjoint copies of a saturating set clearly gives rise to an (m)-saturating set. Then it is not difficult to prove the existence of an (m)-saturating set S in $\mathrm{PG}(2,q)$ with

$$\#S \le m\lfloor 5\sqrt{q\log q}\rfloor. \tag{7.1}$$

7.3 Constructions

Our aim is to construct (m)-saturating sets in the plane $\mathrm{PG}(2,q)$ with size less than $m \cdot \bar{l}(3,2;q)$, where $\bar{l}(3,2;q)$ denotes the size of the smallest known saturating set in $\mathrm{PG}(2,q)$.

7.3.1 q a square

Theorem 7.1 *Let q be a square and let $4 \le s \le \sqrt{q}+1$ be an even integer. Let $\mathcal{L}_1,\dots,\mathcal{L}_s$ be a set of s lines in $\mathrm{PG}(2,\sqrt{q})$ no three of which concurrent. Then the union of these lines is an (m)-saturating set of size $s(\sqrt{q}+2-s)+s(s-1)/2$ in $\mathrm{PG}(2,q)$, with*

$$m = \frac{1}{8}(s^2 - 2s).$$

Theorem 7.2 *Let q be a square and let $3 \le s \le \sqrt{q}+1$ be an integer. Let $\mathcal{L}_1,\dots,\mathcal{L}_s$ be a set of s lines in $\mathrm{PG}(2,\sqrt{q})$ through a common point P. For any other line $\mathcal{L}$ through P choose $s-1$ points $R_{\mathcal{L}}^{(1)},\dots,R_{\mathcal{L}}^{(s-1)}$ in $\mathcal{L}$ distinct from P. Then the union of the lines $\mathcal{L}_1,\dots,\mathcal{L}_s$ and the point set*

$$\bigcup_{\mathcal{L}\ through\ P,\ \mathcal{L}\ne\mathcal{L}_i} \{R_{\mathcal{L}}^{(1)},\dots,R_{\mathcal{L}}^{(s-1)}\}$$

is an (m)-saturating set in $\mathrm{PG}(2,\sqrt{q})$ of size $s\sqrt{q}+(s-1)(\sqrt{q}+1-s)+1$ with $m = \frac{1}{2}(s^2-s)$.

7.3.2 $q = p^r$, $r \ge 3$, r odd

Theorem 7.3 *Let $q = p^r$, and let H be any additive subgroup of $\mathbb{F}_q$ of size p^s, with $2s < r$. Let m be any integer with $1 \le m \le p^{2r-s}$, and let*

$\tau_1, \tau_2, \ldots, \tau_m$ *be a set of distinct non-zero elements in* $\mathbb{F}_q$. *Let* $L_H(X)$ *be as in* (4.5), *and* $\mathcal{M}_H$ *be as in* (4.6). *Then the set*

$$S = \left\{ (L_H(a) : 1 : 1), (L_H(a) : 0 : 1) \mid a \in \mathbb{F}_q \right\}$$
$$\cup \left\{ (\tau_i : b : 1) \mid b \in \mathcal{M}_H, i = 1, \ldots, m \right\}$$
$$\cup \left\{ (1 : \tau_i : 0) \mid i = 1, \ldots, m \right\} \cup \left\{ (1 : 0 : 0) \right\}$$

is an (m)-*saturating set in* $\mathrm{PG}(2,q)$ *of size at most*

$$\frac{2q}{p^s} + m \frac{(p^s - 1)^2}{p - 1} + m.$$

The order of magnitude of the size of S that appears in Theorem 7.3 is $p^{\max\{r-s,\log_p m \cdot (2s-1)\}}$. If s is chosen to be $\lceil r/3 \rceil$, then

$$\#S \leq \begin{cases} 2q^{\frac{2}{3}} + m + m\frac{q^{\frac{2}{3}} - 2q^{\frac{1}{3}} + 1}{p-1} & \text{if } r \equiv 0 \pmod 3 \\[2mm] 2\left(\frac{q}{p}\right)^{\frac{2}{3}} + m + m\frac{p^2\left(\frac{q}{p}\right)^{\frac{2}{3}} - 2p\left(\frac{q}{p}\right)^{\frac{1}{3}} + 1}{p-1} & \text{if } r \equiv 1 \pmod 3 \\[2mm] 2\frac{1}{p}\left(qp\right)^{\frac{2}{3}} + m + m\frac{(qp)^{\frac{2}{3}} - 2(qp)^{\frac{1}{3}} + 1}{p-1} & \text{if } r \equiv 2 \pmod 3. \end{cases}$$

Theorem 7.4 *Let* $q = p^{2s+1}$, *and let* $1 \leq m \leq p$. *Then there exists an* (m)-*saturating set in* $\mathrm{PG}(2,q)$ *of size less than or equal to*

$$\min \left\{ (v+1)p^{s+1} + m \frac{(p^s - 1)^{2v}}{(p-1)^v (p^{(2s+1)} - 1)^{(v-1)}} + 1 + m \right\}$$

where the minimum is taken over all $v \in \{1, \ldots, 2s+1\}$.

For several values of s and p, Theorem 7.4 improves the probabilistic bound (7.1); namely, there exists some integer v such that

$$(v+1)p^{s+1} + m \frac{(p^s - 1)^{2v}}{(p-1)^v (p^{(2s+1)} - 1)^{(v-1)}} + 1 + m < 5m\sqrt{q \log q}.$$

7.3.3 q a prime For q a prime, the smallest known explicitly described saturating sets have about $Cq^{3/4}$ points; see Section 4.2. Here we show that a slight modification of a construction by Bartocci ([6]; see also [86, Example 4.3]) provides (m)-saturating sets of about the same size with $m < \sqrt{q}$.

Theorem 7.5 *Let s be a divisor of $q-1$ and let H_s be the subgroup of $\mathbb{F}_q^*$ of index s. For an integer $m < \frac{q-1}{s}$, let $V_1, \ldots, V_m$ be m disjoint systems of representatives of the cosets of H_s different from H_s. Let*

$$S = \{(t, t^2) \mid x \in H_s\} \cup \{(0, -v) \mid v \in V_1 \cup \ldots \cup V_m\} \cup T,$$

where T is any subset of the ideal line of size $\lceil (1 + \sqrt{1+8m})/2 \rceil$. Then S is an (m)-saturating set in $\mathrm{PG}(2, q)$ of size

$$\frac{q-1}{s} + m(s-1) + \lceil (1 + \sqrt{1+8m})/2 \rceil,$$

provided that

$$q > \left((s-1)^2 + \sqrt{(s-1)^4 + 2ms^2 + 4s} \right)^2. \tag{7.2}$$

We only present the outline of the proof, which once again involves algebraic curves. It is easily seen that an affine point $P = (a, b)$ in $\mathrm{AG}(2, q)$ off the parabola $\mathcal{P}(q)$ is covered by a secant of

$$S' = \{((t, t^2) \mid x \in H_s)\}$$

passing through two distinct points (x^s, x^{2s}), (y^s, y^{2s}) if and only if (x, y) is a suitable $\mathbb{F}_q$-rational point of the curve

$$\mathcal{C}_P : X^s Y^s - a(X^s + Y^s) + b = 0.$$

The curve $\mathcal{C}_P$ is a generalized Fermat curve. Then $\mathcal{C}_P$ is absolutely irreducible, and its genus is $(s-1)^2$; see e.g. [43]. Then by the Hasse-Weil Theorem, together with (7.2), the curve $\mathcal{C}_P$ has more than $2ms^2 + 4s$ $\mathbb{F}_q$-rational points. Then there are at least m distinct secants of S passing through P. Now let $P = (a, a^2)$ be a point of $\mathcal{P}(q)$ not in S. For each $i = 1, \ldots, m$, write $a = v_{i,j} d_i^s$ with $v_i^j \in V_i$. Then (a, a^2) is collinear with (d_i^{-s}, d_i^{-2s}) and $(0, -v_{i,j})$. Finally, if P is an ideal point, then it is covered at least m times by the ideal line.

It is worth noting that, for several primes $q \leq 1217$, the (m)-saturating sets obtained in Theorem 7.5 have size smaller than m copies of the saturating sets described in Theorem 4.7.

7.4   Concatenating construction

The concatenating construction, as introduced in Section 6.3 for saturating sets, exists for (m)-saturating sets with $m > 1$ as well.

The integer $\lceil (1 + \sqrt{1+8m})/2 \rceil$ appears in Theorem 7.6 below since, if it does not exceed q, than it is the smallest size of an (m)-saturating set in $\mathrm{PG}(1, q)$.

Theorem 7.6 *Let r, m be positive integers such that*

$$\lceil (1 + \sqrt{1 + 8m})/2 \rceil \leq \min\{q, q^r + 1 - m\}.$$

Then there exists in $\mathrm{PG}(2r + 1, q)$ *an* (m)-*saturating set of size*

$$q^r \lceil (1 + \sqrt{1 + 8m})/2 \rceil + m \frac{q^r - 1}{q - 1}.$$

Theorem 7.7 *Let S be an* (m)-*saturating set in* $\mathrm{PG}(2, q)$ *of size* n_0, *and let r be a positive integer such that* $n_0 \leq q^r + 1 - m$. *Then there exists an* (m)-*saturating set of size* $q^r n_0 + m \frac{q^r - 1}{q - 1}$ *in* $\mathrm{PG}(2r + 2, q)$.

7.5 Computational results

The results of a computer search on (2)-saturating sets, minimal with respect to set-theoretical inclusion, are reported in Table 1 below.

Table 1: Minimal (2)-saturating sets in $\mathrm{PG}(2, q)$

q	$\bar{l}(3, 2; q)$	Spectrum of sizes
3	4^1	6^4
4	5^1	$6^2 7^5$
5	6^6	$6^1 7^4 8^{18}$
7	6^3	$8^{13} 9^{564} 10^{424}$
8	6^1	$8^2 9^{154} 10^{3372} 11^{611}$
9	6^1	$8^1 9^{57} 10^{12145} 11^{76749} 12^{3049}$
11	7^1	$10^{1348} [11 - 14]$
13	8^2	$10^2 11^{50794} [12 - 16]$
16	9^4	$11^{52} [12 - 19]$
17	10^{3640}	$[12 - 20]$
19	10^{36}	$[13 - 22]$
23	10^1	$[15 - 26]$
25	12	$[17 - 28]$
27	12	$[17 - 30]$
29	13	$[19 - 32]$
31	14	$[19, 21 - 34]$
32	13	$[20 - 35]$
37	15	$[23, 26 - 40]$
41	16	$[25, 29 - 44]$
43	16	$[25, 30 - 46]$
47	18	$[27, 34 - 50]$
49	18	$[29, 34 - 52]$

The symbol $\bar{l}(3,2;q)$ stands for the smallest known cardinality of a saturating set in PG$(2,q)$. For $q \leq 23$, actually $l(3,2;q) = \bar{l}(3,2;q)$ holds. In the third column, some values of n for which minimal (2)-saturating sets in PG$(2,q)$ of size n exist are reported. For $3 \leq q \leq 17$, the complete spectrum of sizes have been determined. In both the second and the third columns, the superscript over a size indicates the number of distinct (2)-saturating sets of that size (up to collineations). For $3 \leq q \leq 9$, the complete classification of distinct minimal (2)-saturating sets in PG$(2,q)$ has been obtained. Some of the values in the spectrum can be obtained from the explicit constructions of Section 7.3.

An ongoing investigation for $m > 2$ has produced so far a full classification of minimal (m)-saturating sets for $q \leq 4$, $m \leq (q+1)\binom{q}{2}$, as well for other sporadic values of q and m; see [10]. Some of the results are summarized in Table 2; again, the superscript over a size indicates the number of distinct (m)-saturating sets of that size (up to collineations).

Table 2: Minimal (m)-saturating sets in PG$(2,q)$

m	3	4	5	6	7	8
$q=3$	$6^1 7^2$	$7^1 8^2$	$8^1 9^1$	9^3	$9^1 10^1$	$9^1 10^1$
$q=4$	$6^1 7^2 8^7$	$8^3 9^{10}$	$9^6 10^7$	$9^2 10^8 11^1$	$10^2 11^{13}$	$11^7 12^5$
$q=5$	$8^5 9^{65}$	$9^7 10^{133}$	$10^{26} 11^{162}$	$11^{121} 12^{102} 11^3 12^{361}$		
$q=7$	$8^1 9^{10} 10^{1506} 11^{10014} 10^2 11^{3167} 12^{67454} 11^2 12^{11301} 13^{239140}$					

Acknowledgements

I am grateful to Gábor Korchmáros and Alexander A. Davydov for a number of valuable comments on an earlier draft.

References

[1] Luca Maria Abatangelo, *Complete caps in a Galois space* PG$(3, q)$, *q even*, Atti Sem. Mat. Fis. Univ. Modena **29** (1980), no. 2, 215–221.

[2] Nurdagül Anbar, Daniele Bartoli, Massimo Giulietti, and Irene Platoni, *Small complete caps from singular cubics*, Submitted, 2012.

[3] Nurdagül Anbar and Massimo Giulietti, *Bicovering arcs and small complete caps from elliptic curves*, Submitted, 2012.

[4] Tsonka S. Baicheva and Evgenia D. Velikova, *Covering radii of ternary linear codes of small dimensions and codimensions*, IEEE Trans. Inform. Theory **43** (1997), no. 6, 2057–2061.

[5] Tsonka S. Baicheva and Evgenia D. Velikova, *Correction to: "Covering radii of ternary linear codes of small dimensions and codimensions" [IEEE Trans. Inform. Theory 43 (1997), no. 6, 2057–2061]*, IEEE Trans. Inform. Theory **44** (1998), no. 5, 2032.

[6] Umberto Bartocci, *Dense k-systems in Galois planes*, Boll. Un. Mat. Ital. D (6) **2** (1983), no. 1, 71–77.

[7] Daniele Bartoli, Private communication, 2012.

[8] Daniele Bartoli, Alexander A. Davydov, Giorgio Faina, Stefano Marcugini, and Fernanda Pambianco, *New upper bounds on the smallest size of a complete arc in the plane* $PG(2, q)$, Proceedings XIII Int. Workshop on Algebraic and Combin. Coding Theory, ACCT2012, June 2012, pp. 60–66.

[9] Daniele Bartoli, Alexander A. Davydov, Giorgio Faina, Stefano Marcugini, and Fernanda Pambianco, *On sizes of complete arcs in* $PG(2, q)$, Discrete Math. **312** (2012), no. 3, 680–698.

[10] Daniele Bartoli, Alexander A. Davydov, Massimo Giulietti, Stefano Marcugini, and Fernanda Pambianco, *Multiple coverings of the farthest-off points*, In preparation, 2012.

[11] Daniele Bartoli, Alexander A. Davydov, Stefano Marcugini, and Fernanda Pambianco, *New type of estimations for the smallest size of complete arcs in* $PG(2, q)$, Proceedings XIII Int. Workshop on Algebraic and Combin. Coding Theory, ACCT2012, June 2012, pp. 67–72.

[12] Daniele Bartoli and Massimo Giulietti, *Small complete caps in* $PG(3, q)$, In preparation, 2012.

[13] Endre Boros, Tamás Szőnyi, and Krisztián Tichler, *On defining sets for projective planes*, Discrete Math. **303** (2005), no. 1–3, 17–31.

[14] Richard A. Brualdi, Simon Litsyn, and Vera S. Pless, *Covering radius*, Handbook of coding theory, Vol. I, II, North-Holland, Amsterdam, 1998, pp. 755–826.

[15] Richard A. Brualdi, Vera S. Pless, and Richard M. Wilson, *Short codes with a given covering radius*, IEEE Trans. Inform. Theory **35** (1989), no. 1, 99–109.

[16] Aiden A. Bruen, James W. P. Hirschfeld, and David L. Wehlau, *Cubic curves, finite geometry and cryptography*, Acta Appl. Math. **115** (2011), no. 3, 265–278.

[17] Robert Calderbank and William M. Kantor, *The geometry of two-weight codes*, Bull. London Math. Soc. **18** (1986), no. 2, 97–122.

[18] Gérard D. Cohen, Iiro Honkala, Simon Litsyn, and Antoine Lobstein, *Covering codes*, North-Holland Mathematical Library, vol. 54, North-Holland Publishing Co., Amsterdam, 1997.

[19] Gérard D. Cohen, Antoine Lobstein, and Neil J. A. Sloane, *Further results on the covering radius of codes*, IEEE Trans. Inform. Theory **32** (1986), no. 5, 680–694.

[20] Alexander A. Davydov, *Construction of linear covering codes*, Problems of Information Transmission **26** (1990), no. 4, 317–331.

[21] Alexander A. Davydov, *Constructions and families of covering codes and saturated sets of points in projective geometry*, IEEE Trans. Inform. Theory **41** (1995), no. 6, part 2, 2071–2080.

[22] Alexander A. Davydov, *Constructions and families of nonbinary linear codes with covering radius 2*, IEEE Trans. Inform. Theory **45** (1999), no. 5, 1679–1686.

[23] Alexander A. Davydov, *New constructions of covering codes*, Des. Codes Cryptogr. **22** (2001), no. 3, 305–316.

[24] Alexander A. Davydov and Anna Yu. Drozhzhina-Labinskaya, *Constructions, families, and tables of binary linear covering codes*, IEEE Trans. Inform. Theory **40** (1994), no. 4, 1270–1279.

[25] Alexander A. Davydov, Giorgio Faina, Stefano Marcugini, and Fernanda Pambianco, *Computer search in projective planes for the sizes of complete arcs*, J. Geom. **82** (2005), no. 1–2, 50–62.

[26] Alexander A. Davydov, Giorgio Faina, Stefano Marcugini, and Fernanda Pambianco, *Locally optimal (nonshortening) linear covering codes and minimal saturating sets in projective spaces*, IEEE Trans. Inform. Theory **51** (2005), no. 12, 4378–4387.

[27] Alexander A. Davydov, Giorgio Faina, Stefano Marcugini, and Fernanda Pambianco, *On sizes of complete caps in projective spaces* PG(n, q) *and arcs in planes* PG(2, q), J. Geom. **94** (2009), no. 1–2, 31–58.

[28] Alexander A. Davydov, Massimo Giulietti, Stefano Marcugini, and Fernanda Pambianco, *New inductive constructions of complete caps in* $\mathrm{PG}(N,q)$, *q even*, J. Combin. Des. **18** (2010), no. 3, 177–201.

[29] Alexander A. Davydov, Massimo Giulietti, Stefano Marcugini, and Fernanda Pambianco, *Linear nonbinary covering codes and saturating sets in projective spaces*, Adv. Math. Commun. **5** (2011), no. 1, 119–147.

[30] Alexander A. Davydov, Stefano Marcugini, and Fernanda Pambianco, *On saturating sets in projective spaces*, J. Combin. Theory Ser. A **103** (2003), no. 1, 1–15.

[31] Alexander A. Davydov, Stefano Marcugini, and Fernanda Pambianco, *Complete caps in projective spaces* $\mathrm{PG}(n,q)$, J. Geom. **80** (2004), no. 1–2, 23–30.

[32] Alexander A. Davydov, Stefano Marcugini, and Fernanda Pambianco, *Linear codes with covering radius 2, 3 and saturating sets in projective geometry*, IEEE Trans. Inform. Theory **50** (2004), no. 3, 537–541.

[33] Alexander A. Davydov and Patric R. J. Östergård, *New linear codes with covering radius 2 and odd basis*, Des. Codes Cryptogr. **16** (1999), no. 1, 29–39.

[34] Alexander A. Davydov and Patric R. J. Östergård, *New quaternary linear codes with covering radius 2*, Finite Fields Appl. **6** (2000), no. 2, 164–174.

[35] Alexander A. Davydov and Patric R. J. Östergård, *On saturating sets in small projective geometries*, European J. Combin. **21** (2000), no. 5, 563–570.

[36] Alexander A. Davydov and Patric R. J. Östergård, *Linear codes with covering radius R = 2, 3 and codimension tR*, IEEE Trans. Inform. Theory **47** (2001), no. 1, 416–421.

[37] Alexander A. Davydov and Patric R. J. Östergård, *Recursive constructions of complete caps*, J. Statist. Plann. Inference **95** (2001), no. 1–2, 167–173, Special issue on design combinatorics: in honor of S. S. Shrikhande.

[38] Alexander A. Davydov and Patric R. J. Östergård, *Linear codes with covering radius 3*, Des. Codes Cryptogr. **54** (2010), no. 3, 253–271.

[39] Tuvi Etzion and Beniamin Mounits, *Quasi-perfect codes with small distance*, IEEE Trans. Inform. Theory **51** (2005), no. 11, 3938–3946.

[40] Giorgio Faina, *Complete caps having less than* $(q^2 + 1)/2$ *points in common with an elliptic quadric of* PG$(3, q)$, q *odd*, Rend. Mat. Appl. (7) **8** (1988), no. 2, 277–281.

[41] Giorgio Faina and Fernanda Pambianco, *A class of complete k-caps in* PG$(3, q)$ *for q an odd prime*, J. Geom. **57** (1996), no. 1-2, 93–105.

[42] Giorgio Faina, Fabio Pasticci, and Lorenzo Schmidt, *Small complete caps in galois spaces*, Ars Combin., to appear.

[43] Stefania Fanali and Massimo Giulietti, *On the number of rational points of generalized Fermat curves over finite fields*, Int. J. Number Theory **8** (2012), no. 4, 1087–1097.

[44] Ernst M. Gabidulin, Alexander A. Davydov, and Leonid M. Tombak, *Linear codes with covering radius 2 and other new covering codes*, IEEE Trans. Inform. Theory **37** (1991), no. 1, 219–224.

[45] Massimo Giulietti, *On small dense sets in Galois planes*, Electron. J. Combin. **14** (2007), no. 1, Research Paper 75.

[46] Massimo Giulietti, *Small complete caps in Galois affine spaces*, J. Algebraic Combin. **25** (2007), no. 2, 149–168.

[47] Massimo Giulietti, *Small complete caps in* PG(N, q), q *even*, J. Combin. Des. **15** (2007), no. 5, 420–436.

[48] Massimo Giulietti and Fernando Torres, *On dense sets related to plane algebraic curves*, Ars Combin. **72** (2004), 33–40.

[49] Heikki O. Hämäläinen, Iiro S. Honkala, Markku K. Kaikkonen, and Simon N. Litsyn, *Bounds for binary multiple covering codes*, Des. Codes Cryptogr. **3** (1993), no. 3, 251–275.

[50] Heikki O. Hämäläinen, Iiro S. Honkala, Simon Litsyn, and Patric R. J. Östergård, *Football pools – a game for mathematicians*, Amer. Math. Monthly **102** (1995), no. 7, 579–588.

[51] Heikki O. Hämäläinen, Iiro S. Honkala, Simon N. Litsyn, and Patric R. J. Östergård, *Bounds for binary codes that are multiple coverings of the farthest-off points*, SIAM J. Discrete Math. **8** (1995), no. 2, 196–207.

[52] Heikki O. Hämäläinen and Seppo Rankinen, *Upper bounds for football pool problems and mixed covering codes*, J. Combin. Theory Ser. A **56** (1991), no. 1, 84–95.

[53] James W. P. Hirschfeld, *Finite projective spaces of three dimensions*, Oxford Mathematical Monographs, The Clarendon Press Oxford University Press, New York, 1985, Oxford Science Publications.

[54] James W. P. Hirschfeld, *Algebraic curves, arcs, and caps over finite fields*, Quaderni del Dipartimento di Matematica dell' Università del Salento 5, Dipartimento di Matematica, Università del Salento, Lecce, 1986.

[55] James W. P. Hirschfeld, *Projective geometries over finite fields*, second ed., Oxford Mathematical Monographs, The Clarendon Press Oxford University Press, New York, 1998.

[56] James W. P. Hirschfeld and Leo Storme, *The packing problem in statistics, coding theory and finite projective spaces*, J. Statist. Plann. Inference **72** (1998), no. 1-2, 355–380, R. C. Bose Memorial Conference (Fort Collins, CO, 1995).

[57] James W. P. Hirschfeld and Leo Storme, *The packing problem in statistics, coding theory and finite projective spaces: update 2001*, Finite geometries, Dev. Math., vol. 3, Kluwer Acad. Publ., Dordrecht, 2001, pp. 201–246.

[58] James W. P. Hirschfeld and Jeff A. Thas, *General Galois geometries*, Oxford Mathematical Monographs, The Clarendon Press Oxford University Press, New York, 1991, Oxford Science Publications.

[59] Iiro S. Honkala, *On the normality of multiple covering codes*, Discrete Math. **125** (1994), no. 1–3, 229–239, 13th British Combinatorial Conference (Guildford, 1991).

[60] Iiro S. Honkala and Simon Litsyn, *Generalizations of the covering radius problem in coding theory*, Bull. Inst. Combin. Appl. **17** (1996), 39–46.

[61] Jeong H. Kim and Van H. Vu, *Small complete arcs in projective planes*, Combinatorica **23** (2003), no. 2, 311–363.

[62] György Kiss, István Kovács, Klavdija Kutnar, János Ruff, and Primož Šparl, *A note on a geometric construction of large Cayley graphs of given degree and diameter*, Stud. Univ. Babeş-Bolyai Math. **54** (2009), no. 3, 77–84.

[63] Gábor Korchmáros, *New examples of complete k-arcs in* PG(2, q), European J. Combin. **4** (1983), no. 4, 329–334.

[64] Sándor J. Kovács, *Small saturated sets in finite projective planes*, Rend. Mat. Appl. (7) **12** (1992), no. 1, 157–164.

[65] Rudolf Lidl and Harald Niederreiter, *Finite fields*, Encyclopedia of Mathematics and its Applications, vol. 20, Addison-Wesley Publishing Company Advanced Book Program, Reading, MA, 1983, With a foreword by P. M. Cohn.

[66] Lucio Lombardo-Radice, *Sul problema dei k-archi completi in* $S_{2,q}$ *(q = p^t, p primo dispari)*, Boll. Un. Mat. Ital. (3) **11** (1956), 178–181.

[67] Mirka Miller and Jozef Siran, *Moore graphs and beyond: A survey of the degree/diameter problem*, Electronic Journal of Combinatorics **Dynamic Survey DS14** (2005), Online journal, print version is known as Journal of Combinatorics (electronic ISSN 1077-8926, print ISSN 1097-1440).

[68] Patric R. J. Östergård and Heikki O. Hämäläinen, *A new table of binary/ternary mixed covering codes*, Des. Codes Cryptogr. **11** (1997), no. 2, 151–178.

[69] Fernanda Pambianco and Leo Storme, Unpublished manuscript, 1995.

[70] Fernanda Pambianco and Leo Storme, *Small complete caps in spaces of even characteristic*, J. Combin. Theory Ser. A **75** (1996), no. 1, 70–84.

[71] Giuseppe Pellegrino, *On complete caps, not ovaloids, in the space* PG(3, q) *with q odd*, Rend. Circ. Mat. Palermo (2) **47** (1998), no. 1, 141–168.

[72] Jörn Quistorff, *On codes with given minimum distance and covering radius*, Beiträge Algebra Geom. **41** (2000), no. 2, 469–478.

[73] Jörn Quistorff, *Correction: "On codes with given minimum distance and covering radius" [Beiträge Algebra Geom. 41 (2000), no. 2, 469–478; MR1801437 (2001i:94082)] by J. Quistorff*, Beiträge Algebra Geom. **42** (2001), no. 2, 601–611.

[74] Hans-Georg Rück, *A note on elliptic curves over finite fields*, Math. Comp. **49** (1987), no. 179, 301–304.

[75] Beniamino Segre, *On complete caps and ovaloids in three-dimensional Galois spaces of characteristic two*, Acta Arith. **5** (1959), 315–332.

[76] Beniamino Segre, *Ovali e curve σ nei piani di Galois di caratteristica due.*, Atti Accad. Naz. Lincei Rend. Cl. Sci. Fis. Mat. Nat. (8) **32** (1962), 785–790.

[77] Beniamino Segre, *Introduction to Galois geometries*, Atti Accad. Naz. Lincei Mem. Cl. Sci. Fis. Mat. Natur. Sez. I (8) **8** (1967), 133–236.

[78] Beniamino Segre, *Proprietà elementari relative ai segmenti ed alle coniche sopra un campo qualsiasi ed una congettura di Seppo Ilkka per il caso dei campi di Galois*, Ann. Mat. Pura Appl. (4) **96** (1972), 289–337.

[79] Beniamino Segre and Umberto Bartocci, *Ovali ed altre curve nei piani di Galois di caratteristica due*, Acta Arith. **18** (1971), 423–449.

[80] Henning Stichtenoth, *Algebraic function fields and codes*, second ed., Graduate Texts in Mathematics, vol. 254, Springer-Verlag, Berlin, 2009.

[81] René Struik, *Covering codes*, Ph.D. thesis, Eindhoven University of Technology, 1994.

[82] Tamás Szőnyi, *Complete arcs in finite projective geometries*, Ph.D. thesis, Univ. L. Eötvös, Budapest, 1984.

[83] Tamás Szőnyi, *Small complete arcs in Galois planes*, Geom. Dedicata **18** (1985), no. 2, 161–172.

[84] Tamás Szőnyi, *Note on the order of magnitude of k for complete k-arcs in* $PG(2, q)$, Discrete Math. **66** (1987), no. 3, 279–282.

[85] Tamás Szőnyi, *Arcs in cubic curves and 3-independent subsets of abelian groups*, Combinatorics (Eger, 1987), Colloq. Math. Soc. János Bolyai, vol. 52, North-Holland, Amsterdam, 1988, pp. 499–508.

[86] Tamás Szőnyi, *Complete arcs in galois planes: a survey*, Quaderni del Seminario di Geometrie Combinatorie 94, Dipartimento di Matematica "G. Castelnuovo", Università degli Studi di Roma "La Sapienza", Roma, January 1989.

[87] Tamás Szőnyi, *Some applications of algebraic curves in finite geometry and combinatorics*, Surveys in combinatorics, 1997 (London), London Math. Soc. Lecture Note Ser., vol. 241, Cambridge Univ. Press, Cambridge, 1997, pp. 197–236.

[88] Jacobus H. van Lint, *Codes*, Handbook of combinatorics, Vol. 1 (R. L. Graham, M. Grötschel, and L. Lovász, eds.), MIT Press, Cambridge, MA, USA, 1995, pp. 773–807.

[89] José Felipe Voloch, *A note on elliptic curves over finite fields*, Bull. Soc. Math. France **116** (1988), no. 4, 455–458 (1989).

[90] José Felipe Voloch, *On the completeness of certain plane arcs. II*, European J. Combin. **11** (1990), no. 5, 491–496.

[91] José Felipe Voloch, Private communication, 2011.

[92] Lawrence C. Washington, *Elliptic curves*, second ed., Discrete Mathematics and its Applications (Boca Raton), Chapman & Hall/CRC, Boca Raton, FL, 2008, Number theory and cryptography.

[93] William C. Waterhouse, *Abelian varieties over finite fields*, Ann. Sci. École Norm. Sup. (4) **2** (1969), 521–560.

Dipartimento di Matematica e Informatica
Università degli Studi di Perugia
Via Vanvitelli, 1 – 06123 – Italy
giuliet@dmi.unipg.it

Bent functions and their connections to combinatorics

Tor Helleseth and Alexander Kholosha

Abstract

Bent functions were first introduced by Rothaus in 1976 as an interesting combinatorial object with the important property of having the maximum distance to all affine functions. Bent functions have many applications to coding theory, cryptography and sequence design. For many years the focus was on the construction of binary bent functions. There are several known examples of binary monomial and binomial bent functions. In 1985, Kumar, Scholtz and Welch generalized bent functions to the case of an arbitrary finite field. In recent years, new results on p-ary bent functions have appeared. This paper gives an updated overview of some of the recent results and open problems on bent functions and their connections. This includes some recent constructions of weakly regular monomial and binomial bent functions useful for constructing certain combinatorial objects such as partial difference sets, strongly regular graphs and association schemes.

1 Introduction

By a p-ary function of n variables we understand a mapping of the Galois field $\mathbb{F}_{p^n}$ (or of the vector space $\mathbb{F}_p^n$ of all p-ary vectors of length n) that takes its values in $\mathbb{F}_p$. In this paper, we shall always endow this vector space with the structure of a field, thanks to the choice of a basis of $\mathbb{F}_{p^n}$ over $\mathbb{F}_p$. The case when $p = 2$ corresponds to Boolean functions that are frequently used in the pseudo-random generators of stream ciphers and play a central role in their security.

Boolean bent functions are interesting combinatorial objects introduced by Rothaus [63] in 1976. These are functions of an even number of variables n, that are maximally nonlinear in the sense that their Walsh transform takes precisely the values $\pm 2^{n/2}$. This corresponds to the fact that their Hamming distance to all affine functions is optimal. Therefore, the construction of Boolean bent functions is heavily related to the covering radius of the first order Reed-Muller code. Nonlinearity of bent functions is extremely useful and important in cryptographic applications where one frequently needs functions which cannot be easily approximated by linear functions. Using bent functions, one can construct families of sequences with optimal cross-correlation properties. This justifies the great interest in bent functions from researchers in the areas of coding theory, cryptography and sequence design.

Kumar, Scholtz and Welch in [47] generalized the notion of Boolean bent functions to the case of functions over an arbitrary finite field. It was discovered recently that p-ary bent functions are useful for constructing certain combinatorial objects such as partial difference sets, strongly regular graphs and association schemes.

Despite their simple and natural definition, bent functions turn out to admit a very complicated structure in general. A complete classification of bent functions looks hopeless even in the binary case. In the case of generalized bent functions, things are naturally much more complicated. However, many explicit methods are known for constructing bent functions either from scratch (primary constructions) or based on other, simpler bent functions (secondary constructions).

The paper is organized as follows. In Section 2, we discuss different polynomial representations of a p-ary function. Further, we define a p-ary bent function in terms of its Walsh transform, and two subclasses of bent functions, weakly and non-weakly regular. We also provide some basic properties of bent functions and define an equivalence relation that preserves bentness. In Section 3, we present the fundamental classes of bent functions and a few univariate and bivariate constructions. In Section 4, connections to coding theory and, in particular, to the first-order Reed-Muller codes are discussed. Section 5 shows that bent functions can be used to discover new classes of such combinatorial objects as strongly regular graphs and association schemes. Finally, in Section 6 we show that bent functions are useful in sequence design.

2   Definitions and basic properties

The *trace function* from a finite field $\mathbb{F}_{p^n}$ to a subfield $\mathbb{F}_{p^k}$ is a mapping $\mathrm{Tr}_k^n : \mathbb{F}_{p^n} \to \mathbb{F}_{p^k}$ defined by

$$\mathrm{Tr}_k^n(x) = \sum_{i=0}^{\frac{n}{k}-1} x^{p^{ki}}.$$

For $k = 1$ we have the absolute trace and use the notation $\mathrm{Tr}_n()$ for $\mathrm{Tr}_1^n()$. Note that $(x,y) \mapsto \mathrm{Tr}_k^n(xy)$ defines a non-degenerate bilinear form.

Let $f : \mathbb{F}_p^n \to \mathbb{F}_p$ be a p-ary function. If we identify $\mathbb{F}_p^n$ with $\mathbb{F}_{p^n}$ (by considering $\mathbb{F}_{p^n}$ as an n-dimensional vector space over $\mathbb{F}_p$), all p-ary functions can be described by $\mathrm{Tr}_n(F(x))$ for some function $F : \mathbb{F}_{p^n} \to \mathbb{F}_{p^n}$ of degree at most $p^n - 1$. This is called the *univariate representation*. If we do not identify $\mathbb{F}_p^n$ with $\mathbb{F}_{p^n}$, the p-ary function has a representation as a multinomial in $x_1, \ldots, x_n$, where the variables x_i occur with exponent at most $p - 1$. This is called the *multivariate representation* or *algebraic normal form (ANF)*. The multivariate representation is unique. The *algebraic degree* of a p-ary function is the degree of the polynomial giving its multivariate representation.

The univariate representation is not unique. A unique univariate form of a Boolean function, called the *trace representation* is the following:

$$f(x) = \sum_{j \in \Gamma_n} \mathrm{Tr}_{o(j)}(A_j x^j) + A_{p^n-1} x^{p^n-1},$$

where Γ_n is the set of integers obtained by choosing the smallest element in each cyclotomic coset modulo $p^n - 1$ (with respect to p), $o(j)$ is the size of the cyclotomic coset containing j, $A_j \in \mathbb{F}_{p^{o(j)}}$ and $A_{p^n-1} \in \mathbb{F}_p$. The algebraic degree of f is equal to $\max_{\{j \mid A_j \neq 0\}}(w_p(j))$, where $w_p(j)$ is the weight of the p-ary expansion of j. Thus, p-ary linear functions are exactly all functions of the form $\mathrm{Tr}_n(ax)$ for some $a \in \mathbb{F}_{p^n}$.

The *bivariate representation* of a p-ary function of even dimension $n = 2m$ is defined as follows: we identify $\mathbb{F}_p^n$ with $\mathbb{F}_{p^m} \times \mathbb{F}_{p^m}$ and consider the argument of f as an ordered pair (x, y) of elements in $\mathbb{F}_{p^m}$. There exists a unique bivariate polynomial $\sum_{0 \leq i,j \leq p^m-1} A_{i,j} x^i y^j$ over $\mathbb{F}_{p^m}$ that represents f. The algebraic degree of f is equal to $\max_{\{(i,j) \mid A_{i,j} \neq 0\}}(w_p(i) + w_p(j))$. Since f takes on its values in $\mathbb{F}_p$ the bivariate representation can be written in the form $f(x, y) = \mathrm{Tr}_m(P(x, y))$, where $P(x, y)$ is some polynomial in two variables over $\mathbb{F}_{p^m}$.

It is always tempting to find functions with desirable properties (for instance, bent functions) that also have a simple representation as a univariate/bivariate polynomial. Having a simple representation is not just an esthetic matter but also a way to prove some facts otherwise intangible. It is natural to take the number of terms in a univariate/bivariate polynomial representation as a measure of its simplicity. However, bent functions with a simple univariate representation tend to have a complex bivariate representation and vice versa. This is illustrated in Subsection 3.2.

For a function $f : \mathbb{F}_{p^n} \to \mathbb{F}_p$ its *Walsh transform* $\hat{f}$ is defined by

$$\hat{f}(y) = \sum_{x \in \mathbb{F}_{p^n}} \zeta_p^{f(x) - \mathrm{Tr}_n(xy)}$$

where $\zeta_p = e^{\frac{2\pi i}{p}}$ is the complex primitive p-th root of unity and elements of $\mathbb{F}_p$ are considered as integers modulo p. The function f can be recovered from the Walsh transform by the inverse transform

$$\zeta_p^{f(x)} = \frac{1}{p^n} \sum_{y \in \mathbb{F}_{p^n}} \hat{f}(y) \zeta_p^{\mathrm{Tr}_n(xy)}.$$

If the function f is given by a multivariate representation then the standard inner product $x \cdot y$ of vectors in $\mathbb{F}_p^n$ is used in place of the bilinear form $\mathrm{Tr}_n(xy)$.

Observe the following useful property of the Walsh transform coefficients:

$$\sum_{b \in \mathbb{F}_{p^n}} |\hat{f}(b)|^2 = \sum_b \sum_x \sum_y \zeta_p^{f(x)-f(y)+\mathrm{Tr}_n(b(y-x))}$$

$$= \sum_x \sum_y \zeta_p^{f(x)-f(y)} \sum_b \zeta_p^{\mathrm{Tr}_n(b(y-x))}$$

$$= p^{2n},$$

where the sums over b, x, y run through $\mathbb{F}_{p^n}$. This implies that the average value of the absolute square of the Walsh transform coefficients is p^n. Functions with all Walsh coefficients having the same absolute square are called *bent*.

Definition 2.1 Let $f : \mathbb{F}_{p^n} \to \mathbb{F}_p$.

(i) f is a *p-ary bent function* if all its Walsh coefficients satisfy $|\hat{f}(y)|^2 = p^n$.

(ii) A bent function f is *regular* if for every $y \in \mathbb{F}_{p^n}$, the normalized Walsh coefficient $p^{-n/2}\hat{f}(y)$ is equal to a complex p-th root of unity, i.e., $p^{-n/2}\hat{f}(y) = \zeta_p^{f^*(y)}$ for some function f^* mapping $\mathbb{F}_{p^n}$ into $\mathbb{F}_p$.

(iii) A bent function f is *weakly regular* if there exists a complex number u having unit magnitude such that $up^{-n/2}\hat{f}(y) = \zeta_p^{f^*(y)}$ for all $y \in \mathbb{F}_{p^n}$.

Hereafter, $p^{n/2}$ with odd n stands for the *positive* square root of p^n. If f is a weakly regular p-ary bent function then f^* is the *dual* of f.

In the Boolean case, the dimension n must obviously be even. Boolean bent functions are regular. If f is a weakly regular bent function and $\hat{f}(y) = u^{-1}p^{n/2}\zeta_p^{f^*(y)}$ for $y \in \mathbb{F}_{p^n}$ then the inverse Walsh transform of f gives

$$up^{n/2}\zeta_p^{f(-x)} = \sum_{y \in \mathbb{F}_{p^n}} \zeta_p^{f^*(y)-\mathrm{Tr}_n(xy)} = \hat{f}_*(x).$$

Thus, the dual of a weakly regular bent function is again a weakly regular bent function.

When p is odd, the Walsh transform coefficients of a p-ary bent function f satisfy

$$p^{-n/2}\hat{f}(y) = \begin{cases} \pm \zeta_p^{f^*(y)}, & \text{if } n \text{ is even or } n \text{ is odd and } p \equiv 1 \pmod 4, \\ \pm i \zeta_p^{f^*(y)}, & \text{if } n \text{ is odd and } p \equiv 3 \pmod 4, \end{cases}$$

(2.1)

where i is a complex primitive 4th root of unity [47]. Therefore, regular bent functions can only be found for even n and for odd n with $p \equiv 1 \pmod 4$. Also, for a weakly regular bent function, the constant u in Definition 2.1 can only be equal to ± 1 or $\pm i$.

For any $j \in \mathbb{F}_p$ and $y \in \mathbb{F}_{p^n}$, we write $N_y(j) = \#\{x \in \mathbb{F}_{p^n} : f(x) - \text{Tr}_n(yx) = j\}$. We see that

$$\hat{f}(y) = N_y(0) + N_y(1)\zeta_p + \cdots + N_y(p-1)\zeta_p^{p-1}.$$

If f is a p-ary bent function of even dimension n then using (2.1) we obtain that for any $y \in \mathbb{F}_{p^n}$

$$\sum_{j=0}^{p-1} N_y(j)\zeta_p^j \mp p^{n/2}\zeta_p^{f^*(y)} = 0,$$

where all the coefficients are integer numbers. It is well known that the polynomial $p(x) = x^{p-1} + \cdots + x^2 + x + 1$ is irreducible over the rational number field, $p(\zeta_p) = 0$ and, thus, $p(x)$ is the minimal polynomial of ζ_p over the rational numbers. Therefore, for a fixed y, the values of $N_y(j)$ are all equal except for $N_y(f^*(y))$ that differs from the rest by $\pm p^{n/2}$. Thus, $N_y(j)$ takes on two different values.

If $n = 2m + 1$ is odd then we use the following well-known identity:

$$\sum_{j=1}^{p-1} \left(\frac{j}{p}\right)\zeta_p^j = \begin{cases} p^{1/2}, & \text{if } p \equiv 1 \pmod 4 \\ ip^{1/2}, & \text{if } p \equiv 3 \pmod 4 \end{cases}$$

where $\left(\frac{j}{p}\right)$ denotes the Legendre symbol that is equal to 0 if p divides j (in particular, $\left(\frac{0}{p}\right) = 0$), equal to 1 if j is a square modulo p and to -1 otherwise. It follows that for a fixed y, the difference between $N_y(j)$ and $N_y(f^*(y))$ is equal to p^m for a half of those j such that $j \neq f^*(y)$ and is equal to $-p^m$ for the remainder. In particular, $N_y(f^*(y)) \geq p^m$. Thus, $N_y(j)$ takes on three different values.

Theorem 2.2 *Let* $n = 2m$ *and* $f : \mathbb{F}_{p^n} \to \mathbb{F}_p$ *be a p-ary bent function. Then*

$$N_0(0) = p^{n-1} - \mu(p^m - p^{m-1}), \quad N_0(1) = \cdots = N_0(p-1) = p^{n-1} + \mu p^{m-1},$$

where $\mu = \pm 1$. *In particular,* $\hat{f}(0) = \pm p^m$.

If f is a p-ary bent function on $\mathbb{F}_{p^n}$ then [63, 42]

$$2 \le \deg f \le \frac{(p-1)n}{2} + 1.$$

Moreover, if f is a weakly regular bent function (that includes the case of Boolean bent functions) with $(p-1)n \ge 4$ then

$$2 \le \deg f \le \frac{(p-1)n}{2}.$$

Theorem 2.3 ([63, 47]) *(The propagation criterion) Let* $f : \mathbb{F}_{p^n} \to \mathbb{F}_p$. *Then* f *is bent if and only if the mappings* $D_a(f) : \mathbb{F}_{p^n} \to \mathbb{F}_p$ *with* $D_a(f)(x) = f(x + a) - f(x)$ *are balanced for all* $a \ne 0$, *i.e., the number of solutions* $f(x + a) - f(x) = b$ *is* p^{n-1} *for all* $a \ne 0$ *and all* $b \in \mathbb{F}_p$.

For p-ary functions, it is natural to consider equivalence relations that preserve some critical properties. The most important equivalence is defined below.

Definition 2.4 Functions $f, g : \mathbb{F}_p^n \to \mathbb{F}_p$ are *extended-affine equivalent* (in brief, EA-equivalent) if there exist affine permutations L_1 of $\mathbb{F}_p$ and L_2 of $\mathbb{F}_p^n$ and an affine function $l : \mathbb{F}_p^n \to \mathbb{F}_p$ such that $g(x) = (L_1 \circ f \circ L_2)(x) + l(x)$. A class of functions is *complete* if it is a union of EA-equivalence classes. The *completed class* is the smallest possible complete class that contains the original one.

If a function f is not affine then any function EA-equivalent to it has the same algebraic degree as f. On the other hand, the function f and its dual f^* do not necessarily have the same degree (this can be seen in many examples from Subsections 3.3 and 3.2). Any function which is EA-equivalent to a bent function is bent. It is common to consider a bent function as "new" if it is not equivalent to any of the known classes.

There also exists another notion of equivalence: the CCZ-equivalence of vectorial functions. However, for bent functions (and for Boolean functions in general), CCZ equivalence coincides with EA equivalence (see [3] and [6, Theorem 3]).

We end this section by defining the following combinatorial objects relevant to our further discussions.

Definition 2.5 Let G be a multiplicative group of order m.

(i) A k-subset D of G is an $(m, k, \lambda; k - \lambda)$ *difference set* in G if every non-identity element in G can be represented as gh^{-1} with $g, h \in D$, in exactly λ ways.

(ii) A k-subset D of G is an (m, k, λ, μ) *partial difference set (PDS)* if each non-identity element in D can be represented as gh^{-1} with $g, h \in D$, in exactly λ ways, and each non-identity element in $G \setminus D$ can be represented as gh^{-1} with $g, h \in D$, in exactly μ ways.

(iii) Let N be a subgroup in G of order n. A k-subset R of G is called an $(m/n, n, k, \lambda)$ *relative difference set (RDS)* in G relative to N if every element $g \in G \setminus N$ can be represented in exactly λ ways in the form gh^{-1} $(g, h \in R, g \neq h)$, and no non-identity element in N has such a representation.

3 Some known classes of bent functions

In this section, we provide some constructions of bent functions both in Boolean and p-ary cases. Our survey of non-Boolean functions is quite comprehensive, but the Boolean functions we present are just the most simple and illustrative. For other results on Boolean bent functions, together with many of the proofs, the reader is referred to [9].

3.1 Fundamental classes of bent functions

The following is a complete list of Boolean bent functions on $\mathbb{F}_2^{2m}$ for $1 \leq m \leq 3$ (up to equivalence). For the functions F_3, F_4, F_5 and F_6, we have $m = 3$.

1. $x_1 x_2$ for $m = 1$,

2. $x_1 x_2 + x_3 x_4$ for $m = 2$,

3. $x_1 x_4 + x_2 x_5 + x_3 x_6 = F_3$,

4. $F_3 + x_1 x_2 x_3 = F_4$,

5. $F_4 + x_2 x_4 x_6 + x_1 x_2 + x_4 x_6 = F_5$,

6. $F_5 + x_3 x_4 x_5 + x_1 x_2 + x_3 x_5 + x_4 x_5 = F_6$.

The corresponding difference sets are inequivalent, but the design corresponding to the difference set from F_3 is equivalent to the design corresponding to F_4. The simplest and best understood bent functions have degree two:

Proposition 3.1 ([53, Chapter 15]) *Suppose that $f = \sum_{i,j} a_{i,j} x_i x_j$ is a quadratic function in $2m$ variables over $\mathbb{F}_2$. Then f is bent if and only if one of the following equivalent conditions is satisfied:*

(i) *The matrix $(a_{i,j} + a_{j,i})_{i,j=1,\ldots,2m} \in \mathbb{F}_2^{(2m,2m)}$ is invertible.*

(ii) *The symplectic form $B(x,y) := f(x+y) + f(x) + f(y)$ is non-degenerate.*

Example 3.2 The matrix corresponding to the Boolean function $f(x) = x_1 x_2 + x_2 x_3 + x_3 x_4$ is $\begin{pmatrix} 0 & 1 & 0 & 0 \\ 1 & 0 & 1 & 0 \\ 0 & 1 & 0 & 1 \\ 0 & 0 & 1 & 0 \end{pmatrix}$ and this is invertible, hence f is bent. Similarly, the function $x_1 x_2 + x_2 x_3 + x_3 x_4 + x_1 x_4$ is not bent, since the corresponding matrix $\begin{pmatrix} 0 & 1 & 0 & 1 \\ 1 & 0 & 1 & 0 \\ 0 & 1 & 0 & 1 \\ 1 & 0 & 1 & 0 \end{pmatrix}$ is singular.

Proposition 3.3 ([34]) *Any quadratic p-ary function $f(x)$ mapping $\mathbb{F}_{p^n}$ to $\mathbb{F}_p$ is bent if and only if the bilinear form $B(x,y) := f(x+y) - f(x) - f(y) + f(0)$ associated with f is nondegenerate. Moreover, all quadratic p-ary bent functions are (weakly) regular.*

Each Boolean quadratic bent function over $\mathbb{F}_2^{2m}$ is EA-equivalent to the function

$$f(x_1, \ldots, x_{2m}) := x_1 x_2 + x_3 x_4 + \cdots + x_{2m-1} x_{2m}.$$

Thus, the class of Boolean quadratic bent functions is complete and consists of a single EA-equivalence class.

There are exactly two inequivalent nondegenerate quadratic forms on $\mathbb{F}_{p^n}$ with p odd:

$$\begin{aligned} f_1 &:= x_1^2 + x_2^2 + \cdots + x_n^2, \\ f_2 &:= x_1^2 + x_2^2 + \cdots + g \cdot x_n^2, \end{aligned} \tag{3.1}$$

where g is a nonsquare in $\mathbb{F}_p$. Here "equivalence" of quadratic forms is the usual linear algebra equivalence. Both functions are bent. If n is odd, then $g \cdot f_2$ is EA-equivalent to f_1, hence there is only one quadratic bent function if n is odd. If n is even, the two quadratic bent functions are EA-inequivalent.

Construction 1 ([20, 47, 54]) *(Maiorana-McFarland) Take any permutation π of $\mathbb{F}_p^m$ and any function $\sigma : \mathbb{F}_p^m \to \mathbb{F}_p$. Then $f : \mathbb{F}_p^m \times \mathbb{F}_p^m \to \mathbb{F}_p$ with $f(x, y) := x \cdot \pi(y) + \sigma(y)$ is a bent function. Moreover, the bijectivity of π is necessary and sufficient for f being bent. Such bent functions are regular and the dual function is equal to $f^*(x, y) = y \cdot \pi^{-1}(x) + \sigma(\pi^{-1}(x))$.*

Example 3.4 The Boolean function $x_1 x_4 + x_2 x_5 + x_3 x_6 + x_4 x_5 x_6$ is a Maiorana-McFarland bent function on 6 variables.

The Maiorana-McFarland construction can be used to construct bent functions of degree $(p-1)m$ by choosing a function σ of degree $(p-1)m$. Frequently, several apparently different constructions of bent functions turn out to be special cases of the Maiorana-McFarland construction. In particular, Boolean quadratic bent functions all belong to the completed Maiorana-McFarland class. On the other hand, for odd p and even n, the quadratic bent function f_2 in (3.1) does not belong to the completed Maiorana-McFarland class [4]. The efficient way to check whether a functions belongs to the completed Maiorana-McFarland class is the use of second-order derivatives as follows.

Theorem 3.5 ([20]) *Let n be even n. A bent function $f : \mathbb{F}_{p^n} \to \mathbb{F}_p$ belongs to the completed Maiorana-McFarland class if and only if there exists an $n/2$-dimensional vector subspace V in $\mathbb{F}_{p^n}$ such that the second order derivatives*

$$D_a D_c f(x) = f(x + a + c) - f(x + a) - f(x + c) + f(x)$$

vanish for any $a, c \in V$.

Construction 2 ([20]) *(Partial spread) Let $V = \mathbb{F}_2^n$ with $n = 2m$. Let U_i, $i \in I$, be a collection of subspaces of dimension m with $U_i \cap U_j = \{0\}$ for all $i \neq j$. If $|I| = 2^{m-1}$, the set $D_I^- = \bigcup_{i \in I} U_i \setminus \{0\}$ is a difference set with parameters $(2^{2m}, 2^{2m-1} - 2^{m-1}, 2^{2m-2} - 2^{m-1}; 2^{2m-2})$. Similarly, if $|I| = 2^{m-1} + 1$, the union $D_I^+ = \bigcup_{i \in I} U_i$ is a difference set with parameters $(2^{2m}, 2^{2m-1} + 2^{m-1}, 2^{2m-2} + 2^{m-1}; 2^{2m-2})$. The functions $f^- : V \to \mathbb{F}_2$ (or f^+) with $f^-(x) = 1$ if and only if $x \in D_I^-$ (or $f^+(x) = 1$ if and only if $x \in D_I^+$) are bent functions.*

These functions are bent functions of *partial spread type*, since a collection of subspaces of dimension m with pairwise trivial intersection is called a partial spread. The functions f^+ are called of type $\mathcal{PS}^+$, the others of type $\mathcal{PS}^-$.

Example 3.6 Let $q = 2^m$, and view $\mathbb{F}_{q^2}$ as a 2-dimensional space over $\mathbb{F}_q$, but also as a $2m$-dimensional vector space over $\mathbb{F}_2$. The 1-dimensional subspaces of $\mathbb{F}_{q^2}$ (viewed as a 2-dimensional vector space) are m-dimensional subspaces over $\mathbb{F}_2$. The $q + 1$ subspaces of dimension 1 over $\mathbb{F}_q$ are

$$U_\alpha := \{\alpha \cdot x : x \in \mathbb{F}_q\}$$

where $\alpha^{q+1} = 1$, i.e., α is in the multiplicative group of order $q+1$ in $\mathbb{F}_{q^2}$. These subspaces intersect pairwise trivially, hence we may take any 2^{m-1} or $2^{m-1} + 1$ of these subspaces to construct f^- or f^+.

The parameters of D_I^+ are complementary to the parameters of D_I^-, but the difference sets are, in general, not complements of each other. All functions in $\mathcal{PS}^-$ have algebraic degree $n/2$. On the contrary, if $n/2$ is even then class $\mathcal{PS}^+$ contains all quadratic bent functions [20]. A partial spread where the union of the subspaces covers the entire vector space is called a *spread*. The spread constructed in Example 3.6 is called the *regular spread*. Spreads of subspaces of dimension m in $2m$-dimensional subspaces can be used to describe translation planes. There are numerous spreads, hence many partial spreads. Many partial spreads are not contained in spreads. For partial spreads contained in spreads, the two constructions in Construction 2 are complements of each other, i.e., for any f^-, there is another partial spread such that $f^- = f^+ + 1$.

Definition 3.7 The class $\mathcal{PS}_{ap}$ is the subclass of $\mathcal{PS}^-$ where a subspread of the regular spread is used.

There also exists a stronger notion of bentness for Boolean functions: a Boolean function f over $\mathbb{F}_{2^n}$ is *hyper-bent* if $f(x^k)$ is bent for any k coprime with $2^n - 1$. All the functions in $\mathcal{PS}_{ap}$ class are hyper-bent [9].

Some of the known constructions of bent functions are direct, that is, do not use previously constructed bent functions as building blocks. We call these *primary constructions*. The others, sometimes leading to recursive constructions, will be called *secondary constructions*. Most of the secondary constructions of Boolean bent functions are not explained here and can be found in [9, Section 6.4.2].

Construction 3 *(A recursive construction) Let $f_1 : \mathbb{F}_p^{m_1} \to \mathbb{F}_p$ and $f_2 : \mathbb{F}_p^{m_2} \to \mathbb{F}_p$ be p-ary bent functions, then $f_1 * f_2 : \mathbb{F}_p^{m_1+m_2} \to \mathbb{F}_p$ with $(f_1 * f_2)(x_1, x_2) = f_1(x_1) + f_2(x_2)$ is p-ary bent.*

This construction was formulated originally in a much more general form using relative difference sets [61]. Since the function $f : \mathbb{F}_p \to \mathbb{F}_p$ with $f(x) = x^2$ is a p-ary bent function if p is odd, Construction 3 yields bent functions $\mathbb{F}_p^n \to \mathbb{F}_p$ for all n (p odd).

3.2 Boolean monomial and Niho bent functions

The following monomial functions $f(x) = \mathrm{Tr}_n(\alpha x^d)$ are bent on $\mathbb{F}_{2^n}$ with $n = 2m$:

1. $d = 2^k + 1$ with $n/\gcd(k, n)$ being even and $\alpha \notin \{y^d : y \in \mathbb{F}_{2^n}\}$;

2. $d = r(2^m - 1)$ with $\gcd(r, 2^m + 1) = 1$ and $\alpha \in \mathbb{F}_{2^m}$ has Kloosterman sum -1 [15, 20] (see the discussion surrounding Theorem 3.14, in particular for the definition of Kloosterman sum);

3. $d = 2^{2k} - 2^k + 1$ with $\gcd(k, n) = 1$ and $\alpha \notin \{y^3 : y \in \mathbb{F}_{2^n}\}$ [21, 49];

4. $d = (2^k + 1)^2$ with $n = 4k$ and k odd, $\alpha \in \omega\mathbb{F}_{2^k}$ with $\omega \in \mathbb{F}_4 \setminus \mathbb{F}_2$ [16, 51];

5. $d = 2^{2k} + 2^k + 1$ with $n = 6k$ and $k > 1$, $\alpha \in \mathbb{F}_{2^{3k}}$ with $\mathrm{Tr}_{\mathbb{F}_{2^{3k}}/\mathbb{F}_{2^k}}(\alpha) = 0$ [8].

These functions are *monomial bent functions*. The functions in Part 1 are quadratic, and those in Parts 4 and 5 belong to the Maiorana-McFarland class. Functions in Part 2 are in the class $\mathcal{PS}_{ap}$. An exhaustive search shows that there are no other monomial bent functions for $n \leq 20$.

A positive integer d (always understood modulo $2^n - 1$ with $n = 2m$) is a *Niho exponent* if $d \equiv 2^j \pmod{2^m - 1}$ for some $j < n$. As we consider $\mathrm{Tr}_n(ax^d)$ with $a \in \mathbb{F}_{2^n}$, without loss of generality, we can assume that d is in normalized form, i.e., with $j = 0$. Then we have a unique representation $d = (2^m - 1)s + 1$ with $2 \leq s \leq 2^m$. The following are examples of bent functions consisting of one or more Niho exponents:

1. The quadratic function $\mathrm{Tr}_m(ax^{2^m+1})$ with $a \in \mathbb{F}_{2^m}^*$. (So here $s = 2^{m-1} + 1$.)

2. Binomials of the form $f(x) = \mathrm{Tr}_n(\alpha_1 x^{d_1} + \alpha_2 x^{d_2})$, where $2d_1 \equiv 2^m + 1 \pmod{2^n - 1}$ and $\alpha_1, \alpha_2 \in \mathbb{F}_{2^n}^*$ are such that $(\alpha_1 + \alpha_1^{2^m})^2 = \alpha_2^{2^m+1}$. Equivalently, writing $a = (\alpha_1 + \alpha_1^{2^m})^2$ and $b = \alpha_2$ we have $a = b^{2^m+1} \in \mathbb{F}_{2^m}^*$ and

$$f(x) = \mathrm{Tr}_m(ax^{2^m+1}) + \mathrm{Tr}_n(bx^{d_2}).$$

We note that if $b = 0$ and $a \neq 0$ then f is a bent function we have already seen in Part 1. The possible values of d_2 are [25, 36]:

$$d_2 = (2^m - 1)3 + 1,$$
$$6d_2 = (2^m - 1) + 6 \quad \text{(with the condition that } m \text{ is even).}$$

These functions have algebraic degree m and do not belong to the completed Maiorana-McFarland class [5].

3. [50] Take $r > 1$ with $\gcd(r, m) = 1$ and define

$$f(x) = \mathrm{Tr}_n\left(ax^{2^m+1} + \sum_{i=1}^{2^{r-1}-1} x^{d_i}\right),$$

where $2^r d_i = (2^m - 1)i + 2^r$ and $a \in \mathbb{F}_{2^n}$ is such that $a + a^{2^m} = 1$. The dual of f, calculated using Proposition 3.11, is equal to

$$f^*(y) = \mathrm{Tr}_m\left((u(1 + y + y^{2^m}) + u^{2^{n-r}} + y^{2^m})(1 + y + y^{2^m})^{1/(2^r-1)}\right),$$

where $u \in \mathbb{F}_{2^n}$ is arbitrary with $u + u^{2^m} = 1$. Moreover, if $d < m$ is a positive integer defined uniquely by $dr \equiv 1 \pmod{m}$ then the algebraic degree of f^* is equal to $d + 1$. Both the function f and its dual belong to the completed Maiorana-McFarland class [12]. On the other hand, f^* is not a Niho bent function.

4. Bent functions in a bivariate representation obtained from the known o-polynomials (see the discussion at the end of this section for details).

Bent functions of Niho type were first defined in the univariate representation. The following result is used for proving the bentness of such functions.

Theorem 3.8 ([25, 50]) *Assume that*

$$d_i = (2^m - 1)s_i + 1 \quad (i = 1, \dots, r)$$

are Niho exponents and

$$f(x) = \mathrm{Tr}_n\left(\sum_{i=1}^{r} \alpha_i x^{d_i}\right)$$

with $\alpha_i \in \mathbb{F}_{2^n}$. Then for every $c \in \mathbb{F}_{2^n}$ we have $\hat{f}(c) = (N(c) - 1)2^m$, where $N(c)$ is the number of $u \in S$ such that

$$cu + \overline{cu} + \sum_{i=1}^{r}(\alpha_i u^{1-2s_i} + \overline{\alpha_i}\,\overline{u}^{1-2s_i}) = 0,$$

where $\overline{x} = x^{2^m}$ and $S = \{u \in \mathbb{F}_{2^n} : u\overline{u} = 1\}$. In particular, f is bent if and only if $N(c) \in \{0, 2\}$.

Example 3.9 To illustrate the use of Theorem 3.8 we prove the bentness of the quadratic monomial function known as the *Kasami* bent function. We have $f(x) = \mathrm{Tr}_n(a^2 x^{2^m+1})$ with $a \in \mathbb{F}_{2^n}$ such that $a + a^{2^m} \neq 0$ (or, equivalently, $f(x) = \mathrm{Tr}_m(bx^{2^m+1})$ with $b = (a + a^{2^m})^2 \in \mathbb{F}_{2^m}^*$). In the notation of Theorem 3.8, $r = 1$, $s_1 = 1/2$ and $\alpha_1 = a^2$. Then $\hat{f}(c) = (N(c) - 1)2^m$, where $N(c)$ is the number of $u \in \mathcal{S}$ such that $cu + \overline{c}\overline{u} + a^2 + \overline{a}^2 = 0$. Since $u\overline{u} = 1$, we may rewrite the last equation as $cu + \overline{c}u^{-1} = (a + \overline{a})^2$, and this equation has 0 or 2 solutions since $a + \overline{a} \neq 0$.

It turns out that Niho bent functions can also be equivalently represented in a bivariate form. It was in the early seventies when Dillon, in his thesis [20], introduced a class of bent functions denoted by H, where bentness is proven under some conditions which were difficult to analyze fully. (In this class, Dillon was able to exhibit only functions belonging, up to the affine equivalence, to the Maiorana-McFarland class.) This problem received attention recently in [13], where the following construction was suggested:

Construction 4 *Define a class $\mathcal{H}$ of functions in their bivariate representation as follows*

$$g(x, y) = \begin{cases} \mathrm{Tr}_m\left(xH(\frac{y}{x})\right), & \text{if } x \neq 0, \\ \mathrm{Tr}_m(\mu y), & \text{if } x = 0, \end{cases} \tag{3.2}$$

where $\mu \in \mathbb{F}_{2^m}$ and H is a mapping from $\mathbb{F}_{2^m}$ to itself with $G(z) := H(z) + \mu z$ satisfying

$$z \to G(z) + \beta z \quad \text{is 2-to-1 on } \mathbb{F}_{2^m} \text{ for any } \beta \in \mathbb{F}_{2^m}^*. \tag{3.3}$$

Condition (3.3) is necessary and sufficient for g to be bent.

The bentness criterion for functions in class $\mathcal{H}$ is proven using the following result, which also gives an indication to how to find the dual of such a function.

Theorem 3.10 ([13]) *Take a function g having the form (3.2) and any $\alpha, \beta \in \mathbb{F}_{2^m}$. Then the Walsh transform coefficients of g are*

$$\hat{g}(\alpha, \beta) = \begin{cases} 2^m N_{\alpha, \beta} & \text{if } \beta = \mu \\ 2^m (N_{\alpha, \beta} - 1) & \text{if } \beta \neq \mu \end{cases}$$

where $N_{\alpha, \beta} = |\{z \in \mathbb{F}_{2^m} \mid H(z) + \beta z + \alpha = 0\}|$.

Corollary 3.11 *Let g be a bent function having the form* (3.2). *Then its dual function* g^*, *represented in its bivariate form, satisfies* $g^*(\alpha, \beta) = 1$ *if and only if the equation* $H(z) + \beta z = \alpha$ *has no solution in* $\mathbb{F}_{2^m}$.

Note that Dillon's class H and the class $\mathcal{H}$ are the same up to the addition of a linear term. Any mapping G on $\mathbb{F}_{2^m}$ that satisfies (3.3) is called an *o-polynomial*. Any o-polynomial defines a hyperoval in $\mathrm{PG}(2, 2^m)$. Using Construction 4, every o-polynomial results in a bent function in class $\mathcal{H}$. The following is the list of the monomial o-polynomials due to Segre (1962) and Glynn (1983). The remaining four known classes of o-polynomials (including Subiaco and Adelaide o-polynomials) are listed in [13].

1. $G(z) = z^{2^i}$ with $\gcd(i, m) = 1$.

2. $G(z) = z^6$ with m odd.

3. $G(z) = z^{3 \cdot 2^k + 4}$ with $m = 2k - 1$.

4. $G(z) = z^{2^k + 2^{2k}}$ with $m = 4k - 1$.

5. $G(z) = z^{2^{2k+1} + 2^{3k+1}}$ with $m = 4k + 1$.

For example, from the o-polynomial z^6 and using (3.2) we obtain the bent function $g(x, y) = \mathrm{Tr}_m(x^{-5} y^6)$ for odd m that has a simple bivariate representation. On the contrary, its univariate trace representation is complex.

A bent function $\mathbb{F}_{2^n} \to \mathbb{F}_2$ with $n = 2m$ belongs to class $\mathcal{H}$ if and only if its restriction to each coset $u\mathbb{F}_{2^m}$ with $u \in \mathbb{F}_{2^n}^*$ is linear. Thus, Niho bent functions are just functions of class $\mathcal{H}$ viewed in their univariate representation. In particular, binomial Niho bent functions with $d_2 = (2^m - 1)3 + 1$ correspond to Subiaco hyperovals [36], functions with $6d_2 = (2^m - 1) + 6$ correspond to Adelaide hyperovals and the functions listed under 3 above (consisting of 2^r terms) are obtained from Frobenius map $G(z) = z^{2^{m-r}}$ (i.e., translation hyperovals) [12].

3.3 p-ary bent functions in univariate form

Definition 3.12 A function $F(x)$ over $\mathbb{F}_{p^n}$ is *planar* if for any nonzero $c \in \mathbb{F}_{p^n}$ the mapping $F(x + c) - F(x)$ is a bijection on $\mathbb{F}_{p^n}$.

Except for one class, all known planar functions are quadratic [6] which means that they can be represented by so called Dembowski-Ostrom polynomials (see [18]). The only known examples of nonquadratic planar functions are the family over $\mathbb{F}_{3^n}$ defined by $F(x) = x^{\frac{3^k+1}{2}}$ with $\gcd(k, n) = 1$

and odd k; the members of this family are known as *Coulter-Matthews* functions [18]. The following theorem shows that every planar function gives a family of p-ary bent functions.

Theorem 3.13 ([11]) *A function F mapping $\mathbb{F}_{p^n}$ to itself is planar if and only if for every nonzero $a \in \mathbb{F}_{p^n}$ the function $\mathrm{Tr}_n(aF)$ is bent.*

The p-ary bent functions $\mathrm{Tr}_n(aF)$ obtained from Dembowski-Ostrom polynomials are quadratic, hence they are weakly regular (see Proposition 3.3). It was shown in [66, 26] that the bent functions coming from the Coulter-Matthews planar functions are also weakly regular.

For any $a \in \mathbb{F}_{p^n}$, the Kloosterman sum is defined by

$$K(a) = \sum_{c \in \mathbb{F}_{p^n}^*} \zeta_p^{\mathrm{Tr}(c + ac^{-1})}.$$

Theorem 3.14 ([34]) *Let $n = 2m$ and $t > 0$ be an arbitrary integer with $\gcd(t, p^m + 1) = 1$ for an odd prime p. For any nonzero $a \in \mathbb{F}_{p^n}$, define the following p-ary function mapping $\mathbb{F}_{p^n}$ to $\mathbb{F}_p$*

$$f(x) = \mathrm{Tr}_n\left(ax^{t(p^m-1)}\right). \tag{3.4}$$

Then for any $y \in \mathbb{F}_{p^n}^*$, the corresponding Walsh transform coefficient of f is equal to

$$\hat{f}(y) = 1 + K\left(a^{p^m+1}\right) + p^m \zeta_p^{-\mathrm{Tr}_n(a^{p^m} y^{t(p^m-1)})} \quad and$$
$$\hat{f}(0) = 1 - (p^m - 1)K\left(a^{p^m+1}\right).$$

Assuming $p^m > 3$, then f is bent if and only if $K\left(a^{p^m+1}\right) = -1$. Moreover, if the latter holds then f is a regular bent function of degree $(p-1)m$.

Take $p = 2$ and without loss of generality assume $a \in \mathbb{F}_{2^m}$. Then exactly the same result as in Theorem 3.14 holds for any m in the binary case giving the so-called *Dillon class* of bent functions [20, 51, 15] (see the second class of functions listed at the beginning of Subsection 3.2). Moreover, the Kloosterman sum over $\mathbb{F}_{2^m}$ takes on *all* the integer values in the closed range $[-2\sqrt{2^m}, 2\sqrt{2^m}]$ that are equal to -1 modulo 4 (see [48, Theorem 3.4]). This means that binary Dillon bent functions exist.

According to the result of Katz and Livné [44] (see also [55, Theorem 6.4]), as c runs over $\mathbb{F}_{3^m}^*$ the Kloosterman sum $K(c)$ takes on *all* the integer values in the range $(-2\sqrt{3^m}, 2\sqrt{3^m})$ that are equal to -1 modulo 3. In particular, there exists at least one $a \in \mathbb{F}_{3^n}$ such that $K\left(a^{3^m+1}\right) = -1$.

This means that in the ternary case (i.e., when $p = 3$), if the conditions of Theorem 3.14 are satisfied, there exists at least one $a \in \mathbb{F}_{3^n}$ such that function (3.4) is bent. Moreover, there are no bent functions having the form of (3.4) when $p > 3$ since in this case, the Kloosterman sum never takes on the value -1 as shown in [45]. Here we want to mention the following crosscorrelation conjecture of Helleseth (see Section 6 for the definitions) which has a connection with Kloosterman sums:

Conjecture 3.15 ([31]) *If $d \equiv 1 \pmod{p-1}$ then the periodic correlation of an m-sequence and its d-decimation contains the value -1.*

The conjecture is not true in the opposite direction in general. Note that Kloosterman sum values make up the periodic correlation of an m-sequence and its reverse (so $d = -1$). In the binary and ternary cases, $d = -1 \equiv 1 \pmod{p-1}$ and it is known that the Kloosterman sum always takes on the value -1 for these values of p. The fact that the Kloosterman sum never is equal to -1 in a non-binary and non-ternary case means that when the periodic correlation of an m-sequence and its d-decimation with $d = -1$ contains the value -1 then $d \equiv 1 \pmod{p-1}$.

Take any $m > 0$ and select all integers in the range $\{0, \ldots, 3^m - 1\}$ that do not contain 2-digits (digits equal to 2) in their ternary expansion and none of their 1-digits are adjacent (the least significant digit is cyclically linked with the most significant). Further, split this set into cyclotomic cosets modulo $3^m - 1$, take coset leaders and denote this subset I_m. So $I_1 = \{0\}$, $I_2 = I_3 = \{0, 1\}$, $I_4 = I_5 = \{0, 1, 10\}$, $I_6 = I_7 = \{0, 1, 10, 28, 91\}$, $I_8 = \{0, 1, 10, 28, 82, 91, 253, 820\}$, $I_9 = \{0, 1, 10, 28, 82, 91, 253, 271, 757, 820\}$, for example. Obviously, $I_t \subseteq I_{t+1}$. In combinatorics, the corresponding binary strings are known as *binary necklaces* of length m with the forbidden subsequence '11'. The cardinalities of I_m form the sequence A000358 of [1].

Theorem 3.16 ([34, 33, 30]) *Let $n = 2m$ with m odd. The ternary function $f : \mathbb{F}_{3^n} \to \mathbb{F}_3$ given by*

$$f(x) = \mathrm{Tr}_n \left(a x^{\frac{3^n-1}{4} + 3^m + 1} \right)$$

is a weakly regular bent function of degree n if $a = \xi^{\frac{3^m+1}{4}}$ and ξ is a primitive element of $\mathbb{F}_{3^n}$. Moreover, for any $y \in \mathbb{F}_{3^n}$ the corresponding Walsh transform coefficient of f is equal to $\hat{f}(y) = -3^m \zeta_3^{g(y)}$ with

$$g(y) = -\mathrm{Tr}_n \left(a^{-1} y^{\frac{3^n-1}{4} + 3^m + 1} \right) \sum_{t \in I_m} \mathrm{Tr}_{o(t)} \left((a y^{-2})^{t(3^m+1)} \right),$$

where $o(t)$ is the size of the cyclotomic coset modulo $3^m - 1$ that contains t.

In particular, for $k = 3$, the dual function has just two terms in the trace representation and

$$g(y) = \text{Tr}_6(a^{97}y^{14} + a^3 y^{98}).$$

The open problem remaining is to find the trace representation of g in general.

In the following theorem, we describe the class of bent functions consisting of two terms (so called binomial functions). This is the only infinite class of nonquadratic p-ary functions, in a univariate representation over fields of arbitrary odd characteristic, that has been proven to be bent.

Theorem 3.17 ([35]) *Let $n = 4k$. The p-ary function $f : \mathbb{F}_{p^n} \to \mathbb{F}_p$ given by*

$$f(x) = \text{Tr}_n \left(x^{p^{3k}+p^{2k}-p^k+1} + x^2 \right)$$

is a weakly regular bent function of degree $(p-1)k+2$. Moreover, for any $y \in \mathbb{F}_{p^n}$ the corresponding Walsh transform coefficient of f is equal to

$$\hat{f}(y) = -p^{2k} \zeta_p^{\text{Tr}_k(x_0)/4},$$

where x_0 is a unique root in $\mathbb{F}_{p^k}$ of the polynomial

$$y^{p^{2k}+1} + (y^2 + X)^{(p^{2k}+1)/2} + y^{p^k(p^{2k}+1)} + (y^2 + X)^{p^k(p^{2k}+1)/2}. \quad (3.5)$$

In particular, if $y^2 \in \mathbb{F}_{p^{2k}}$ then $x_0 = -\text{Tr}_{\mathbb{F}_{p^{2k}}/\mathbb{F}_{p^k}} (y^2)$.

Note that the polynomial (3.5) gives an interesting description of the dual function to the function f in Theorem 3.17. It is hard to believe that such a bent function would ever be found if it was not a dual of a bent function with a simple binomial formula. It is also interesting that in the binary case when $p = 2$, the decimation $2^{3k} - 2^{2k} + 2^k + 1$ which is cyclotomic equivalent to the exponent in the first term of the above bent function, was studied by Niho in [57, Theorem 3–7] and Helleseth in [32]. They proved that the cross-correlation function between two binary m-sequences that differ by this decimation is four-valued and found the distribution (see Theorem 6.6 (ii)).

All known univariate polynomials representing infinite classes of p-ary bent functions are summarized in Table 1. Here ξ denotes a primitive element of $\mathbb{F}_{3^n}$, "r" and "wr" refer to regular and weakly regular bent

Table 1: Nonquadratic p-ary Bent Functions

n	d or $F(x)$	a	deg	Remarks
	$\frac{3^k+1}{2}$, $\gcd(k,n)=1$, k-odd	$a \neq 0$	$k+1$	tern. r, wr
$2m$	$t(3^m-1)$, $\gcd(t,3^m+1)=1$	$K(a^{\frac{3^m+1}{}})=0$	n	tern. r
$2m$	$\frac{3^n-1}{4}+3^m+1$, m-odd	$\xi^{\frac{3^m+1}{4}}$	n	tern. wr
$4k$	$x^{p^{3k}+p^{2k}-p^k+1}+x^2$		$(p-1)k+2$	wr

functions respectively. The first three families in the table are monomials of the form $\mathrm{Tr}_n(ax^d)$ while the last one is a binomial bent function in the form $\mathrm{Tr}_n(F(x))$. When the value of n is not specified in the table it means that n is arbitrary. Naturally, all the exponents d and coefficients a can be replaced with their cyclotomic equivalents. The table does not include numerous examples of binomial ternary bent functions consisting of two Dillon type exponents (see Theorem 3.14). Using Theorem 3.5, it was proved in [4] that all bent functions in Table 1 possibly except for those of Dillon type, do not belong to the completed Maiorana-McFarland class.

It was long believed that all p-ary bent functions are weakly regular. However, some counterexamples were found recently. In particular, the ternary function f mapping $\mathbb{F}_{3^6}$ to $\mathbb{F}_3$ and given by

$$f(x) = \mathrm{Tr}_6\left(\xi^7 x^{98}\right)$$

where ξ is a primitive element of $\mathbb{F}_{3^6}$, is bent but not weakly regular. Further, the following table gives known examples of ternary binomial bent functions of the form $\mathrm{Tr}_n(a_1 x^{d_1} + a_2 x^{d_2})$ that are not weakly regular, obtained from a computer search. Here ξ denotes a primitive element of $\mathbb{F}_{3^n}$.

n	a_1	d_1	a_2	d_2
3	1	8	1	14
4	1	4	ξ^{10}	22
6	ξ^7	14	ξ^{35}	70
6	ξ	20	ξ^{41}	92

Just one representative of each EA-equivalence class is listed. An interesting open problem is to find an infinite class of non-weakly regular bent functions in a univariate representation.

4 Connections to coding theory

In coding theory bent functions were first indirectly studied by Helle-seth, Kløve and Mykkeltveit [37] in their investigations of the covering radius of the first-order Reed-Muller codes. This problem was first stud-ied in the 1970s by coding theorists without knowledge of bent functions, that were discovered just a few years earlier. To review some of the results on the covering radius problem and its connections to bent functions some notation from coding theory are needed.

The *Hamming distance* between binary vectors $a = (a_1, a_2, \cdots, a_n)$ and $b = (b_1, b_2, \cdots, b_n)$ is the number of coordinates where they differ, i.e.,

$$d(a, b) = |\{i \mid a_i \neq b_i, i \leq i \leq n\}|.$$

The *sphere* of radius r around a is the set of vectors of distance $\leq r$ from a.

Definition 4.1 An $[n, k, d]$ *code* C is a k-dimensional subspace of $\mathbb{F}_2^n$ such that the minimum Hamming distance between any two distinct codewords equals d.

The *dual code* of C is an $[n, n - k, d^\perp]$ code defined by

$$C^\perp = \{a \mid a \cdot c = 0 \text{ for all } c \in C\}.$$

A *generator matrix* G of a linear $[n, k, d]$ code is a $k \times n$ matrix whose row space generates all codewords in the code.

Definition 4.2 The *covering radius* of a code is the smallest integer ρ such that the spheres of radius ρ around the codewords cover the complete space.

Finding the covering radius of a code is equivalent to finding the largest Hamming distance of any n-dimensional vector from the code. An impor-tant but challenging problem in coding theory is find the covering radius of a code. In particular, we will discuss the status of this problem for the first-order Reed-Muller code defined next.

Let $x = (x_1, x_2, \cdots, x_n)$ and let v_f be a vector of length 2^n, where the coordinates are indexed by the elements of $\mathbb{F}_2^n$. The value of v_f in position x is obtained by evaluating $f(x)$ in $x \in \mathbb{F}_2^n$, so

$$v_f = (f(x))_{x \in \mathbb{F}_2^n}.$$

The *first order Reed-Muller code* of length 2^n, is obtained using all binary affine polynomials $f(x) = \sum_{i=1}^n b_i x_i + b_0$, i.e.,

$$RM(1, n) = \{v_f \mid deg(f) \leq 1\}.$$

The parameters of $RM(1, n)$ are $[2^n, n + 1, 2^{n-1}]$. Except for the all-one vector and the all-zero vector, all codewords in the code have Hamming weight 2^{n-1}.

Example 4.3 For $n = 3$ the following is a generator matrix of $RM(1, 3)$,

$$G = \begin{bmatrix} 1 & 1 & 1 & 1 & 1 & 1 & 1 & 1 \\ 1 & 1 & 1 & 1 & 0 & 0 & 0 & 0 \\ 1 & 1 & 0 & 0 & 1 & 1 & 0 & 0 \\ 1 & 0 & 1 & 0 & 1 & 0 & 1 & 0 \end{bmatrix}.$$

Each affine function $f(x) = b_1 x_1 + b_2 x_2 + b_3 x_3 + b_0$ gives a vector $v_f = (b_0, b_1, b_2, b_3)G$. In the special case when $f(x) = x_1 + x_3 + 1$ we obtain the codeword $v_f = (1101)G = (10100101)$.

The distance from an arbitrary vector $v_f = (f(x))_{x \in \mathbb{F}_2^n}$ to a codeword $c = b \cdot x + a \in RM(1, n)$ can be found via the Walsh transform by

$$
\begin{aligned}
(-1)^a \hat{f}(b) &= \sum_{x \in \mathbb{F}_2^n} (-1)^{f(x)+b \cdot x+a} \\
&= (2^n - d(v_f, c)) - d(v_f, c) \\
&= 2^n - 2d(v_f, c),
\end{aligned}
$$

where $d(v_f, c)$ denotes the Hamming distance between the binary vectors v_f and c.

Since the average value of $|\hat{f}(b)|$ is $2^{n/2}$, the covering radius ρ_n of $RM(1, n)$ satisfies

$$\rho_n \leq 2^{n-1} - 2^{\frac{n}{2}-1}.$$

Vectors v_f where equality holds correspond to bent functions $f(x)$.

Theorem 4.4 ([38]) *The function $f : \mathbb{F}_2^n \to \mathbb{F}_2$ is bent if and only if*

$$\#\{x \in \mathbb{F}_2^n : f(x) \neq l(x) + \epsilon\} = 2^{n-1} \pm 2^{\frac{n}{2}-1}$$

for all linear mappings $l : \mathbb{F}_2^n \to \mathbb{F}_2$ and all $\epsilon \in \{0, 1\}$.

Theorem 4.4 shows that f has the largest distance to all affine functions, hence bent functions solve the covering radius problem for first order Reed-Muller codes of length 2^n with n even. In other words, bent functions are *maximum nonlinear functions* $\mathbb{F}_2^n \to \mathbb{F}_2$ if n is even, and therefore the covering radius of the first order Reed-Muller code of even dimension equals

$$\rho_n = 2^{n-1} - 2^{\frac{n}{2}-1}.$$

For odd values of n it appears to be a very difficult and challenging problem to compute the covering radius of the first order Reed-Muller code of period 2^n. However, one can use the covering radius of $RM(1, n-1)$ and the inductive structure of the Reed-Muller code to establish the following simple bounds for the covering radius (see [37]):

$$2^{n-1} - 2^{\frac{n-1}{2}} \le \rho_n \le 2^{n-1} - 2^{\frac{n}{2}-1} \text{ for odd } n.$$

Other bounds on the covering radius of the first order Reed-Muller codes can be found via the analysis of the existence (or nonexistence) of certain linear self-complementary codes. To see this, we can argue as follows.

Let v be a vector of distance ρ_n to $RM(1, n)$. Since the code is linear, we can assume without loss of generality that the distance to the all-zero codeword is ρ_n. Consider the distance between v and a codeword c, different from the all-one codeword, in the Reed-Muller code,

$$
\begin{array}{ccccc}
v & 111\ldots111 & 111\ldots111 & 000\ldots000 & 000\ldots000 \\
c & 111\ldots111 & 000\ldots000 & 111\ldots111 & 000\ldots000
\end{array}
$$
$$\underbrace{}_{w} \quad \underbrace{}_{\rho_n-w} \quad \underbrace{}_{2^{n-1}-w}$$

Since $d(v, c) = \rho_n - w + 2^{n-1} - w \ge \rho_n$ we have $w \le 2^{n-2}$. Consider the code C obtained by restricting the codewords in $RM(1, n)$ to the positions where v has a 1. Since the all-one vector belongs to C, the code is self-complementary and its parameters are

$$[\rho_n, n+1, \rho_n - 2^{n-2}]$$

and with dual minimum distance $d^\perp \ge 4$. Thus studying whether codes with such parameters exist or not we can obtain bounds on ρ_n.

The covering radius for $RM(1, 5)$ was determined by Berlekamp and Welch [2] to be $\rho_5 = 12$. Their investigations were motivated by the fact that this code was used in the Mariner '69 space mission. The largest odd value of n where ρ_n is known is $n = 7$; here $\rho_7 = 56$. This was proved by Mykkeltveit [56] by showing that a $[57, 8, 25]$ self-complementary code with $d^\perp \ge 4$ does not exist. An alternative proof was given in 1996 by Hou [41].

The next odd case of $n = 9$ has been open for more than 30 years. The best lower bound known for this case is $\rho_9 \ge 242$ due to Kavut and Yucel [64]. To settle the first open case one has to decide whether certain self-complementary codes with one of the parameters $[243, 10, 115]$ or $[244, 10, 116]$ and $d^\perp \ge 4$ do exist.

Kavut and Yucel [64] also have the best lower bounds for $n = 11$, where $\rho_{11} \geq 996$ and $n = 13$, where $\rho_{13} \geq 4040$. For $n = 15$ the best lower bound is by Paterson and Wiedemann [59, 60] who also showed that

$$\rho_n \geq 2^{n-1} - \frac{27}{32} 2^{\lceil \frac{n}{2} \rceil - 1} \text{ for all odd } n \geq 15.$$

For the first few values of the covering radius of $RM(1, n)$ the following bounds are therefore known:

n	3	4	5	6	7	8
ρ_n	2	6	12	28	56	120
n	9	10	11	12	13	
ρ_n	$242 - 244$	496	$996 - 1001$	2016	$4040 - 4050$	

The upper bound on the covering radius of the Reed-Muller code is a special case of a more general bound for the covering radius of any code C.

Definition 4.5 A code C (possibly nonlinear) has *strength* $s(C) = 2$ if all pairs occur equally often in each pair of coordinates.

Theorem 4.6 ([37]) *Let C be a code of length n and strength $s(C) = 2$. Then the covering radius ρ of C obeys the inequality*

$$\rho \leq \frac{n - \sqrt{n}}{2}.$$

For a linear code, $d^{\perp} = s(C) + 1$. Thus since the first-order Reed-Muller has strength $s = 3 = d^{\perp} - 1$ and length 2^n the covering radius ρ_n of the code satisfies $\rho_n \leq 2^{n-1} - 2^{\frac{n}{2}-1}$.

The result of the theorem follows rather elementarily by considering the coset $v + C$ where $d(v, C) = \rho$ and using well known expressions for the sum of the squares of the weights of all binary vectors in the coset. The bound above, frequently called the Norse bound, can be improved for codes of strength $s > 2$. However, the main problem still remains:

Open problem Find the covering radius of the first-order Reed-Muller code for all odd values of $m \geq 9$.

5 Bent functions and other combinatorial objects

Weakly regular bent functions are useful for constructing certain combinatorial objects such as Hadamard matrices, partial difference sets, strongly regular graphs and association schemes as illustrated in this section. Difference sets and partial difference sets are defined in Section 2.

One class of combinatorial objects associated with partial difference sets are strongly regular graphs: A graph Γ with ν vertices is called a (ν, k, λ, μ) *strongly regular graph (SRG)* if each vertex is adjacent to exactly k other vertices, any two adjacent vertices have exactly λ common neighbors, and two nonadjacent vertices have exactly μ common neighbors.

Given a group G of order ν and a k-subset D of G, the graph $\Gamma = (V, E)$ defined as follows is called the *Cayley graph* generated by D:

1. the vertex set V is G;

2. two vertices g, h are joined by an edge if and only if $gh^{-1} \in D$.

The Cayley graph generated by a k-subset D of a finite multiplicative group G is a (ν, k, λ, μ) SRG if and only if D is a (ν, k, λ, μ) PDS with the identity element of G not contained in D and $\{d^{-1} \mid d \in D\} = D$ (see [52]).

Theorem 5.1 *A Boolean function* $f : \mathbb{F}_{2^n} \to \mathbb{F}_2$ *is bent if and only if the matrix of size* 2^n *consisting of elements* $\frac{1}{2^{n/2}} \hat{f}(u + v)$ *with* $u, v \in \mathbb{F}_{2^n}$ *is a Hadamard matrix.*

Theorem 5.2 ([20]) *A function* $f : \mathbb{F}_2^n \to \mathbb{F}_2$ *is bent if and only if the set* $\{x \in \mathbb{F}_2^n : f(x) = 1\}$ *is a difference set in* $\mathbb{F}_2^n$ *with parameters* $(2^n, 2^{n-1} \pm 2^{(n/2)-1}, 2^{n-2} \pm 2^{(n/2)-1}; 2^{n-2})$. *Equivalent difference sets give rise to EA-equivalent bent functions, but EA-equivalent bent functions need not necessarily give rise to equivalent difference sets; see Example 5.3.*

Example 5.3 If f and g are EA-equivalent Boolean bent functions, the corresponding difference sets need not be equivalent. One reason is that the complemented function f, i.e., $f + 1$ describes the complementary difference set. But, more seriously, adding a linear mapping l to f may not preserve equivalence of the corresponding difference sets. Using the bivariate notation introduced earlier, the functions $f(x, y) = \mathrm{Tr}_4(x \cdot y^7)$ and $g(x, y) = f(x, y) + \mathrm{Tr}_4(x)$ with $x, y \in \mathbb{F}_{2^4}$, are EA equivalent. Both functions describe difference sets with parameters $(256, 120, 56; 64)$. These difference sets are inequivalent since the corresponding designs are not isomorphic.

Theorem 5.2 does not hold for p-ary bent functions with p odd. Instead, we have the following:

Theorem 5.4 ([61]) *The set* $R_f := \{(x, f(x)) : x \in \mathbb{F}_p^n\} \subset \mathbb{F}_p^{n+1}$ *is a* (p^n, p, p^n, p^{n-1}) *RDS in an elementary abelian group if and only if the function* f *is p-ary bent.*

The following result provides a characterization of ternary weakly regular bent function via partial difference sets. Generalization to the case of arbitrary odd p is also possible (see [17, 27]).

Theorem 5.5 ([65]) *Take* $m \geq 2$ *and let* $f : \mathbb{F}_{3^{2m}} \to \mathbb{F}_3$ *be a ternary bent function satisfying* $f(-x) = f(x)$ *and* $f(0) = 0$. *Define* $D_i := \{x \in \mathbb{F}_{3^{2m}} : f(x) = i\}$ *for each* $0 \leq i \leq 2$. *Then the following hold:*

(i) f *is weakly regular if and only if* D_1 *and* D_2 *are both*

$$(3^{2m}, 3^{2m-1} + \epsilon 3^{m-1}, 3^{2m-2}, 3^{2m-2} + \epsilon 3^{m-1}) PDS,$$

where $\epsilon = \pm 1$ *(the choice of* ϵ *for* D_1 *and* D_2 *should be the same).*

(ii) *if* f *is weakly regular then* $D := D_0 \setminus \{0\}$ *is a*

$$(3^{2m}, 3^{2m-1} - 2\epsilon 3^{m-1} - 1, 3^{2m-2} - 2\epsilon 3^{m-1} - 2, 3^{2m-2} - \epsilon 3^{m-1}) PDS,$$

where ϵ *is the same as in the first part of the theorem.*

When $\epsilon = 1$, the partial difference sets D_0, D_1 and D_2 (and the associated SRG) are of *negative Latin square type*, and if $\epsilon = -1$ they are of *Latin square type*. The number of constructions for SRG of negative Latin square type is substantially less than for their counterparts. That is why it seems particularly interesting that some SRG of negative Latin square type obtained from D_1 and D_2 using ternary bent functions in Theorems 3.16, 3.17 and those obtained from Coulter-Matthews planar functions are inequivalent to any of the known constructions. Also, for $p > 3$, some bent functions in Theorem 3.17 result in new SRG (see [17]). This has been checked for a few particular constructions in small dimensions; obtaining a general result in this direction would be important.

Note that nonsingular quadratic forms on $\mathbb{F}_q$ with q odd (these are exactly the p-ary quadratic bent functions, see Proposition 3.3) result in the SRG that are known as *affine polar graphs*.

Finally, we remark that weakly regular bent functions can be used to construct association schemes by the use of Schur rings [62, 27].

6   Crosscorrelation and bent sequences

The linear recursion of degree n over $\mathbb{F}_p$ defined by

$$\sum_{i=0}^{n} c_i s_{t+i} = 0 \quad \text{where } c_0, c_n \neq 0$$

with initial state $(s_0, s_1, \ldots, s_{n-1})$ generates a periodic sequence $\{s_t\}$. The *characteristic polynomial* of the linear recursion is

$$f(x) = \sum_{i=0}^{n} c_i x^i.$$

Since n consecutive elements of the sequence determine the sequence completely, the maximal period for a recursion of degree n is $p^n - 1$. In the case when $f(x) \in \mathbb{F}_p[x]$ is a primitive polynomial, the recursion is known to generate a *maximal linear sequence* (or an m-sequence) of period $p^n - 1$.

Some important and well-known properties of m-sequences are:

- The m-sequence has period $p^n - 1$, each nonzero element occurs p^{n-1} times and the zero element occurs $p^{n-1} - 1$ times.

- For any m-sequence $\{s_t\}$ and $\tau \neq 0$ (mod $p^n - 1$) the sequence $\{s_{t+\tau} - s_t\}$ is also an m-sequence.

Let $\{u_t\}$ and $\{v_t\}$ be two p-ary sequences of period ε. The *crosscorrelation* between the two sequences at shift τ for $0 \leq \tau < \varepsilon$, is

$$C_{u,v} = \sum_{t=0}^{\varepsilon-1} \zeta_p^{u_{t+\tau} - v_t},$$

where ζ_p is a complex p-th root of unity. If the two sequences $\{u_t\}$ and $\{v_t\}$ are equal we use the term *autocorrelation* instead of crosscorrelation. A useful property of m-sequences is their optimal two-level autocorrelation, which is important for synchronization purposes in many communication systems:

Lemma 6.1 *The autocorrelation function $C_{s,s}(\tau)$ of the m-sequence $\{s_t\}$ satisfies*

$$C_{s,s}(\tau) = \begin{cases} p^n - 1 \ \textit{if } \tau = 0 \pmod{p^n - 1} \\ -1 \ \textit{if } \tau \neq 0 \pmod{p^n - 1}. \end{cases}$$

This result is an easy consequence of the balanced distribution of the elements in an m-sequence and the fact that $\{s_{t+\tau} - s_t\}$ is an m-sequence for $\tau \neq 0$ (mod $p^n - 1$).

The m-sequence $\{s_t\}$ can (after a suitable cyclic shift) be described simply by the trace function

$$s_t = \mathrm{Tr}_n(\xi^t),$$

where ξ is a zero of the characteristic polynomial $f(x)$. The different shifts of the m-sequence are obtained by

$$s_{t+\tau} = \text{Tr}_n(c\xi^t),$$

where $c = \xi^\tau \in \mathbb{F}_{p^n}^* = \mathbb{F}_{p^n} \setminus \{0\}$. Two m-sequences $\{u_t\}$ and $\{v_t\}$ of the same period $p^n - 1$ are related by a decimation d such that $u_t = v_{dt+\tau}$ and $\gcd(d, p^n - 1) = 1$.

The crosscorrelation between any two m-sequences of the same period $\varepsilon = p^n - 1$ that differ by a decimation d can be described by the following exponential sum by using the trace function representation

$$C_d(\tau) = \sum_{t=0}^{\varepsilon-1} \zeta_p^{s_{t+\tau} - s_{dt}} = \sum_{t=0}^{\varepsilon-1} \zeta_p^{\text{Tr}_n(\xi^{t+\tau} - \xi^{dt})} = \sum_{x \in \mathbb{F}_{p^n}^*} \zeta_p^{\text{Tr}_n(cx - x^d)}$$

where $c = \xi^\tau$. Determining the values and the number of occurrences of each value in the crosscorrelation function $C_d(\tau)$ when τ runs through $\{0, 1, \cdots, p^n - 2\}$ is equivalent to finding the distribution of this exponential sum for any $c \neq 0$.

The following result was stated for $p = 2$ by Golomb [29] without proof and first proved and generalized to any odd p by Helleseth [31].

Theorem 6.2 *If $d \notin \{1, p, \cdots, p^{n-1}\}$ (so the two m-sequences are cyclically distinct), then $C_d(\tau)$ is at least three-valued when $\tau = 0, 1, \cdots, p^n - 2$.*

Therefore, the crosscorrelation between m-sequences takes on at least three values. For binary sequences of length $2^n - 1$, the following is a complete list of all decimations known to give three-valued crosscorrelation. It is a challenging and open problem to decide whether this list is complete.

Theorem 6.3 *The crosscorrelation $C_d(\tau)$ is three-valued and the complete correlation distribution is known for the following values of d:*

(i) $d = 2^k + 1$ with $\frac{n}{\gcd(k,n)}$ odd [28].

(ii) $d = 2^{2k} - 2^k + 1$ with $\frac{n}{\gcd(k,n)}$ odd [43].

(iii) $d = 2^{\frac{n}{2}} + 2^{\frac{n+2}{4}} + 1$ with $n \equiv 2 \pmod 4$ [19].

(iv) $d = 2^{\frac{n+2}{2}} + 3$ with $n \equiv 2 \pmod 4$ [19].

(v) $d = 2^{\frac{n-1}{2}} + 3$ with n odd [7].

(vi) *[23, 40]*

$$d = \begin{cases} 2^{\frac{n-1}{2}} + 2^{\frac{n-1}{4}} - 1, & n \equiv 1 \pmod{4} \\ 2^{\frac{n-1}{2}} + 2^{\frac{3n-1}{4}} - 1, & n \equiv 3 \pmod{4}. \end{cases}$$

We now define some important classes of vectorial Boolean functions.

Definition 6.4 Let $F : \mathbb{F}_{2^n} \to \mathbb{F}_{2^n}$. For an odd n, if the Walsh transform coefficients

$$\hat{f}(u, v) = \sum_{x \in \mathbb{F}_{2^n}} (-1)^{\mathrm{Tr}_n(vF(x)+ux)}$$

take on the values $\{0, \pm 2^{\frac{n+1}{2}}\}$ for any $u, v \in \mathbb{F}_{2^n}$ then F is called *Almost Bent (AB)*. If for any $a \neq 0$ and $b \in \mathbb{F}_{2^n}$, the equation

$$F(x + a) + F(x) = b$$

has at most two solutions in $\mathbb{F}_{2^n}$ then F is called *Almost Perfect Nonlinear (APN)*.

It was shown by Chabaud and Vaudenay [14] that an AB function is APN. Note that any AB power function x^d with $\gcd(d, 2^n - 1) = 1$ corresponds to m-sequences having 3-valued crosscorrelation with values $\{0, \pm 2^{\frac{n+1}{2}}\}$. The four known AB power exponents are Cases (i), (ii) (both with n odd and $\gcd(k, n) = 1$), (v) and (vi) in Theorem 6.3.

For a mapping $F : \mathbb{F}_{2^n} \to \mathbb{F}_{2^n}$ define a Boolean function γ_F on $\mathbb{F}_{2^{2n}}$ by

$$\gamma_F(a, b) = \begin{cases} 1 & \text{if } a \neq 0 \text{ and } F(x + a) + F(x) = b \text{ has solutions,} \\ 0 & \text{otherwise.} \end{cases}$$

It was showed by Carlet, Charpin and Zinoviev [10] that F is an AB function if and only if γ_F is bent. We illustrate this construction using the Welch function (Theorem 6.3 Case (v)) that also as a bonus provides a nice connection between a permutation polynomial and a bent function.

Example 6.5 Let $n = 2m + 1$ and $F(x) = x^d$ where $d = 2^m + 3$. This exponent gives a 3-valued crosscorrelation and defines an AB function [7]. Furthermore, Dobbertin [23] showed that

$$F(x + 1) + F(x) + 1 = Q(x^{2^m} + x),$$

where

$$Q(x) = x^{2^{m+1}+1} + x^3 + x$$

is a permutation polynomial. Since $F(x + a) + F(x) = b$ is equivalent to $F(x + 1) + F(x) = b/a^d$ it therefore follows that $Q(x^{2^m} + x) = b/a^d + 1$ can be considered instead. Hence,

$$\gamma_F(a, b) = \begin{cases} \text{Tr}_n\left(Q^{-1}(b/a^d + 1)\right) + 1 & \text{if } a \neq 0, \\ 0 & \text{otherwise} \end{cases}$$

is a bent function.

For $p = 2$, all the known infinite families of decimations that give a four-valued crosscorrelation are of the Niho type. The corresponding exponents are given in the following theorem.

Theorem 6.6 *Let $e_2(i)$ be the highest power of 2 dividing the integer i and assume $n = 2m$. The crosscorrelation $C_d(\tau)$ is four-valued and the complete correlation distribution is known for the following values of d:*

(i) $d = \frac{2^{(m+1)r} - 1}{2^r - 1}$ *for $0 < r < m$ and $\gcd(r, n) = 1$ with $n \equiv 0$ (mod 4) [22].*

(ii) $d = (2^m - 1)\frac{2^m - 1}{2^r - 1} + 2$ *with $2r$ dividing m and $n \equiv 0$ (mod 4) [39].*

(iii) $d = (2^m - 1)s + 1$ *with $s \equiv 2^r(2^r \pm 1)^{-1}$ (mod $2^m + 1$) and $e_2(r) < e_2(m)$ [24].*

The particular cases of Theorem 6.6 (i) with $r = 1$ and Theorem 6.6 (ii) with $r = m/2$ are due to Niho [57]. Also observe that the decimation in Theorem 6.6 (ii) with $r = m/2$ gives interesting p-ary bent functions if $p = 2$ is replaced by an odd prime p (see Theorem 3.17).

In Code-Division Multiple-Access (CDMA) applications a very important problem in sequence design is to find large families of sequences such that the maximum absolute value of the nontrivial auto- and cross-correlation between all sequences in the family is low. Let

$$\mathcal{F} = \{\{s_i(t)\} \mid 1 \leq i \leq M\}$$

be a family of M sequences of period n with elements from $\mathbb{F}_p$. Let θ_{ij} denote the cross correlation between the i-th and the j-th sequence in the family at shift τ, i.e.,

$$\theta_{ij}(\tau) = \sum_{t=0}^{n-1} \zeta_p^{s_i(t+\tau) - s_j(t)}.$$

The main goal is to minimize

$$\theta_{\max} = \max\{|\theta_{ij}(\tau)| \mid \text{either } i \neq j \text{ or } \tau \neq 0\}$$

for given values of n and M.

The following family of sequences that has optimal values of $\theta_{\max}$ and several other desirable properties is constructed from bent functions:

Definition 6.7 Let $n = 2m$ when p is odd and let $n = 2m = 4k$ when $p = 2$. Let $h : \mathbb{F}_p^m \to \mathbb{F}_p$ be a bent function and ζ_p be a primitive p-th root of unity. Let $\{\gamma_1, \ldots, \gamma_m\}$ be a basis for the subfield $\mathbb{F}_{p^m}$ over $\mathbb{F}_p$ and let $\sigma \in \mathbb{F}_{p^n} \setminus \mathbb{F}_{p^m}$. Define $f_i : \mathbb{F}_{p^n} \to \mathbb{F}_p$ by

$$f_i(x) = h_i(\mathbf{y}) + \mathrm{Tr}_n(\sigma x),$$

where $\mathbf{y} = (y_1, \ldots, y_m)$ with $y_i = \mathrm{Tr}_n(\gamma_i x)$ and $h_i(\mathbf{y}) = h(\mathbf{y}) + \mathbf{u}_i \cdot \mathbf{y}$ with $\mathbf{u}_i$ ranging over all of $\mathbb{F}_p^m$ as i varies between 1 and p^m.

A family of *bent sequences* is a family of the form

$$\mathcal{F} = \{\{s_i(t)\} \mid 1 \leq i \leq p^m\}$$

where $s_i(t) = f_i(\alpha^t)$ for some fixed primitive element α in $\mathbb{F}_{p^n}$.

Theorem 6.8 ([58, 46]) *Let $\mathcal{F}$ be a family of bent sequences. Then*

(i) *Each sequence $\{s_i(t)\} \in \mathcal{F}$ has period $p^n - 1$.*

(ii) *Let N_b be the number of times the element b occurs in a period of $\{s_i(t)\}$. Then*

$$N_b = \begin{cases} p^{n-1} & \text{if } b \in \mathbb{F}_p^* \\ p^{n-1} - 1 & \text{if } b = 0, \end{cases}$$

i.e., the sequences in $\mathcal{F}$ are balanced.

(iii) *The maximum correlation of $\mathcal{F}$ is $\theta_{\max} \leq p^m + 1$.*

7 Conclusions

A survey of some recent constructions of bent functions and their connections to several areas and objects of combinatorics such as coding theory, strongly regular graphs and difference sets has been given. There are still many open problems on the construction and analysis of bent functions. In particular, it would be interesting to understand the connections between a bent function and its dual bent function. Why does the dual of a simple bent function sometimes appear to be very complicated? Do there exist other simpler descriptions of the dual of a bent function that may lead to other nice combinatorial structures?

References

[1] OEIS Foundation Inc. (2011), The On-Line Encyclopedia of Integer Sequences, http://oeis.org.

[2] Elwyn R. Berlekamp and Lloyd R. Welch, *Weight distributions of the cosets of the* (32, 6) *Reed-Muller code*, IEEE Trans. Inf. Theory **18** (1972), no. 1, 203–207.

[3] Lilya Budaghyan and Claude Carlet, *CCZ-equivalence of single and multi output Boolean functions*, Finite Fields: Theory and Applications (Providence, Rhode Island) (Gary McGuire, Gary L. Mullen, Daniel Panario, and Igor E. Shparlinski, eds.), Contemporary Mathematics, vol. 518, American Mathematical Society, 2010, pp. 43–54.

[4] Lilya Budaghyan, Claude Carlet, Tor Helleseth, and Alexander Kholosha, *Generalized bent functions and their relation to Maiorana-Mcfarland class*, Proceedings of the 2012 IEEE International Symposium on Information Theory, IEEE, July 2012, pp. 1217–1220.

[5] Lilya Budaghyan, Claude Carlet, Tor Helleseth, Alexander Kholosha, and Sihem Mesnager, *Further results on Niho bent functions*, IEEE Trans. Inf. Theory (2012), accepted.

[6] Lilya Budaghyan and Tor Helleseth, *New commutative semifields defined by new PN multinomials*, Cryptography and Communications **3** (2011), no. 1, 1–16.

[7] Anne Canteaut, Pascale Charpin, and Hans Dobbertin, *Binary m-sequences with three-valued crosscorrelation: A proof of Welch's conjecture*, IEEE Trans. Inf. Theory, **46** (2000), no. 1, 4–8.

[8] Anne Canteaut, Pascale Charpin, and Gohar M. Kyureghyan, *A new class of monomial bent functions*, Finite Fields Appl. **14** (2008), no. 1, 221–241.

[9] Claude Carlet, *Boolean functions for cryptography and error-correcting codes*, Boolean Models and Methods in Mathematics, Computer Science, and Engineering (Yves Crama and Peter L. Hammer, eds.), Encyclopedia of Mathematics and its Applications, vol. 134, Cambridge University Press, Cambridge, 2010, pp. 257–397.

[10] Claude Carlet, Pascale Charpin, and Victor A. Zinoviev, *Codes, bent functions and permutations suitable for DES-like cryptosystems*, Des. Codes Cryptogr. **15** (1998), no. 2, 125–156.

[11] Claude Carlet and Sylvie Dubuc, *On generalized bent and q-ary perfect nonlinear functions*, Finite Fields and Applications: Proceedings of the Fifth International Conference (Berlin) (Dieter Jungnickel and Harald Niederreiter, eds.), Springer-Verlag, 2001, pp. 81–94.

[12] Claude Carlet, Tor Helleseth, Alexander Kholosha, and Sihem Mesnager, *On the dual of bent functions with 2^r Niho exponents*, Proceedings of the 2011 IEEE International Symposium on Information Theory, IEEE, July/August 2011, pp. 657–661.

[13] Claude Carlet and Sihem Mesnager, *On Dillon's class H of bent functions, Niho bent functions and o-polynomials*, J. Combin. Theory Ser. A **118** (2011), no. 8, 2392–2410.

[14] Florent Chabaud and Serge Vaudenay, *Links between differential and linear cryptanalysis*, Advances in Cryptology - EuroCrypt '94 (Berlin) (Alfredo De Santis, ed.), Lecture Notes in Computer Science, vol. 950, Springer-Verlag, 1995, pp. 356–365.

[15] Pascale Charpin and Guang Gong, *Hyperbent functions, Kloosterman sums, and Dickson polynomials*, IEEE Trans. Inf. Theory **54** (2008), no. 9, 4230–4238.

[16] Pascale Charpin and Gohar M. Kyureghyan, *Cubic monomial bent functions: A subclass of M*, SIAM J. Discrete Math. **22** (2008), no. 2, 650–665.

[17] Yeow Meng Chee, Yin Tan, and Xian De Zhang, *Strongly regular graphs constructed from p-ary bent functions*, J. Algebraic Combin. **34** (2011), no. 2, 251–266.

[18] Robert S. Coulter and Rex W. Matthews, *Planar functions and planes of Lenz-Barlotti class II*, Des. Codes Cryptogr. **10** (1997), no. 2, 167–184.

[19] Thomas W. Cusick and Hans Dobbertin, *Some new three-valued crosscorrelation functions for binary m-sequences*, IEEE Trans. Inf. Theory **42** (1996), no. 4, 1238–1240.

[20] John F. Dillon, *Elementary Hadamard difference sets*, Ph.D. thesis, University of Maryland, 1974.

[21] John F. Dillon and Hans Dobbertin, *New cyclic difference sets with Singer parameters*, Finite Fields Appl. **10** (2004), no. 3, 342–389.

[22] Hans Dobbertin, *One-to-one highly nonlinear power functions on* GF(2^n), Appl. Algebra Engrg. Comm. Comput. **9** (1998), no. 2, 139–152.

[23] Hans Dobbertin, *Almost perfect nonlinear power functions on* GF(2^n): *The Niho case*, Inform. and Comput. **151** (1999), no. 1–2, 57–72.

[24] Hans Dobbertin, Patrick Felke, Tor Helleseth, and Petri Rosendahl, *Niho type cross-correlation functions via Dickson polynomials and Kloosterman sums*, IEEE Trans. Inf. Theory **52** (2006), no. 2, 613–627.

[25] Hans Dobbertin, Gregor Leander, Anne Canteaut, Claude Carlet, Patrick Felke, and Philippe Gaborit, *Construction of bent functions via Niho power functions*, J. Combin. Theory Ser. A **113** (2006), no. 5, 779–798.

[26] Keqin Feng and Jinquan Luo, *Value distributions of exponential sums from perfect nonlinear functions and their applications*, IEEE Trans. Inf. Theory **53** (2007), no. 9, 3035–3041.

[27] Tao Feng, Bin Wen, Qing Xiang, and Jianxing Yin, *Partial difference sets from quadratic forms and p-ary weakly regular bent functions*, arXiv:1002.2797v2, 2011.

[28] Robert Gold, *Maximal recursive sequences with 3-valued recursive cross-correlation functions*, IEEE Trans. Inf. Theory **14** (1968), no. 1, 154–156.

[29] Solomon W. Golomb, *Theory of transformation groups of polynomials over* GF(2) *with applications to linear shift register sequences*, Information Sci. **1** (1968), no. 1, 87–109.

[30] Guang Gong, Tor Helleseth, Honggang Hu, and Alexander Kholosha, *On the dual of certain ternary weakly regular bent functions*, IEEE Trans. Inf. Theory **58** (2012), no. 4, 2237–2243.

[31] Tor Helleseth, *Some results about the cross-correlation function between two maximal linear sequences*, Discrete Math. **16** (1976), no. 3, 209–232.

[32] Tor Helleseth, *A note on the cross-correlation function between two binary maximal length linear sequences*, Discrete Math. **23** (1978), no. 3, 301–307.

[33] Tor Helleseth, Henk D. L. Hollmann, Alexander Kholosha, Zeying Wang, and Qing Xiang, *Proofs of two conjectures on ternary weakly regular bent functions*, IEEE Trans. Inf. Theory, **55** (2009), no. 11, 5272–5283.

[34] Tor Helleseth and Alexander Kholosha, *Monomial and quadratic bent functions over the finite fields of odd characteristic*, IEEE. Trans. Inf. Theory, **52** (2006), no. 5, 2018–2032.

[35] Tor Helleseth and Alexander Kholosha, *New binomial bent functions over the finite fields of odd characteristic*, IEEE Trans. Inf. Theory, **56** (2010), no. 9, 4646–4652.

[36] Tor Helleseth, Alexander Kholosha, and Sihem Mesnager, *Niho bent functions and Subiaco hyperovals*, Theory and Applications of Finite Fields (Providence, Rhode Island) (Michel Lavrauw, Gary L. Mullen, Svetla Nikova, Daniel Panario, and Leo Storme, eds.), Contemporary Mathematics, vol. 579, American Mathematical Society, 2012, pp. 91–101.

[37] Tor Helleseth, Torleiv Kløve, and Johannes J. Mykkeltveit, *On the covering radius of binary codes*, IEEE Trans. Inf. Theory **24** (1978), no. 5, 627–628.

[38] Tor Helleseth and P. Vijay Kumar, *Sequences with low correlation*, Handbook in Coding Theory (Vera S. Pless and W. Cary Huffman, eds.), vol. II, chapter 21 Elsevier Science B.V., Amsterdam, 1998, pp. 1765–1853.

[39] Tor Helleseth and Petri Rosendahl, *New pairs of m-sequences with 4-level cross-correlation*, Finite Fields Appl. **11** (2005), no. 4, 674–683.

[40] Henk D. L. Hollmann and Qing Xiang, *A proof of the Welch and Niho conjectures on cross-correlations of binary m-sequences*, Finite Fields Appl. **7** (2001), no. 2, 253–286.

[41] Xiang-Dong Hou, *Covering radius of the Reed-Muller code R(1, 7) – A simpler proof*, J. Combin. Theory Ser. A **74** (1996), no. 2, 337–341.

[42] Xiang-Dong Hou, *p-ary and q-ary versions of certain results about bent functions and resilient functions*, Finite Fields Appl. **10** (2004), no. 4, 566–582.

[43] Tadao Kasami, *The weight enumerators for several classes of subcodes of the 2nd order binary Reed-Muller codes*, Inform. and Control **18** (1971), no. 4, 369–394.

[44] Nicholas M. Katz and Ron Livné, *Sommes de Kloosterman et courbes elliptiques universelles en caractéristiques 2 et 3*, Comptes Rendus de l'Académie des Sciences Paris, Série I – Mathematique **309** (1989), no. 11, 723–726.

[45] Keijo Petteri Kononen, Marko Juhani Rinta-aho, and Keijo O. Väänänen, *On integer values of Kloosterman sums*, IEEE Trans. Inf. Theory **56** (2010), no. 8, 4011–4013.

[46] P. Vijay Kumar, *On bent sequences and generalized bent functions*, Ph.D. thesis, University of Southern California, 2004.

[47] P. Vijay Kumar, Robert A. Scholtz, and Lloyd R. Welch, *Generalized bent functions and their properties*, J. Combin. Theory Ser. A **40** (1985), no. 1, 90–107.

[48] Gilles Lachaud and Jacques Wolfmann, *The weights of the orthogonals of the extended quadratic binary Goppa codes*, IEEE Trans. Inf. Theory **36** (1990), no. 3, 686–692.

[49] Philippe Langevin and Gregor Leander, *Monomial bent functions and Stickelberger's theorem*, Finite Fields Appl. **14** (2008), no. 3, 727–742.

[50] Gregor Leander and Alexander Kholosha, *Bent functions with 2^r Niho exponents*, IEEE Trans. Inf. Theory **52** (2006), no. 12, 5529–5532.

[51] Nils Gregor Leander, *Monomial bent functions*, IEEE Trans. Inf. Theory **52** (2006), no. 2, 738–743.

[52] S.L. Ma, *A survey of partial difference sets*, Des. Codes Cryptogr. **4** (1994), no. 3, 221–261.

[53] Florence J. MacWilliams and Neil James A. Sloane, *The Theory of Error-Correcting Codes*, North-Holland Mathematical Library, vol. 16, North-Holland, Amsterdam, 1996, Ninth impression.

[54] Robert L. McFarland, *A family of difference sets in non-cyclic groups*, J. Combin. Theory Ser. A **15** (1973), no. 3, 1–10.

[55] Marko Moisio, *Kloosterman sums, elliptic curves, and irreducible polynomials with prescribed trace and norm*, Acta Arith. **132** (2008), no. 4, 329–350.

[56] Johannes J. Mykkeltveit, *The covering radius of the* (128,8) *Reed-Muller code is* 56, IEEE Trans. Inf. Theory **26** (1980), no. 3, 359–362.

[57] Yoji Niho, *Multi-valued cross-correlation functions between two maximal linear recursive sequences*, Ph.D. thesis, University of Southern California, Los Angeles, 1972.

[58] John D. Olsen, Robert A. Scholtz, and Lloyd R. Welch, *Bent-function sequences*, IEEE Trans. Inf. Theory **28** (1982), no. 6, 858–864.

[59] Nick J. Patterson and Douglas H. Wiedemann, *The covering radius of the $(2^{15}, 16)$ Reed-Muller code is at least 16276*, IEEE Trans. Inf. Theory **29** (1983), no. 3, 354–356.

[60] Nick J. Patterson and Douglas H. Wiedemann, *Corrections to "The covering radius of the $(2^{15}, 16)$ Reed-Muller code is at least 16276"*, IEEE Trans. Inf. Theory **36** (1990), no. 2, 443.

[61] Alexander Pott, *Finite Geometry and Character Theory*, Lecture Notes in Mathematics, vol. 1601, Springer-Verlag, Berlin, 1995.

[62] Alexander Pott, Yin Tan, Tao Feng, and San Ling, *Association schemes arising from bent functions*, Des. Codes Cryptogr. **59** (2011), no. 1–3, 319–331.

[63] Oscar S. Rothaus, *On "bent" functions*, J. Combin. Theory Ser. A **20** (1976), no. 3, 300–305.

[64] Melek Diker Yücel Selçuk Kavut, *9-variable Boolean functions with nonlinearity 242 in the generalized rotation symmetric class*, Inform. and Comput. **208** (2010), no. 4, 341–350.

[65] Yin Tan, Alexander Pott, and Tao Feng, *Strongly regular graphs associated with ternary bent functions*, J. Combin. Theory Ser. A **117** (2010), no. 6, 668–682.

[66] Jin Yuan, Claude Carlet, and Cunsheng Ding, *The weight distribution of a class of linear codes from perfect nonlinear functions*, IEEE Trans. Inf. Theory **52** (2006), no. 2, 712–717.

The Selmer Center
Department of Informatics, University of Bergen, P.O. Box 7800
N-5020 Bergen, Norway
{Tor.Helleseth, Alexander.Kholosha}@ii.uib.no

The complexity of change

Jan van den Heuvel

Abstract

Many combinatorial problems can be formulated as "Can I transform configuration 1 into configuration 2, if only certain transformations are allowed?". An example of such a question is: given two k-colourings of a graph, can I transform the first k-colouring into the second one, by recolouring one vertex at a time, and always maintaining a proper k-colouring? Another example is: given two solutions of a SAT-instance, can I transform the first solution into the second one, by changing the truth value one variable at a time, and always maintaining a solution of the SAT-instance? Other examples can be found in many classical puzzles, such as the 15-Puzzle and Rubik's Cube.

In this survey we shall give an overview of some older and some more recent work on this type of problem. The emphasis will be on the computational complexity of the problems: how hard is it to decide if a certain transformation is possible or not?

1   Introduction

Reconfiguration problems are combinatorial problems in which we are given a collection of configurations, together with some transformation rule(s) that allows us to change one configuration to another. A classic example is the so-called *15-puzzle* (see Figure 1): 15 tiles are arranged on a 4×4 grid, with one empty square; neighbouring tiles can be moved to the empty slot. The normal aim is, given an initial configuration, to move the tiles to the position with all numbers in order (right-hand picture in Figure 1). Readers of a certain age may remember Rubik's cube and its relatives as examples of reconfiguration puzzles (see Figure 2).

More abstract kinds of reconfiguration problems abound in graph theory. For instance, suppose we are given a planar graph and two 4-colourings of that graph. Is it possible to transform the first 4-colouring into the second one, by recolouring one vertex at a time, and never using more than 4 colours? Taking any two different 4-colourings of the complete graph K_4 shows that the answer is not always yes. But what would happen if we allowed a fifth colour? And whereas it is easy to see what the situation is with two 4-colourings of K_4, how hard is it to decide in general if two given 4-colourings of some planar graph can be transformed from one to the other by recolouring one vertex at a time?

Figure 1: Two configurations of the 15-puzzle (left picture © 2008 Theon; used under CC BY-SA 3.0 license, right picture © 2006 Booyabazooka; public domain; both from Wikimedia Commons).

Figure 2: Rubik's cube (© 2006 Booyabazooka; used under CC BY-SA 3.0 license; from Wikimedia Commons).

As a final (class of) examples in this introduction, we mention reconfiguration problems on satisfiability problems. Given some Boolean formula and two satisfying assignments of its variables, is it possible to transform one assignment into the other by changing the value of one variable at a time, but so that the formula remains TRUE during the whole sequence of transformations?

In this survey we concentrate on *complexity* considerations of transformation problems. In other words, we are interested in knowing *how hard* it is computationally to decide if the answer to some problem involving transformation is "yes" or "no". More specifically, we will look at two types of those complexity question, which very roughly can be described as follows:

A-TO-B-PATH

Instance: Description of a collection of feasible configurations; description of one or more transformations changing one configuration to another; description of two feasible configurations A, B.

Question: Is it possible to change configuration A into configuration B by a sequence of transformations in which each intermediate configuration is a feasible configuration as well?

PATH-BETWEEN-ALL-PAIRS

Instance: Description of a collection of feasible configurations; description of one or more transformations changing one configuration to another.

Question: Is it possible, given any two feasible configurations A, B, to change configuration A into configuration B by a sequence of transformations in which each intermediate configuration is a feasible configuration as well?

Of course, there are many other questions that can be asked: how many steps does it take to go from one configuration to another? Which two configurations are furthest apart? Etc., etc. Many of these questions have been considered for particular problems, and where opportune we shall mention some of this work.

An alternative way to formulate this type of problem is by using the concept of a *configuration graph*. This is the graph that has as vertex set the collection of all possible feasible configurations, and where two configurations are connected by an edge if there is a transformation changing one to the other. Note that nothing that we have said so far rules out the possibility that the transformation goes one way only, but in general we will assume that we can always go back and forth between configurations. This means the configuration graph can be taken to be an undirected graph.

Using the language of configuration graphs, the two general decision problems above can be rephrased as follows. PATH-BETWEEN-ALL-PAIRS: is the configuration graph connected? A-TO-B-PATH: given two vertices (configurations) in the configuration graph, are they in the same component?

In most of this survey we will use fairly informal language. So we may use "step" or "move" instead of "transformation" (a one-step change). On the other hand, "transform configuration A to configuration B", "move

from A to B" or "go from A to B" usually indicate a sequence of transformations.

1.1   A little bit on computational complexity

This survey cannot give a full definition of the complexity classes we
will encounter, and we only give a general, intuitive, description of some of
them. The interested reader can find all details in appropriate textbooks,
such as Garey & Johnson [25] and Papadimitriou [45].

We assume the reader is familiar with the concept of decision problems
(problems that have as answer either "yes" or "no") and the complexity classes **P**, **NP** and **coNP**. We will also regularly encounter the class
PSPACE. A decision problem is in **PSPACE**, or *can be solved in polynomial space*, if there exists an algorithm that solves the problem and that
uses an amount of memory that is polynomial in the size of the input.
The related non-deterministic complexity class **NPSPACE** is similarly
defined as the class of decision problems for which there exists a non-
deterministic algorithm that can recognise "yes"-instances of the problem
using an amount of memory that is polynomial in the size of the input.
For a non-deterministic algorithm we mean by *recognising "yes"-instances*
that for every "yes"-instance (but for none of the "no"-instances) there is
a possible run of the algorithm that finishes in finite time with a "yes"
answer.

We obviously have $\mathbf{P} \subseteq \mathbf{NP} \cap \mathbf{coNP}$ and $\mathbf{PSPACE} \subseteq \mathbf{NPSPACE}$,
and a little bit of thought should convince the reader that we also have
$\mathbf{NP} \cup \mathbf{coNP} \subseteq \mathbf{PSPACE}$. (Trial and error of all possible solutions of a
problem in **NP** or **coNP** can be done in polynomial space.) For most of
these inclusions it is unknown if they are proper inclusions or if the classes
are in fact the same, leading to some of the most important problems in
computer science (settling whether or not $\mathbf{P} = \mathbf{NP}$ is worth a million
dollars [16]). The one exception is that we know that **PSPACE** and
NPSPACE are in fact the same, by the celebrated theorem of Savitch [47].

Within each complexity class we can define so-called *complete* problems. Again, we refer to the appropriate textbooks for the precise definition; for us it is enough intuitively to assume that these are the "most
difficult" problems in their class.

1.2   Computational complexity of reconfiguration problems

In order to be able to ask sensible questions (and obtain sensible an-
swers) about the complexity of reconfiguration problems, we will
make some assumptions regarding their properties. In particular, when

describing the possible configurations, we will assume that these are not given as a full set of all configurations, but by some compact description. Otherwise, if the set of all possible configurations was part of the input, most decision problems about those configurations would trivially be possible in polynomial time because the input would be very large.

More precisely, we assume that an instance of the input contains an algorithm to decide if a candidate configuration really is feasible or not. Similarly, we are in general not interested in problems where the collection of possible transformations needs to be given in the form of an exhaustive list of all pairs that are related by the transformation. Instead, we assume that the input contains an algorithm to decide, given two configurations, whether or not we can get the second configuration from the first by a single transformation.

Regarding these algorithmic issues of the description of an instance of a configuration problem, we make the following assumptions:

A1: Deciding if a given possible configuration is a feasible configuration can be done in polynomial time.

A2: Given two feasible configurations, deciding if there is a transformation from the first to the second can be done in polynomial time.

Note that these assumptions guarantee that both of our general reconfiguration problems are in **NPSPACE** (hence in **PSPACE** by Savitch's Theorem). The following is a non-deterministic algorithm for A-TO-B-PATH that would work in polynomial space, when required to decide if it is possible to have a sequence of transformations from configuration A to configuration B:

1: Given the initial configuration A, "guess" a next configuration A_1. Check that A_1 is indeed a feasible configuration and that there is transformation from A to A_1. If A_1 is a valid next configuration, "forget" the initial configuration A and replace it by A_1.

2: Repeat the step above until the target configuration B has been reached.

If there is indeed a way to go from configuration A to configuration B, then a sequence of correct guesses in the algorithm above will indeed recognise that, using a polynomial amount of space; while if there is no sequence of transformations from A to B, then the algorithm will never finish. To extend the algorithm to the PATH-BETWEEN-ALL-PAIRS problem, we just need to repeat this task for all possible pairs. This means systematically generating all pairs of candidate configurations, and testing those. Since each candidate configuration has a size that is polynomial in the size of the original input (as it can be tested in polynomial time whether or not a candidate configuration is feasible), this brute-force generation of all possible pairs of configurations and testing whether or not they are feasible and connected can be done in a polynomial amount of memory as well.

The fact that all problems we consider are "automatically" in **PSPACE** means that we are in particular interested in determining if a particular variant is in a more restricted class (**P**, **NP**), or if it is in fact **PSPACE**-complete.

A final property that all examples we look at will have is that the transformations are symmetric: if we can transform one configuration into another, then we can also go the other way round. There is no real reason why this symmetry should always be the case, it is just that most reconfiguration problems considered in the literature have this feature. In particular, the sliding token problems we look at in Section 4 could just as well be formulated for directed graphs, leading in general to directed reconfiguration problems.

2   Reconfiguration of satisfiability problems

A collection of interesting results regarding the reconfiguration of solutions of a given Boolean formula were obtained by Gopalan et al. [27, 28]. They considered the following general set-up. Given a Boolean formula φ with n Boolean variables, the feasible configurations are those assignments from $\{T, F\}^n$ that satisfy φ (i.e., for which φ gives the value TRUE); the allowed transformation is changing the value of exactly one of the variables.

The collection of all possible assignments, together with edges added between those pairs that differ in exactly one variable, gives us the structure of the (graph of the) n-dimensional hypercube. This means that the configuration graph for a Boolean formula reconfiguration problem is the induced subgraph of the hypercube induced by the satisfying assignments. It is this additional structure that should give hope of a better understanding of this type of reconfiguration problem.

The first to analyse the connectivity properties of the configuration graphs of this type of problem were Gopalan et al. [27, 28]. To describe their results, we need a few more definitions.

A *logical relation* R is a subset of $\{T, F\}^k$, where k is the *arity* of R. For instance, if $R_{1/3} = \{TFF, FTF, FFT\}$, then $R_{1/3}(x_1, x_2, x_3)$ is TRUE if and only if exactly one of x_1, x_2, x_3 is T. For S a finite set of logical relations, a CNF(S)-*formula* over a set of variables $V = \{x_1, x_2, \ldots, x_n\}$ is a finite conjunction $C_1 \wedge C_2 \wedge \cdots \wedge C_m$ of clauses built using relations from S, variables from V, and the constants T and F. Hence each C_i is an expression of the form $R(\xi_1, \xi_2, \ldots, \xi_k)$, where R is a relation of arity k, and each ξ_j is a variable from V or one of the Boolean constants T, F. The satisfiability problem $\textsc{Sat}(\mathcal{S})$ associated with a finite set of logical relations $\mathcal{S}$ asks: given a CNF$(\mathcal{S})$-formula φ, is it satisfiable?

As an example, if we use the relation $R_{1/3}$ as above, and set $\mathcal{S}_{1/3} = \{R_{1/3}\}$, then CNF$(\mathcal{S}_{1/3})$ consists of Boolean expressions of the form

$$\varphi = (x_i \vee x_j \vee x_k) \wedge (x_{i'} \vee x_{j'} \vee x_{k'}) \wedge \cdots .$$

Finally, such an expression φ is true for some assignment from $\{T, F\}$ to the variables if and only if each clause $x_i \vee x_j \vee x_k$ has exactly one variable that is T. The satisfiability problem $\textsc{Sat}(\mathcal{S}_{1/3})$ is known as $\textsc{Positive-1-In-3-Sat}$; a decision problem that is **NP**-complete [48].

Another, better known, example is obtained by taking

$$\mathcal{S}_2 = \{\{TF, FT, TT\}, \{FF, FT, TT\}, \{FF, TF, FF\}\}.$$

Here the relation $R = \{TF, FT, TT\}$ indicates that $R(x_1, x_2)$ is TRUE when at least one of x_1, x_2 is T, hence it represents clauses of the form $x_1 \vee x_2$. Similarly, $\{FF, FT, TT\}$ represents clauses $\neg x_1 \vee x_2$, and $\{FF, TF, FT\}$ represents clauses $\neg x_1 \vee \neg x_2$. We see that CNF$(\mathcal{S}_2)$ is exactly the set of Boolean expressions that can be formulated with clauses that are disjunctions of two literals. (A *literal* is one of x_1 or $\neg x_i$ for a variable $x_i \in V$.) We call such a formula a *2-CNF-formula*.

Similarly, the *3-CNF-formulas* are exactly those formulas whose set of relations is $\mathcal{S}_3 = \{R_0, R_1, R_2, R_3\}$, where

$$R_0 = \{T, F\}^3 \setminus \{FFF\}, \quad R_1 = \{T, F\}^3 \setminus \{TFF\},$$
$$R_2 = \{T, F\}^3 \setminus \{TTF\}, \quad R_3 = \{T, F\}^3 \setminus \{TTT\}.$$

Note that this also means that $\textsc{Sat}(\mathcal{S}_2)$ and $\textsc{Sat}(\mathcal{S}_3)$ are equivalent to the well-known 2-SAT and 3-SAT decision problems, respectively.

Schaefer [48] proved a celebrated dichotomy theorem about the complexity of $\textsc{Sat}(\mathcal{S})$: for certain sets $\mathcal{S}$ (nowadays called *Schaefer sets*), $\textsc{Sat}(\mathcal{S})$ is solvable in polynomial time; while for all other sets $\mathcal{S}$ the problem is **NP**-complete.

In [27, 28], the following two decision problems are considered for given $\mathcal{S}$.

$\textsc{st-Conn}(\mathcal{S})$

Instance: A CNF$(\mathcal{S})$-formula φ, and two satisfying assignments **s** and **t** of φ.

Question: Is there a path between **s** and **t** in the configuration graph of solutions of φ?

Conn($\mathcal{S}$)

Instance: A CNF($\mathcal{S}$)-formula φ.

Question: Is the configuration graph of solutions of φ connected?

The key concept for these problems appears to be that of a *tight* set of relations $\mathcal{S}$ — see [27, 28] for a precise definition of this concept. Here we only note that every Schaefer set is tight.

Theorem 2.1 (Gopalan et al. [27, 28]) *Let $\mathcal{S}$ be a finite set of logical relations.*

(a)   *If $\mathcal{S}$ is tight, then* ST-Conn($\mathcal{S}$) *is in* **P**; *otherwise,* ST-Conn($\mathcal{S}$) *is* **PSPACE**-*complete.*

(b)   *If $\mathcal{S}$ is tight, then* Conn($\mathcal{S}$) *is in* **coNP**; *if $\mathcal{S}$ is tight but not Schaefer, then* Conn($\mathcal{S}$) *is* **coNP**-*complete; otherwise,* Conn($\mathcal{S}$) *is* **PSPACE**-*complete.*

(c)   *If every relation R in $\mathcal{S}$ is the set of solutions of a 2-CNF-formula, then* Conn($\mathcal{S}$) *is in* **P**.

Major parts of the proof of Theorem 2.1 in [27, 28] follow a similar strategy to the proof of Schaefer's Theorem in [48]. Given a set of relations $\mathcal{S}$, a k-ary relation R is *expressible from* S if there is a CNF($\mathcal{S}$)-formula $\varphi(x_1, \ldots, x_k, z_1, \ldots, z_m)$ such that R coincides with the set of all assignments to $x_1, \ldots, x_k$ that satisfy

$$(\exists z_1) \cdots (\exists z_m)\, \varphi(x_1, \ldots, x_k, z_1, \ldots, z_m).$$

Then the essential part of the proof of Schaefer's Theorem is that if a set $\mathcal{S}$ of relations is not-Schaefer, then every logical relation is expressible from $\mathcal{S}$.

The authors in [27, 28] extend the concept of expressibility to *structural expressibility*. Informally, a relation R is structurally expressible from a set of relations $\mathcal{S}$, if R is expressible using some CNF($\mathcal{S}$)-formula $\varphi(x_1, \ldots, x_k, z_1, \ldots, z_m)$ and if the subgraph of the hypercube formed by the satisfying assignments of φ has components that 'resemble' the components formed by the subgraph of the hypercube of the relations in R. The crucial result is that if a set of relations $\mathcal{S}$ is not tight, then every logical relation is structurally expressible from $\mathcal{S}$. Although the outline of this (part of the) proof is very similar to the outline of the corresponding part of Schaefer's Theorem, the actual proof is considerably more involved.

Additionally, the following results on the structure of the configuration graphs of the solutions of different CNF($\mathcal{S}$)-formula are obtained in [27, 28].

Theorem 2.2 (Gopalan et al. [27, 28]) *Let S be a finite set of logical relations.*

(a)   *If S is tight, then for any CNF(S)-formula φ with n variables, if two satisfying assignments s and t of φ are connected by a path, then the number of transformations needed to go from s to t is $O(n)$.*

(b)   *If S is not tight, then there exists an exponential function $f(n)$ such that for every n_0 there exists a CNF(S)-formula φ with $n \geq n_0$ variables and two satisfying assignments s and t of φ that are connected by a path, but where the number of transformations needed to go from s to t is at least $f(n)$.*

Regarding the result in Theorem 2.1(b), in their original paper [27] the authors in fact conjectured a trichotomy for the complexity of $\mathrm{CONN}(S)$, conjecturing that if S is Schaefer, then $\mathrm{CONN}(S)$ is actually in **P**. They showed this conjecture to be true for particular types of Schaefer set (see Theorem 2.1(b) for one example). The conjecture was disproved (assuming **P** $\neq$ **NP**) by Makino et al. [41], who found a set of Schaefer relations for which $\mathrm{CONN}(S)$ remains **coNP**-complete. In the updated version [28], a modified trichotomy conjecture for the complexity of $\mathrm{CONN}(S)$ is formulated.

3   Reconfiguration of graph colourings

Reconfiguration of different kinds of graph colourings is probably one of the best studied examples of reconfiguration problems. We will look at some particular variants. The required background in the basics of graph theory can be found in any textbook on graph theory, such as Bondy & Murty [7] or Diestel [21].

3.1   Single-vertex recolouring of vertex colourings

Recall that a *k-colouring* of a graph $G = (V, E)$ is an assignment $\varphi : V \rightarrow \{1, \ldots, k\}$ such that $\varphi(u) \neq \varphi(v)$ for every edge $uv \in E$. A graph is *k-colourable* if it has a k-colouring.

We start by considering the case where we are allowed to recolour one vertex at a time, while always maintaining a valid k-colouring. We immediately get the following two reconfiguration problems, for a fixed positive integer k.

k-Colour-Path

Instance: A graph G together with two k-colourings α and β.

Question: Is it possible to transform the first colouring α into the second
colouring β by recolouring one vertex at a time, while always
maintaining a valid k-colouring?

k-Colour-Mixing

Instance: A graph G.

Question: Is it possible, for any two k-colourings of G, to transform the
first one into the second one by recolouring one vertex at a
time, while always maintaining a valid k-colouring?

Let us call a graph G *k-mixing* if the answer to the second decision problem
is yes. The use of the work "mixing" in this context derives from its relation with work on rapid mixing of Markov chains to sample combinatorial
configurations; we say more about this in the final Section 5.

A variant of graph colouring is *list-colouring*. Here we assume that each
vertex v of a graph G has its own *list* $L(v)$ of colours. An *L-colouring* of
the vertices is an assignment $\varphi : V \longrightarrow \bigcup_{v \in V} L(v)$ such that $\varphi(v) \in L(v)$
for each vertex $v \in V$, and $\varphi(u) \neq \varphi(v)$ for every edge $uv \in E$. We
call such a colouring a *k-list-colouring* if each list $L(v)$ contains at most k
colours.

Regarding a transformation of a list-colouring, we use the obvious
choice: recolour one vertex at a time, where the new colour must come
from the list for that vertex list. With all this, we say that a graph G
with list assignments L is *L-list-mixing* if for every two L-colourings α
and β, we can transform α into β by recolouring one vertex at a time,
while always maintaining a valid L-colouring.

We also have two related decision problems.

k-List-Colour-Path

Instance: A graph G, list assignments $L(v)$ with $|L(v)| \leq k$ for each
vertex $v \in V$, and two L-colourings α and β.

Question: Is it possible to transform the first colouring α into the second
colouring β by recolouring one vertex at a time, while always
maintaining a valid L-colouring?

k-List-Colour-Mixing

Instance: A graph G and list assignments $L(v)$ with $|L(v)| \leq k$ for each
vertex $v \in V$.

Question: Is G L-list-mixing?

Before we look in some detail at what is known about the decision problems defined in this subsection, we give some other results on (list-)mixing. The following result has been obtained independently several times, but the first instance appears to be in (preliminary versions) of Dyer et al. [22]. The *degeneracy* $\deg(G)$ of a graph G is the minimum integer d so that every subgraph of G has a vertex of degree at most d. In other words, $\deg(G)$ is the maximum, over all subgraphs H of G, of the minimum degree of H.

Theorem 3.1 (Dyer et al. [22]) *For any graph G, if $k \geq \deg(G) + 2$, then G is k-mixing. In fact, G is L-list-mixing for any list assignment L such that $|L(v)| \geq \deg(G) + 2$ for all $v \in V$.*

Proof We only prove the k-mixing statement; the proof of the list version is essentially the same.

We use induction on the number of vertices of G. The result is obviously true for the graph with one vertex, so suppose G has at least two vertices. Let v be a vertex with degree at most $\deg(G)$, and consider $G' = G - v$. Note that $\deg(G') \leq \deg(G)$, hence we also have $k \geq \deg(G') + 2$, and by induction we can assume that G' is k-mixing.

Take two k-colourings α and β of G, and let α', β' be the k-colourings of G' induced by α, β. Since G' is k-mixing, there exists a sequence $\alpha' = \gamma'_0, \gamma'_1, \ldots, \gamma'_N = \beta'$ of k-colourings of G' so that two consecutive colourings γ'_{i-1}, γ'_i, $i = 1, \ldots, N$, differ in the colour of one vertex, say v_i. Set $c_i = \gamma'_i(v_i)$, the colour of v_i after recolouring. We now try to take the same recolouring steps to recolour G, starting from α. If for some i it is not possible to recolour v_i, this must be because v_i is adjacent to v and v at that moment has colour c_i. But because v has degree at most $\deg(G) \leq k - 2$, there is a colour $c \neq c_i$ that does not appear on any of the neighbours of v. Hence we can first recolour v to c, then recolour v_i to c_i, and continue.

In this way we find a sequence of k-colourings of G, starting at α, and ending in a colouring in which all the vertices except possibly v will have the same colour as in β. But then, if necessary, we can do a final recolouring of v to give it the colour from β, completing the proof. □

Theorem 3.1 is best possible, as can be seen, for example, by the complete graphs K_n and trees. (For a tree T we have $\deg(T) = 1$, while it is trivial to check that a graph with at least one edge is never 2-mixing.) Constructions of graphs that are k-mixing for specific values of k, and not for other values, can be found in Cereceda et al. [12].

Since the degeneracy $\deg(G)$ of a graph H is clearly at most the maximum degree $\Delta(G)$, Theorem 3.1 immediately implies that for $k \geq \Delta(G) + 2$, G is k-mixing, as already noted by Jerrum [34, 35].

An interesting result that is related to Theorem 3.1 was proved by Choo & MacGillivray [15]. They proved that if $k \geq \deg(G) + 3$, then the configuration graph (formed by all k-colourings with edges between colourings that differ in the colour on one vertex) is Hamiltonian. In other words, for those k we can start at any k-colouring of G and then there is a sequence of single-vertex recolourings so that every other k-colouring appears exactly once, ending with the original starting colouring.

When $k \geq \deg(G) + 2$, the proof of Theorem 3.1 provides an algorithm to find a sequence of transformations between any two k-colourings of G. But the best upper bound on the number of steps that can be obtained from the proof is exponential in the number of vertices of G. No graph is known for which such an exponential number of steps is necessary. In fact the following is conjectured in Cereceda [11].

Conjecture 3.2 (Cereceda [11]) *For a graph G on n vertices and integer $k \geq \deg(G) + 2$, any two k-colourings of G can be transformed from one into the other using $O(n^2)$ single-vertex recolouring steps.*

If true, the value $O(n^2)$ in Conjecture 3.2 would be best possible. Some weaker versions of the conjecture were proved in [11].

Theorem 3.3 (Cereceda [11]) *Conjecture 3.2 is true under the stronger assumptions $k \geq 2\deg(G) + 1$ or $k \geq \Delta(G) + 2$.*

Theorem 3.3 means that Conjecture 3.2 is true if G is a tree (then $\deg(G) + 2 = 3 = 2\deg(G) + 1$) or if G is regular (in which case $\deg(G) = \Delta(G)$).

Note that Theorem 3.1 has algorithmic consequences if we restrict the decision problems to classes of graph in which each graph has degeneracy at most some fixed upper bound. For instance, as planar graphs have degeneracy at most 5, we obtain that 7-COLOUR-MIXING and 7-COLOUR-PATH restricted to planar graphs are trivially in **P**, as the answer is always "yes".

We now return to the recolouring complexity problems introduced earlier in this subsection. The following are some results that are known about those problems.

Theorem 3.4

(a)   If $k = 2$, then k-COLOUR-PATH and k-COLOUR-MIXING are in **P**.

(b)   If $k = 3$, then k-COLOUR-PATH is in **P**; while k-COLOUR-MIXING is **coNP**-complete (Cereceda et al. [13, 14]).

(c)   For all $k \geq 4$, k-COLOUR-PATH is **PSPACE**-complete (Bonsma and Cereceda [8]).

Theorem 3.5

(a)   If $k = 2$, then k-LIST-COLOUR-PATH and k-LIST-COLOUR-MIXING are in **P**.

(b)   For all $k \geq 3$, k-LIST-COLOUR-PATH is **PSPACE**-complete (Bonsma and Cereceda [8]).

We already noted that the claims in Theorem 3.4(a) are trivial: if we have only two colours, then the end-vertices of any edge can never be recoloured.

The results in Theorem 3.4(b) are clearly the odd ones among the list of results above. The proof in [13] of the **coNP**-completeness of 3-COLOUR-MIXING uses the concept of *folding*: given two non-adjacent vertices u and v that have a common neighbour, a *fold on u and v* is the identification of u and v (together with removal of any double edges produced). We say a graph G is *foldable* to H if there exists a sequence of folds that transforms G to H. Folding of graphs, and its relation to vertex colouring, has been studied before, see for instance [17].

Combining observations from [12] for non-bipartite 3-colourable graphs, with the structural characterisation of bipartite graphs that are not 3-mixing in [13], gives the following.

Theorem 3.6 (Cereceda et al. [12, 13]) *Let G be a connected 3-colourable graph. Then G is not 3-mixing if and only if G is foldable to the 3-cycle C_3 or the 6-cycle C_6.*

It is easy to see that every non-bipartite 3-colourable connected graph can be folded to C_3, so the interesting part of this theorem is the characterisation of bipartite non-3-mixing graphs as being foldable to C_6. In a sense, the theorem shows that C_3 and C_6 are the 'minimal' graphs that are not 3-mixing.

Theorem 3.5(b) is not explicitly given in Bonsma and Cereceda [8], but follows from the proof of Theorem 3.4(c) in that paper.

The results that 2-LIST-COLOUR-PATH and 2-LIST-COLOUR-MIXING can be done in polynomial time can be proved directly with some effort. But it is easy to see that in fact any 2-list-colouring problem can be reduced to a Boolean 2-CNF-formula. For each vertex v and colour $c \in L(v)$, introduce a Boolean variable $x_{v,c}$. Then for each vertex v with $L(v) = \{c, d\}$ we add a clause $x_{v,c} \vee x_{v,d}$; while for each edge uv and colour $c \in L(u) \cap L(v)$ we add a clause $(\neg x_{u,c}) \vee (\neg x_{v,c})$. We can now use the results that checking the connectivity of the solution space of a 2-CNF-formula is in **P**, see Theorem 2.1(c).

As well as the computational complexity of some of the recolouring problems, we also know something about the number of recolourings we might need.

Theorem 3.7

(a)   *For a graph G on n vertices, if two 3-colourings of G can be connected by a sequence of single-vertex recolourings, then this can be done in $O(n^2)$ steps (Cereceda et al. [14]).*

(b)   *For any $k \geq 3$ there exists an exponential function $f(n)$ such that for every n_0, there exists a graph G on $n \geq n_0$ vertices, an assignment L of lists of size k to the vertices of G, and two L-colourings α and β of G, such that we can transform α into β by recolouring one vertex at a time, but where the number of recolourings required is at least $f(n)$ (Bonsma and Cereceda [8]).*

(c)   *For any $k \geq 4$, the result in (b) also holds for ordinary k-colouring recolourings.*

The bound $O(n^2)$ in Theorem 3.7(a) is best possible.

Two remaining questions are the complexity class of k-COLOUR-MIXING for $k \geq 4$ and k-LIST-COLOUR-MIXING for $k \geq 3$. In view of the fact that k-COLOUR-PATH and k-LIST-COLOUR-PATH for those values of k are **PSPACE**-complete, one would expect the mixing variants to have a similar complexity. On the other hand, since the **coNP**-completeness of 3-COLOUR-MIXING is obtained by a particular structure that needs to be present in a graph to fail to be 3-mixing, a similar graph-structural condition for mixing with more colours might well be possible.

In view of the results that many of the recolouring problems are not in **P** (assuming that $\mathbf{P} \neq \mathbf{NP}$), it is interesting to find restricted instances for which the recognition problems are in **P**. A natural choice for a graph class where one expects this to happen is the class of bipartite graphs, since many colouring problems are trivial in that class. Surprisingly, restricting the input of the decision problems to just bipartite graphs does not change any of the results in Theorems 3.4–3.7. A restriction to planar graphs has more surprising results, as expressed in the final result of this subsection.

Theorem 3.8

(a)   *When restricted to planar graphs, 3-COLOUR-MIXING becomes polynomial (Cereceda et al. [13]).*

(b)   *When restricted to planar graphs, k-COLOUR-PATH for $4 \leq k \leq 6$ and k-LIST-COLOUR-PATH for $3 \leq k \leq 6$ remain **PSPACE**-complete (Bonsma and Cereceda [8]). Both decision problems are in **P** for $k \geq 7$ when restricted to planar graphs.*

(c)   *When restricted to bipartite planar graphs, k-COLOUR-PATH for $k = 4$ and k-LIST-COLOUR-PATH for $3 \leq k \leq 4$ remain **PSPACE**-complete (Bonsma and Cereceda [8]). Both decision problems are in **P** for $k \geq 5$ when restricted to bipartite planar graphs.*

The results that k-COLOUR-PATH and k-LIST-COLOUR-PATH are polynomial for planar or bipartite planar graphs and larger k, follows directly from upper bounds for the degeneracy for those graphs and Theorem 3.1.

3.2 Kempe chain recolouring

Given a k-colouring φ of a graph G, a *Kempe chain* is a connected component of the subgraph of G induced by the vertices coloured with one of two given colours. In other words, if c_1, c_2 are two different colours, and $W \subseteq V$ is the collection of vertices coloured either c_1 or c_2, then a Kempe chain is a connected component of the induced subgraph of G with vertex set W. By a *Kempe chain recolouring* we mean switching the two colours on a Kempe chain. Kempe chains and Kempe chain recolouring have been essential concepts in the proofs of many classical results on colouring, such as the Four Colour Theorem from Appel & Haken [1, 3, 2] and Vizing's Edge-Colouring Theorem [53].

Notice that a Kempe chain recolouring is a generalisation of the single-vertex recolouring transformation from the previous subsection, since such a recolouring corresponds to a Kempe chain recolouring on a Kempe chain consisting of just one vertex. Let us call a graph G k-*Kempe-mixing* if it is possible, for any two k-colourings of G, to transform the first one into the second one by a sequence of Kempe chain recolourings.

From the observation above, we see that if a graph is k-mixing, then it certainly is k-Kempe-mixing. But the reverse need not be true. For instance, it has been observed many times that a bipartite graph is k-Kempe-mixing for any $k \geq 2$ [10, 24, 44], whereas for any $k \geq 2$ there exist bipartite graphs that are not k-mixing [12]. Furthermore, a simple modification of the proof of Theorem 3.1 shows that for any graph G, if $k \geq \deg(G) + 1$, then G is k-Kempe-mixing, as was already proved by Las Vergnas & Meyniel [38].

Very little is known about the complexity of determining if a graph is k-Kempe-mixing. The same holds for the 'path' version of the problem (determining if two given k-colourings can be transformed into one another by a sequence of Kempe chain recolourings). Intuitively, there appear to be at least two reasons why Kempe chain recolouring is so much harder to analyse than single-vertex recolouring. Firstly, for any k-colouring of a graph it is always possible to perform Kempe chain recolourings. This is different from single-vertex recolouring, where it might not be possible to recolour any vertex at all (if all vertices have all colours different from their own appearing on a neighbour). This kind of 'frozen' colourings is essential in many of the analyses and results on single-vertex recolourings.

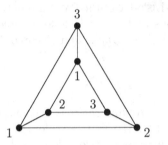 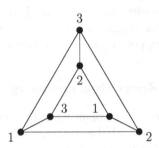

Figure 3: The 3-prism with two 3-colourings that are not related by Kempe chains.

Secondly, whereas a single-vertex recolouring has only a 'local' effect, a Kempe chain can affect many vertices throughout the graph.

The following are some results for planar graphs.

Theorem 3.9
(a) *Every planar graph is 5-Kempe-mixing (Meyniel [43]).*
(b) *If G is a 3-colourable planar graph, then G is 4-Kempe-mixing (Mohar [44]).*

The results in the theorem are best possible in the sense that the number of colours cannot be reduced in either statement [44].

As we have observed earlier, for the single-vertex recolouring problem, the smallest graph that is not 3-mixing is the triangle C_3, while the smallest *bipartite* graph that is not 3-mixing is the 6-cycle C_6. These graphs are essential in the proof that 3-COLOUR-MIXING is **coNP**-complete. The smallest graph that is not 3-Kempe-mixing is the 3-prism $K_3 \square K_2$ (see Figure 3).

It is easy to check that any Kempe chain recolouring in these two colourings will only result in renaming the two colours involved in the Kempe chain, but never changes the structure. It is unknown if the 3-prism is in some way a 'minimal' graph that is not 3-Kempe mixing, or if it is the only one. Neither is it obvious what subgraph relation we should use (like 'foldable' for the single-vertex recolouring problem) when talking about 'minimal' for Kempe-mixing.

Kempe chains have also been used extensively in the analysis of edge-colourings of graphs. Recall that a *k-edge-colouring* of a graph $G = (V, E)$ is an assignment $\varphi : E \to \{1, \ldots, k\}$ such that $\varphi(e) \neq \varphi(f)$ for any two edges e, f that share a common end-vertex. Similar to vertex-colourings, a Kempe chain in an edge-coloured graph is a component of the subgraph formed by the edges coloured with one of two given colours. Note that in this case every Kempe chain is a path or an even length cycle, and a recolouring is again just switching the colours on the chain. Call a graph G k-*Kempe-edge-mixing* if it is possible, given any two k-edge-colourings of G, to transform the first one into the second one by a sequence of Kempe chain recolourings.

Kempe chains on edge-colourings are instrumental in most (if not all) proofs of Vizing's Theorem [53] that a simple graph with maximum degree Δ has an edge-colouring using at most $\Delta + 1$ colours. Hence it is not surprising that results on k-Kempe-edge mixing are related to this constant as well.

Theorem 3.10 (Mohar [44])

(a)   *If a simple graph G can be edge-coloured with k colours, then G is $(k + 2)$-Kempe-edge-mixing.*

(b)   *If G is a simple bipartite graph with maximum degree Δ, then G is $(\Delta + 1)$-Kempe-edge-mixing.*

It is unknown if Theorem 3.10(a) is best possible, nor if the condition that the graph be bipartite in part (b) is necessary. A strongest possible result would be for any simple graph with maximum degree Δ to be $(\Delta + 1)$-Kempe-edge-mixing.

4   Moving tokens on graphs

The 15-puzzle can be considered as a problem involving moving tokens around a given graph, where a token can be moved along an edge to an empty vertex. So the two configurations in Figure 1 can also be drawn as in Figure 4.

Looking at the 15-puzzle in this way immediately suggests all kind of generalisations. An obvious generalisation is to play the game on different graphs. But we can also change the number of tokens, or the way the tokens are labelled. In this section we consider some of the variants that have been studied in the literature.

4.1   Labelled tokens without restrictions

There is an obvious generalisation of the 15-puzzle. For a given graph on n vertices, place $n - 1$ tokens labelled 1 to $n - 1$ on different vertices. The allowed moves are "sliding" a token along an edge onto the unoccupied vertex. The central question is if each of the $n!$ possible token

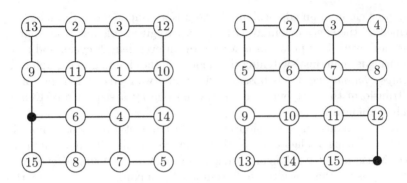

Figure 4: Two configurations of the 15-puzzle on the 4×4 grid.

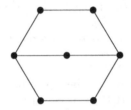

Figure 5: The exceptional graph θ_0.

configurations can be obtained from one another by a sequence of token moves. A complete answer to this was given in Wilson [56]. For a given graph G on n vertices, he defines the *puzzle graph* puz(G) as the graph that has as vertex set all possible placements of the $n-1$ tokens on G, and two configurations are adjacent if they can be obtained from one another by a single move.

Theorem 4.1 (Wilson [56]) *Let G be a 2-connected graph on $n \geq 3$ vertices. Then* puz(G) *is connected, except in the following cases:*

(a) *G is a cycle on $n \geq 4$ vertices (in which case* puz(G) *has $(n-2)!$ components);*

(b) *G is a bipartite graph different from a cycle (then* puz(G) *has two components);*

(c) *G is the graph θ_0 in Figure 5 (then* puz(G) *has six components).*

The condition in Wilson's theorem that the graph G is 2-connected is necessary. It is obvious that for a non-connected graph G, puz(G) is never connected; while if G has a cut-vertex v, then a token can never be moved from one component of $G - v$ to another component.

The proof of Theorem 4.1 in [56] is quite algebraic in nature. This is not surprising, since each token configuration can be considered as a permutation of n labels (with the unoccupied vertex having the label 'empty'). Within that context, a move is just a particular type of transposition involving two labels (one of them always being the 'empty' label). Although Wilson's theorem is not formulated in algorithmic terms, it is easy to derive from it a polynomial time algorithm to decide if puz(G) is connected for a given input graph G.

Since Wilson's work (and often independent of it), many generalisations have been considered in the literature. To describe these in some detail, we need some further notation. Instead of assuming that all tokens are different, we will assume that some tokens can be identical. So tokens come in certain *types* (other authors use *colours* for this), where tokens of the same type are considered indistinguishable (and hence swapping tokens of the same type will not lead to a different configuration). A collection of tokens can have k_1 tokens of type 1, k_2 tokens of type 2, etc. We denote such a typed set by $(k_1, k_2, \ldots, k_p)$, so that $k_1 + \cdots + k_p$ is the total number of tokens. A repeated sequence of p ones can be denoted as $1^{(p)}$.

Given a graph G and token set $(k_1, \ldots, k_p)$, the *puzzle graph*, denoted puz($G; k_1, \ldots, k_p$), is the graph that has as vertex set all possible token placements on G of k_1 tokens of type 1, k_2 tokens of type 2, etc., and two configurations are adjacent if they can be obtained from one another by a single move of a token to a neighbouring empty vertex. This means that if G is a graph on n vertices, then puz(G) $\cong$ puz($G; 1^{(n-1)}$). We will always assume that if G has n vertices then $k_1 + \cdots + k_p \leq n$ and $k_1 \geq k_2 \geq \cdots \geq k_p \geq 1$.

A first generalisation of Wilson's work, in which there may be fewer than $n - 1$ tokens, was considered by Kornhauser et al. [37]. They showed that if G is a graph on n vertices, then for any two configurations from puz($G, 1^{(p)}$), it can be decided in polynomial time if these two configurations are in the same component, i.e., if one configuration can be obtained from the other by a sequence of token moves. Additionally, they showed that if such a transformation is possible, the number of moves required is at most $O(n^3)$, and the order of this bound is best possible.

This work was further extended to token configurations with types as above; first to trees by Auletta et al. [4], and later to general graphs by Goraly & Hassin [29]. Their results prove that for any graph G and typed token set $(k_1, \ldots, k_p)$, given two configurations from puz($G; k_1, \ldots, k_p$), it can be decided in linear time if one configuration can be obtained from the other. Notice that by the result for all tokens being different mentioned earlier, we immediately have that for a graph on n vertices, more than $O(n^3)$ token moves are never needed to move between two configurations.

The work mentioned in the previous paragraphs does not give an explicit characterisation of the puzzle graphs $\text{puz}(G; k_1, \ldots, k_p)$ that are connected (i.e., where any two token configurations of the right type can be obtained from one another by a sequence of token moves). In order to describe such a characterisation, we need some further terminology regarding specific vertex-cut-sets in graphs. For a connected graph G, a *separating path of size one* in G is a cut-vertex. A *separating path of size two* is a cut-edge $e = v_1 v_2$ so that both components of $G - e$ have at least two vertices. Finally, for $\ell \geq 3$, a *separating path of size ℓ* is a path $P = v_1 v_2 \ldots v_\ell$ in G, such that the vertices $v_2, \ldots, v_{\ell-1}$ have degree two, $G - \{v_2, \ldots, v_{\ell-1}\}$ has exactly two components, one containing v_1 and one containing v_ℓ, and where both components have at least two vertices.

Theorem 4.2 (Brightwell et al. [9]) *Let G be a graph on $n \geq 3$ vertices and $(k_1, \ldots, k_p)$ be a token set, with $k_1 + \cdots + k_p \leq n$ and $k_1 \geq k_2 \geq \cdots \geq k_p \geq 1$. Then $\text{puz}(G; k_1, \ldots, k_p)$ is disconnected if and only if at least one of the following cases holds:*

(a)   *G is disconnected, and $p \geq 2$ or $k_1 \leq n - 1$;*

(b)   *$p \geq 2$ and $k_1 + \cdots + k_p = n$;*

(c)   *G is a path and $p \geq 2$;*

(d)   *G is a cycle, $p = 2$ and $k_2 \geq 2$; or G is a cycle and $p \geq 3$;*

(e)   *G is a 2-connected bipartite graph and the token set is $(1^{(n-1)})$;*

(f)   *G is the graph θ_0 in Figure 5 and the token set is one of $(2, 2, 2)$, $(2, 2, 1, 1)$, $(2, 1, 1, 1, 1)$, $(1^{(6)})$;*

(g)   *G has connectivity one and contains a separating path of size at least $n - (k_1 + k_2 + \cdots + k_p)$.*

Note in particular that if G is a 2-connected graph on n vertices different from a cycle, and $(k_1, \ldots, k_p)$ is a token set with $k_1 + \cdots + k_p \leq n - 2$, then $\text{puz}(G; k_1, \ldots, k_p)$ is always connected.

It is possible to extend this theorem to a full characterisation of any two token configurations from any puzzle graph $\text{puz}(G; k_1, \ldots, k_p)$ that are in the same component (hence extending the algorithmic results from Goraly & Hassin [29]). This rather technical and long result can also be found in [9].

The results mentioned above mean that it is quite straightforward to check if one can go from any given token configuration to any other one. So a next natural question is to ask if it is possible to find the shortest path, i.e., to find the minimum number of token moves required between two given token configurations in the same component of $\mathrm{puz}(G; k_1, \ldots, k_p)$. This leads to the following decision problem.

SHORTEST-TOKEN-MOVES-SEQUENCE

Instance: A graph G, a token set $(k_1, \ldots, k_p)$, two token configurations α and β on G of type $(k_1, \ldots, k_p)$, and a positive integer N.

Question: Is it possible to transform configuration α into configuration β
using at most N token moves?

Theorem 4.3 *Restricted to the case that the token sets are (k) (i.e., all tokens are the same),* SHORTEST-TOKEN-MOVES-SEQUENCE *is in* **P**.

Proof   We can assume that the given graph G is connected. (Since two configurations can be transformed into one another if and only if this can be done for the configurations restricted to the components of the graph.) Given two token configurations α and β of k identical tokens on G, let $U = \{u_1, \ldots, u_k\}$ be the set of vertices containing a token in α, and $V = \{v_1, \ldots, v_k\}$ be the same for β.

Form a complete bipartite graph $K_{k,k}$ with parts U and V. For each edge $e_{ij} = u_i v_j$, define the *weight* w_{ij} of e_{ij} as the length of the shortest path from u_i to v_j in G (and denote by P_{ij} such a shortest path in G). It is well-known that a minimum weight perfect matching in a weighted balanced complete bipartite graph can be found in polynomial time (for instance using the Hungarian method, see, e.g., Schrijver [49, Section 17.2]); let M be such a minimum weight perfect matching.

We can assume that $M = \{u_1 v_1, \ldots, u_k v_k\}$. Let W be the total weight in M, i.e., the sum of the lengths of the paths P_{ii}, $i = 1, \ldots, k$. It is obvious that any way to move the tokens from U to V will use at least W steps. We will prove that in fact it is possible to do so using exactly W steps. We use induction on W, observing that if $W = 0$, then $U = V$, so $\alpha \equiv \beta$, and no tokens have to be moved.

If $W > 0$, then at least one element of V, say v_1, has no token on it in α. If $V(P_{11}) \cap U = \{u_1\}$, then we can just move the token from u_1 along P_{11} to v_1, and are done by induction. So assume that P_{11} contains some other elements from U. Take u_i to be the element from $V(P_{11}) \cap U$ nearest to v_1 on P_{11}. Define new paths P'_{1i} and P'_{i1} as follows. Let P'_{1i} be the path formed by going from u_1 along P_{11} to u_i and then continue along P_{ii} to v_i; while P'_{i1} is just the path from u_i along P_{11} to v_1. It is clear that the sum of the lengths of P'_{1i} and P'_{i1} is the same as that sum for P_{11} and P_{ii}, so we can replace P_{11} and P_{ii} by P'_{1i} and P'_{i1} to get another set of paths from U to V of minimum total length. But in this new collection of paths, we can just move u_i along P'_{i1} to v_1, and then continue by induction.   □

It was proved by Goldreich [26][1] that the problem SHORTEST-TOKEN-MOVES-SEQUENCE is **NP**-complete for the case Wilson considered, i.e., if all tokens are different. So somewhere between all tokens the same and all tokens different, the problem switches from being in **P** to being **NP**-complete. In fact, the change-over happens as soon as not all tokens are identical.

Theorem 4.4
Restricted to the case that the token sets are $(k-1,1)$ (i.e., there is one special token and all others are identical), SHORTEST-TOKEN-MOVES-SEQUENCE *is* **NP**-*complete.*

It is possible to prove this using most of the ideas from the proof in Papadimitriou et al. [46] that 'GRAPH-MOTION-PLANNING-WITH-ONE-ROBOT' is **NP**-complete. Motion planning of robot(s) on graphs is very closely related to transformations between token configurations on graphs. Except now there are some special tokens, the 'robots', that have to be moved from an initial position to a specific final position, while all other tokens are just 'obstacles', and their final position is not relevant. The full details of the proof of Theorem 4.4 will appear in Trakultraipruk [52].

4.2 Unlabelled Tokens with Restrictions

If we consider the token problems in the previous subsection for the case that all tokens are identical, then there is very little to prove. The puzzle graph $\text{puz}(G; k)$ (with $k \le |V(G)|$) is connected if and only if $k = |V(G)|$ or G is connected. More specifically, two token configurations are in the same component of $\text{puz}(G; k)$ if and only if they have the same number of tokens on each component of G. Even finding the minimum number of steps to go from one given configuration to another can be done in polynomial time. (Of course, this does not mean that questions about other properties of this kind of reconfiguration graphs cannot be interesting; see for instance Fabila-Monroy et al. [23].)

But the situation changes drastically if only certain positions of tokens are allowed. The following problem was studied in Hearn & Demaine [32]. Recall that a *stable set* in a graph is a set of vertices so that no two in the set are adjacent.

---

[1] Although [26] was published in 2011, it is remarked in it that the work was already completed in 1984, and appeared as a technical report from the Technion in 1993.

STABLE-SLIDING-TOKEN-CONFIGURATIONS
Instance: A graph G, and two token configurations on G using identical
              tokens so that the set of occupied vertices for both configu-
              rations forms a stable set in G.

Question: Is it possible to transform the first given configuration into
              the second one by a sequence of moves of one token along an
              edge, and such that in every intermediate configuration the
              set of occupied vertices is a stable set?

Theorem 4.5 (Hearn & Demaine [32])   *The problem* STABLE-SLIDING-
TOKEN-CONFIGURATIONS *is* **PSPACE**-*complete, even when restricted to
planar graphs with maximum degree three.*

The proof of this theorem in [32] (and many other results in that paper)
rely on a powerful general type of problem that seems to be very suitable
for complexity theoretical reductions. A *non-deterministic constraint logic
machine (NCL machine)* consists of an undirected graph, together with
assignments of non-negative integer weights to its edges and its vertices.
A feasible configuration of an NCL machine is an orientation of the edges
such that the sum of incoming edge-weights at each vertex is at least
the weight of that vertex. A move is nothing other than reversing the
orientation of one edge, guaranteeing that the resulting orientation is still
a feasible configuration.

The following is a natural reconfiguration question for NCL machines.

NCL-CONFIGURATION-TO-EDGES
Instance: An NCL machine, a feasible configuration on that machine,
              and a specific edge of the underlying graph.

Question: Is there a sequence of moves such that all intermediate con-
              figurations are feasible, and ending in a feasible configuration
              in which the specified edge has its orientation reversed?

Theorem 4.6 (Hearn & Demaine [32])
The problem NCL-CONFIGURATION-TO-EDGES *is* **PSPACE**-*complete,
even when restricted to NCL machines in which the underlying graph is
planar, all vertices have degree three, all edge-weights are 1 or 2, and all
vertex weights are 2.*

We return to moving tokens configuration problems. Note that in the
STABLE-SLIDING-TOKEN-CONFIGURATIONS problem, the graph has a 'double' role: it determines both the allowed configurations (stable vertex sets)
and the allowed moves (sliding along an edge). A natural next question

would be what happens when one of these constraints imposed by the
graph is removed. We have already seen that if we remove the constraint
that the configurations must be stable sets, then the problem becomes
easy. But the situation is different if we remove the constraint that token
movement must happen along an edge.

STABLE-SET-RECONFIGURATION

Instance: A graph G, and two token configurations on G using identical
tokens so that the set of occupied vertices for both configurations forms a stable set in G.

Question: Is it possible to transform the first given configuration into
the second one by a sequence of moves of one token at each
step, where a token can move from any vertex to any other
vacant one, and so that in every intermediate configuration
the set of occupied vertices is a stable set?

Theorem 4.7 (Ito et al. [33]) *The problem* STABLE-SET-RECONFIGUR-
ATION *is* **PSPACE**-*complete, even when restricted to planar graphs with
maximum degree three.*

Since independent set problems are easily reduced to problems about
cliques, vertex covers, etc., reconfiguration problems where the vertices
occupied by a token form sets of this type are easily seen to be **PSPACE**-
complete as well. See Ito et al. [33] for more details.

For some other types of sets formed by occupied vertices, the corresponding reconfiguration problems can become polynomial. A classical
example of this is the following.

Theorem 4.8 (Cummins [19]) *Let G be a connected graph with positive
weights on its edges. Then any minimum spanning tree of G can be transformed into any other minimum spanning tree by exchanging one edge at a
time, so that each intermediate configuration is a minimum spanning tree
as well.*

Note that the reconfiguration in Theorem 4.8 can be seen as a token reconfiguration problem by playing on the line graph of G. Similarly, the
following problem is essentially STABLE-SET-RECONFIGURATION played
on line graphs.

MATCHING-RECONFIGURATION

Instance: A graph G, and two matchings of G (subgraphs of degree at
most one).

Question: Is it possible to transform the first matching into the second one by a sequence of moves of one edge at a time, so that in every intermediate configuration the set of chosen edges forms a matching as well?

Theorem 4.9 (Ito et al. [33]) MATCHING-RECONFIGURATION *is in* **P**.

Comparing the reconfiguration problems seen in this subsection that are **PSPACE**-complete with those that are in **P**, it is tempting to conjecture that if the related *decision problem* is **NP**-complete, then the reconfiguration problem is **PSPACE**-complete; whereas if the related decision problem is in **P**, then so is the reconfiguration problem. Such a connection is alluded to in Ito et al. [33]. Nevertheless, in earlier sections we have seen some examples that shows that such a direct connection is not true. For instance, it is **NP**-complete to decide if a graph is 3-colourable, but the single-vertex recolouring reconfiguration problem is in **P**, Theorem 3.4(b).

We close this section with a simplified version of a question from Ito et al. [33]: is the HAMILTON-CYCLE-RECONFIGURATION problem (where two cycles are adjacent if they differ in two edges) **PSPACE**-complete?

5 Applications

Most reconfiguration problems are interesting enough for their own sake, and do not really need an application to justify their study. Nevertheless, many reconfiguration problems have applications or are inspired by problems in related areas. In this section we look at some of those applications and connections.

5.1 Sampling and counting

Randomness plays an important role in many parts of combinatorics and theoretical computer science. Indeed, results from probability theory have led to major developments in both fields. It is therefore unsurprising that researchers are often interested in obtaining random samples of particular combinatorial structures. For example, much attention has been devoted to the problem of sampling from an exponential number of structures (exponential in the size of the object over which the structures are defined) in time polynomial in this quantity. One of the reasons for this is that being able to sample almost uniformly from a set of combinatorial structures is enough to be able to approximately count such structures. See Jerrum [34] for an example illustrating the method in the context of graph colourings, and Jerrum [35] and Jerrum et al. [36] for full details.

   The question of when the configuration graph of a reconfiguration problem is connected is quite old. In particular the configuration graph of the single-vertex recolouring method has been looked at, as a subsidiary issue, by researchers in the statistical physics community studying the 'Glauber dynamics of an anti-ferromagnetic Potts model at zero temperature'. (See Sokal [50] for an introduction to the Potts model and its many relations to graph theory.) Associated with that research is the work on rapid mixing of Markov chains used to obtain efficient algorithms for almost uniform sampling of k-colourings of a given graph. We give a brief description of the basic ideas involved in these areas of research.

   Quite often, the sampling is done via the simulation of an appropriately defined Markov chain. Here the important point is that the Markov chain should be rapidly mixing. This means, loosely speaking, that it should converge to a close approximation of the stationary distribution in time polynomial in the size of the problem instance. For a precise description of this concept and further details, see [35] again.

   In the context of the particular Markov chain used for sampling k-colourings of a graph known as *Glauber dynamics* (originally defined for the *anti-ferromagnetic Potts model at zero temperature*) we have the following. For a particular graph G and value of k, let us denote the Glauber dynamics for the k-colourings of G by $\mathcal{M}_k(G) = (X_t)_{t=0}^{\infty}$. The state space of $\mathcal{M}_k(G)$ is the set of all k-colourings of G, the initial state X_0 is an arbitrary colouring, and its transition probabilities are determined by the following procedure.

   1. Select a vertex v of G uniformly at random.
   2. Select a colour $c \in \{1, 2, \ldots, k\}$ uniformly at random.
   3. If recolouring vertex v with colour c yields a proper colouring, then set X_{t+1} to be this new colouring; otherwise, set $X_{t+1} = X_t$.

   The relation between $\mathcal{M}_k(G)$ and the single-vertex recolouring transformations is immediate. In particular, to be sure that every k-colouring can appear as some state of the Markov chain, we need that G is k-mixing. Thus the fact that a graph is k-mixing is a necessary condition for its Glauber dynamics Markov chain to be rapidly mixing. (This explains the choice of terminology in Section 3.) On the other hand, if a graph is k-mixing it does not mean that its Glauber dynamics Markov chain is rapidly mixing. An example showing this is given by the stars $K_{1,m}$, which are k-mixing for any $k \geq 3$ (see Theorem 3.1) but whose Glauber dynamics is not rapidly mixing for $k \leq m^{1-\varepsilon}$, for fixed $\varepsilon > 0$ (Łuczak & Vigoda [40]).

   Let us point out that much of the work on rapid mixing of the Glauber dynamics Markov chain (as well as that on its many generalisations and variants) has concentrated on specific graphs, or on values of k so large that k-mixing is guaranteed. Many applications in theoretical physics related to the Potts model are of particular interest for crystalline structures, leading to the many studies of the Glauber dynamics and its generalisations on very regular and highly symmetric graphs such as integer grids.

Similar to the single-vertex recolouring method, the Kempe chain re-colouring method (see Subsection 3.2) has also been used to define a Markov chain on the set of all k-colourings of some graph. The correspond-ing approximate sampling algorithm is known as the *Wang-Swendsen-Kotecký dynamics*; see [54, 55].

Many reconfiguration problems we have considered so far can be used to define a Markov chain similar to the ones for vertex-colouring defined above. Again, for such a Markov chain to be a useful tool for almost uni-form sampling and approximate counting, it is necessary that the config-uration graph is connected, leading to questions considered in this survey.

5.2   Puzzles and games

We introduced the study of token configurations on graphs by looking at the classical 15-puzzle. But in fact, many puzzles and games can be described as reconfiguration problems. Following Demaine and Hearn [20] we use the term *(combinatorial) puzzle* if there is only one player, and use *(combinatorial) game* if there are two players. (So we ignore games with more than two players, or with no players (like Conway's Game of Life).)

The puzzles we are interested in are of the following type: "Given some initial configuration and a collection of allowed moves, can some prescribed final configuration (or a final configuration from a prescribed set) be reached in a finite number of moves?" For a game the situation is somewhat more difficult, and several different variants can be described. A quite general one is: "Given some initial configuration, a starting player, a collection of allowed moves which the two players have to play alternately, and a collection of winning configurations for player 1, can player 1 force the game to always reach a winning configuration in a finite number of moves, no matter the moves player 2 chooses?" Another way to describe this question is: "Given the setup of the game and the initial situation, is there a winning strategy for player 1?".

With these descriptions there is an obvious relation between the type of reconfiguration problem we considered and puzzles and games. For many puzzles and games, both existing and specially invented, the complexity of answering the questions above have been considered. A good start to find the relevant results and literature in this area is the extensive survey of Demaine and Hearn [20].

5.3 Other applications

Some reconfiguration problems have more practical applications (leaving aside if "solving puzzles" is really a practical application). In particular, graph recolouring problems can be seen as abstract versions of several real-life problems. One example of this is as a modelling tool for the assignment of frequencies in radio-communication networks. The basic aim of the *Frequency Assignment Problem (FAP)* is to assign frequencies to users of a wireless network, minimising the interference between them and taking care to use the smallest possible range of frequencies. Because the radio spectrum is a naturally limited resource with a constantly growing demand for the services that rely on it, it has become increasingly important to use it as efficiently as possible. As a result, and because of the inherent difficulty of the problem, the subject is huge. For an introduction and survey of different approaches and results we refer the reader to Leese & Hurley [39].

The FAP was first defined as a graph colouring problem by Hale [30]. In this setting, we think of the available frequencies (discretised and appropriately spaced in the spectrum) as colours, transmitters as vertices of a graph, and we add edges between transmitters that must be assigned different frequencies. In order to better capture the subtleties of the 'real-world' problem, this basic model has been generalised in a multitude of different ways. Typically this might involve taking into account the fact that radio waves decay with distance obeying an inverse-square law. For instance, numerical weights can be placed on the edges of the graph to indicate that frequencies assigned to the end-vertices of an edge must differ by at least the amount given by the particular edge-weight.

One of the major factors contributing to the rise in demand for use of the radio spectrum has been the dramatic growth, both in number and in size, of mobile telecommunication systems. In such systems, where new transmitters are continually added to meet increases in demand, an optimal or near-optimal assignment of frequencies will in general not remain so for long. On the other hand, it might be the case that, because of the difficulty of finding optimal assignments, a sub-optimal assignment is to be replaced with a recently-found better one. It thus becomes necessary to think of the assignment of frequencies as a dynamic process, where one assignment is to be replaced with another. In order to avoid interruptions to the running of the system, it is desirable to avoid a complete re-setting of the frequencies used on the whole network. In a graph colouring framework, this leads naturally to the graph-recolouring problems we looked at in Section 3.

Not much attention seems yet to have been devoted to the problem of reassigning frequencies in a network. Some first results can be found in [5, 6, 31, 42]. Most of the work in that literature describes specific heuristic approaches to the problem, often accompanied by some computational simulations.

As hinted at already in Section 4, moving-token puzzles are related to questions on movements of robots. A simple abstraction and discretisation is to assume that one or more robots move along the edges of a graph. There might be additional objects placed on the vertices, playing the role of 'obstacles'. The robots have to move from an initial configuration to some target configuration. In order to pass vertices occupied by obstacles, these obstacles have to be moved out of the way, also along edges.

In general it is not hard to decide whether or not the robots can actually move from their initial to their target configuration. But for practical applications, limiting the number of steps required is also important, leading to problems that are much harder to answer, see e.g. Papadimitriou et al. [46]. A multi-robot motion planning problem in which robots are partitioned in groups such that robots in the same group are interchangeable, comparable to the sliding token problem with different types of tokens, has recently been studied in Solovey & Halperin [51].

The robot motion problem on graphs is closely related to certain puzzles as well. A well-known example of such a puzzle is *Sokoban*, which is played on a square grid where certain squares contain immovable walls or other obstacles. There is also a single 'pusher' who can move certain blocks from one square to a neighbouring one, but only in the direction the pusher can 'push'. Moreover, the pusher can only move along unoccupied squares. The goal of the game is to push the movable blocks to their prescribed final position. Deciding if a given Sokoban configuration can be solved is known to be **PSPACE**-complete, as was proved in Culberson [18].

6   Open problem

It would be easy to end this survey with a long list of open problems: what is the complexity of deciding the following reconfiguration problems: ... ? But a more fundamental, and probably more interesting, problem is to try to find a connection between the complexity of reconfiguration problems and the complexity of the decision problem on the existence of configurations of a particular kind (related to the reconfiguration problem under consideration).

Such connections are regularly alluded to in the literature. For instance, with regard to the complexity of satisfiability reconfiguration problems, Gopalan et al. [27] conjectured that if $\mathcal{S}$ is Schaefer, then CONN($\mathcal{S}$) is in **P**. (This has been disproved since then; see Section 2.) Similarly, Ito et al. [33] write "There is a wealth of reconfiguration versions of NP-complete problems which can be shown PSPACE-complete via extensions, often quite sophisticated, of the original NP-completeness proofs; ... ", as if there is a general connection between the complexity of these two types of decision problem.

But the connection must be more subtle that just "**NP**-completeness of decision problem implies **PSPACE**-completeness of corresponding reconfiguration problem". For instance, it is well known that deciding if a graph is k-colourable is **NP**-complete for any fixed $k \geq 3$. But deciding if two given 3-colourings of a graph are connected via a sequence of single-vertex recolourings is in **P** for $k = 3$ and **PSPACE**-complete for $k \geq 4$; see Theorem 3.4.

Nevertheless, it might be possible to say more about the connection between the complexity of certain decision problems and the complexity of the corresponding reconfiguration problem. In particular for problems that involve labelling of certain objects under constraints (such as satisfiability and graph colouring), and where the allowed transformation is the relabelling of a single object, such a connection might be identifiable. If this is indeed the case, it might give us a better understanding of both the original decision problem and the reconfiguration problem.

Acknowledgements

The author likes to thank the anonymous referee for very careful reading and for suggestions that greatly improved the presentation in the survey.

References

[1] K. Appel and W. Haken, Every planar map is four colourable. I. Discharging, *Illinois J. Math.* **21** (1977), 429–490.

[2] K. Appel and W. Haken, Every planar map is four colourable, *Contemporary Mathematics* **98** (1989), American Mathematical Society, Providence, RI.

[3] K. Appel, W. Haken, and J. Koch, Every planar map is four colourable. II. Reducibility, *Illinois J. Math.* **21** (1977), 491–567.

[4] V. Auletta, A. Monti, M. Parente, and P. Persiano, A linear time algorithm for the feasibility of pebble motion on trees, *Algorithmica* **23** (1999), 223–245.

[5] V. Barbéra and B. Jaumard, Design of an efficient channel block retuning, *Mobile Netw. Appl.* **6** (2001), 501–510.

[6] J. Billingham, R. A. Leese, and H. Rajaniemi, Frequency reassignment in cellular phone networks, *Smith Institute Study Group Report*, available from www.smithinst.ac.uk/Projects/ESGI53/ESGI53-Motorola/Report, (2005).

[7] J. A. Bondy and U. S. R. Murty, *Graph Theory*, Springer, New York (2008).

[8] P. Bonsma and L. Cereceda, Finding paths between graph colourings: PSPACE-completeness and superpolynomial distances, *Theoret. Comput. Sci.* **410** (2009), 5215–5226.

[9] G. Brightwell, J. van den Heuvel, and S. Trakultraipruk, Connectedness of token graphs with labelled tokens. In preparation.

[10] J. K. Burton Jr. and C. L. Henley, A constrained Potts antiferromagnet model with an interface representation, *J. Phys. A* **30** (1997), 8385–8413.

[11] L. Cereceda, Mixing Graph Colourings, PhD Thesis, London School of Economics, (2007).

[12] L. Cereceda, J. van den Heuvel, and M. Johnson, Connectedness of the graph of vertex-colourings, *Discrete Math.* **308** (2008), 913–919.

[13] L. Cereceda, J. van den Heuvel, and M. Johnson, Mixing 3-colourings in bipartite graphs, *European J. Combin.* **30** (2009), 1593–1606.

[14] L. Cereceda, J. van den Heuvel, and M. Johnson, Finding paths between 3-colorings, *J. Graph Theory* **67** (2011), 69–82.

[15] K. Choo and G. MacGillivray, Gray code numbers for graphs, *Ars Math. Contemp.* **4** (2011), 125–139.

[16] Clay Mathematical Institute, The Millennium Prize Problems. www.claymath.org/millennium/.

[17] C. R. Cook and A. B. Evans, Graph folding, in *Proceedings of the 10th Southeastern Conference on Combinatorics, Graph Theory and Computing, Congress. Numer.* **XXIII–XXIV** (1979), pp. 305–314.

[18] J. C. Culberson, Sokoban is PSPACE-complete, Technical Report **TR 97-02**, Department of Computing Science, University of Alberta, available via citeseerx.ist.psu.edu/viewdoc/summary?doi=10.1.1.52.41, (1997).

[19] R. L. Cummins, Hamilton circuits in tree graphs, *IEEE Trans. Circuit Theory* **CT-13** (1966), 82–90.

[20] E. D. Demaine and R. A. Hearn, Playing games with algorithms: Algorithmic combinatorial game theory, arXiv:cs/0106019v2, (2008).

[21] R. Diestel, *Graph Theory*, Springer, Heidelberg (2010).

[22] M. Dyer, A. Flaxman, A. Frieze, and E. Vigoda, Randomly colouring sparse random graphs with fewer colours than the maximum degree, *Random Structures Algorithms* **29** (2006), 450–465.

[23] R. Fabila-Monroy, D. Flores-Peñaloza, C. Huemer, F. Hurtado, J. Urrutia, and D. R. Wood, Token graphs, *Graphs Combin.* **28** (2012), 365–380.

[24] S. J. Ferreira and A. D. Sokal, Antiferromagnetic Potts models on the square lattice: A high-precision Monte Carlo study, *J. Statist. Phys.* **96** (1999), 461–530.

[25] M. R. Garey and D. S. Johnson, *Computers and Intractability: A Guide to the Theory of NP-completeness*, Freeman, New York (1979).

[26] O. Goldreich, Finding the shortest move-sequence in the graph-generalized 15-puzzle is NP-hard, in *Studies in Complexity and Cryptography; Miscellanea on the Interplay between Randomness and Computation*, Lect. Notes Comput. Sci., **6650** (2011), pp. 1–5.

[27] P. Gopalan, P. G. Kolaitis, E. Maneva, and C. H. Papadimitriou, The connectivity of Boolean satisfiability: Computational and structural dichotomies, in *Proceedings of Automata, Languages and Programming, 33rd International Colloquium*, Lect. Notes Comput. Sci., **4051** (2006), pp. 346–357.

[28] P. Gopalan, P. G. Kolaitis, E. Maneva, and C. H. Papadimitriou, The connectivity of Boolean satisfiability: Computational and structural dichotomies, *SIAM J. Comput.* **38** (2009), 2330–2355.

[29] G. Goraly and R. Hassin, Multi-color pebble motion on graphs, *Algorithmica* **58** (2010), 610–636.

[30] W. K. Hale, Frequency assignment: Theory and applications, *Proc. IEEE* **68** (1980), 1497–1514.

[31] J. Han, Frequency reassignment problem in mobile communication networks, *Comput. Oper. Res.* **34** (2007), 2939–2948.

[32] R. A. Hearn and E. D. Demaine, PSPACE-completeness of sliding-block puzzles and other problems through the nondeterministic constraint logic model of computation, *Theoret. Comput. Sci.* **343** (2005), 72–96.

[33] T. Ito, E. D. Demaine, N. J. A. Harvey, C. H. Papadimitriou, M. Sideri, R. Uehara, and Y. Uno, On the complexity of reconfiguration problems, *Theoret. Comput. Sci.* **412** (2011), 1054–1065.

[34] M. Jerrum, A very simple algorithm for estimating the number of k-colourings of a low degree graph, *Random Structures Algorithms* **7** (1995), 157–165.

[35] M. Jerrum, *Counting, Sampling and Integrating: Algorithms and Complexity*, Birkhäuser Verlag, Basel (2003).

[36] M. R. Jerrum, L. G. Valiant, and V. V. Vazirani, Random generation of combinatorial structures from a uniform distribution, *Theoret. Comput. Sci.* **43** (1986), 169–188.

[37] D. Kornhauser, G. Miller, and P. Spirakis, Coordinating pebble motion on graphs, the diameter of permutation groups, and applications, in *Proceedings of the 25th Annual Symposium on Foundations of Computer Science* (1984), pp. 241–250.

[38] M. Las Vergnas and H. Meyniel, Kempe classes and the Hadwiger Conjecture, *J. Combin. Theory Ser. B* **31** (1981), 95–104.

[39] R. A. Leese and S. Hurley (eds.), *Methods and Algorithms for Radio Channel Assignment*, Oxford Univ. Press, Oxford (2003).

[40] T. Łuczak and E. Vigoda, Torpid mixing of the Wang-Swendsen-Kotecký algorithm for sampling colorings, *J. Discrete Alg.* **3** (2005), 92–100.

[41] K. Makino, S. Tamaki, and M. Yamamoto, On the Boolean connectivity problem for Horn relations, *Discrete Appl. Math.* **158** (2010), 2024–2030.

[42] O. Marcotte and P. Hansen, The height and length of colour switching, in *Graph Colouring and Applications* (eds. P. Hansen and O. Marcotte), AMS, Providence (1999), pp. 101–110.

[43] H. Meyniel, Les 5-colorations d'un graphe planaire forment une classe de commutation unique, *J. Combin. Theory Ser. B* **24** (1978), 251–257.

[44] B. Mohar, Kempe equivalence of colorings, in *Graph theory in Paris* (eds. J.A. Bondy, J. Fonlupt, J.-L. Fouquet, J.-C. Fournier, and J.L. Ramírez Alfonsín), Birkhäuser Verlag, Basel (2007), 287–297.

[45] C. H. Papadimitriou, *Computational Complexity*, Addison-Wesley, Boston (1994).

[46] C. H. Papadimitriou, P. Raghavan, M. Sudan, and H. Tamaki, Motion planning on a graph, in *Proceedings of the 35th Annual Symposium on Foundations of Computer Science* (1994), pp. 511–520. Long version available online at: `people.csail.mit.edu/madhu/papers/1994/robot-full.pdf`.

[47] W. J. Savitch, Relationships between nondeterministic and deterministic tape complexities, *J. Comput. System Sci.* **4** (1970), 177–192.

[48] T. Schaefer, The complexity of satisfiability problems, in *Proceedings of the 10th Annual ACM Symposium on Theory of Computing* (1978), pp. 216–226.

[49] A. Schrijver, *Combinatorial Optimization; Polyhedra and Efficiency*, Springer-Verlag, Berlin (2003).

[50] A. D. Sokal, The multivariate Tutte polynomial (alias Potts model) for graphs and matroids, in *Surveys in Combinatorics 2005* Cambridge Univ. Press, Cambridge (2005), pp. 173–226.

[51] K. Solovey and D. Halperin, k-Color multi-robot motion planning, `arXiv:1202.6174v2`, (2012).

[52] S. Trakultraipruk, Connectivity Properties of Some Transformation Graphs, PhD Thesis, London School of Economics, in preparation, (2013).

[53] V. G. Vizing, On an estimate of the chromatic class of a p-graph (in Russian), *Metody Diskret. Analiz.* **3** (1964), 25–30.

[54] J. S. Wang, R. H. Swendsen, R. Kotecký, Antiferromagnetic Potts models, *Phys. Rev. Lett.* **63** (1989), 109–112.

[55] J. S. Wang, R. H. Swendsen, R. Kotecký, Three-state antiferromagnetic Potts models: A Monte Carlo study, *Phys. Rev. B* **42** (1990), 2465–2474.

[56] R. M. Wilson, Graph puzzles, homotopy, and the alternating group, *J. Combin. Theory Ser. B* **16** (1974), 86–96.

Department of Mathematics
London School of Economics
London WC2A 2AE, UK
j.van-den-heuvel@lse.ac.uk

How symmetric can maps on surfaces be?

Jozef Širáň

Abstract

A map, that is, a cellular embedding of a graph on a surface, may admit symmetries such as rotations and reflections. Prominent examples of maps with a 'high level of symmetry' come from Platonic and Archimedean solids. The theory of maps and their symmetries is surprisingly rich and interacts with other disciplines in mathematics such as algebraic topology, group theory, hyperbolic geometry, the theory of Riemann surfaces and Galois theory.

In the first half of the paper we outline the fundamentals of the algebraic theory of regular and orientably regular maps. The second half of the article is a survey of the state-of-the-art with respect to the classification of such maps by their automorphism groups, underlying graphs, and supporting surfaces. We conclude by introducing the notion of 'external symmetries' of regular maps, going well beyond automorphisms, and discuss the corresponding 'super-symmetric' maps.

1   Introduction

Groups are often studied in terms of their action on the elements of a set or on particular objects within a structure. Examples of such situations are abundant and we mention here just a few. Since Cayley's time we know that every group can be viewed as a group of permutations on a set. The study of group actions on vector spaces gave rise to the vast area of representation theory. Investigation of automorphism groups of field extensions generated challenges such as the Inverse Galois Problem. In low-dimensional topology, group actions on trees and on graphs in general led to important findings regarding growth of groups.

The aim of this article is to survey yet another direction of investigation of group actions. Namely, we will be interested in the automorphism groups of maps, that is, cellular decompositions of two-dimensional surfaces, having the 'highest level of symmetry'. The fundamentals of algebraic and topological theory of such maps on orientable surfaces were laid by Jones and Singerman in [69] and a corresponding theory for all surfaces (including those with boundary) was developed by Bryant and Singerman

[15]. By the nature of the subject, the study of symmetry groups of such maps is restricted to groups generated by three involutions two of which commute, or by two elements one of which is an involution. Despite this restriction the corresponding theory is extremely rich and extends to areas which are rather unexpected at a first glance. Examples include actions of such groups on Riemann surfaces and the study of the absolute Galois group by its action on maps.

By a 'symmetry' of an object one usually understands an automorphism, that is, a structure-preserving bijection. We will extend our study also to 'external symmetries', which are not automorphisms since they destroy some of the structure but still return an object isomorphic to the original one. This leads to the question in the title of this article, to which we still do not have a satisfactory answer.

This survey is by no means claiming any completeness. What the reader will find here is our own, often informal, view of building a theory of highly symmetric maps on surfaces, highlighting the most important connections to group theory and the theory of Riemann surfaces, but almost entirely skipping the very well developed algebraic and topological theory of general maps on surfaces. The survey continues by presenting the most important classification results for maps with the highest level of symmetry and concludes with a discussion of external symmetries and remarks on generalisations.

2   Regular maps – algebra and geometry

The study of maps on surfaces and their symmetries is a relatively new branch of mathematics called *topological graph theory*. Its origins, however, go way back into the 19th century to the time of Heawood and his influential Map Colour Problem. One may stretch this even further and refer to the work of Kepler related to his symmetric polyhedra, or even to the school of Plato, whose members were fascinated by the five regular spherical polyhedra, now known as Platonic solids. In fact, our objects of interest, regular maps, are a generalisation of Platonic solids to arbitrary surfaces.

Informally, a map is a cellular decomposition of a connected surface; its 0-cells, 1-cells and 2-cells are the vertices, edges, and faces of the map. Equivalently, a map can be viewed as a cellular embedding of a graph on a surface. The incidence of vertices, edges and faces is understood in the usual way. Still informally, a symmetry of a map may be regarded as a bijective assignment of vertices to vertices, edges to edges, and faces to faces, preserving incidence between these elements. Formal definitions will be given in what follows.

2.1 Maps in terms of flags

To give a formal definition of a map we need to introduce a few concepts.

A graph we will be identified with a 1-dimensional CW-complex. In particular, we will assume that every edge is homeomorphic to the closed unit interval and every loop is homeomorphic to the unit circle. Graphs considered in this article may be infinite and have multiple edges and loops. Except stated otherwise, however, we will assume throughout that *our graphs do not contain semiedges*, that is, edges with two ends just one of which is a vertex.

By a *surface* we will mean a connected Hausdorff topological space in which every point has an open neighbourhood homeomorphic to an open disc; no preliminary assumptions on compactness or orientability are made.

Let Γ be a graph and let $\mathcal{S}$ be a surface. An *embedding* of Γ on $\mathcal{S}$ is a continuous one-to-one function $\vartheta : \Gamma \to \mathcal{S}$. The embedding ϑ is *cellular* if every connected component of $\mathcal{S} \backslash \vartheta(\Gamma)$ is homeomorphic to an open disc. If ϑ is a cellular embedding, the pair $(\mathcal{S}, \vartheta(\Gamma))$ is called a *map*. The graph Γ and the surface $\mathcal{S}$ are the *underlying graph* and the *supporting surface* of the map. In such a case every component of $\mathcal{S} \backslash \vartheta(\Gamma)$ is called a *face* of the map.

To avoid entanglement in the notation, in a map $M = (\mathcal{S}, \vartheta(\Gamma))$ we will identify the graph Γ with its embedded image $\vartheta(\Gamma)$. On an informal level we may think of a map M as a 'crossing-free drawing' of the underlying graph Γ on the supporting surface $\mathcal{S}$, such that when one 'cuts the surface open' along all edges of Γ drawn on the surface, every component one obtains is homeomorphic to a disc (i.e., a '2-cell'). This yields the cellular decomposition mentioned in our informal description of a map given in the preamble to Section 2.

We will give a formal definition of a symmetry as well but this requires equipping a map with a somewhat refined structure, which we shall do next.

Let M be a map on a surface $\mathcal{S}$. We will construct a map M^b from M on the same surface as follows. First, subdivide every edge $e = uv$ of

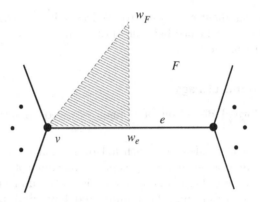

Figure 1: A flag of a map.

M by a new vertex w_e, so that $e = uv$ becomes a walk uw_ev. Let M^s be the resulting map; note that the faces of M and M^s have the same interior. In the interior of every face F of M^s we place a new vertex w_F. To facilitate our description, assume for a moment that the interior of F is identified with the open unit disc centered at w_F. For every presence of a vertex v of M and for every presence of a vertex w_e of M^s appearing on the boundary of F we draw an edge from w_F to v and from w_F to w_e as a straight line segment on the disc. Having done this with every face of M one obtains the *barycentric subdivision* M^b of M, supported by the same surface S. Every face of M^b is called a *flag* of the original map M. From our description it follows that every flag is incident with three vertices of the form v, w_e and w_F, where (v, e, F) is a mutually incident vertex-edge-face triple of M. The boundary of such a flag is formed by the three edges vw_e, w_ew_F and w_Fv; see Fig. 1.

Informally, one speaks of flags of M as a 'triangle' of the barycentric subdivision M^b. It will be useful to 'colour' the vertices v, w_e and w_F by labels 0, 1 and 2, respectively. The labels reflect the dimension of the corresponding elements – a vertex, an edge, and a face *associated* with the flag. Two distinct flags of M are *adjacent* if they share an edge in M^b, and we will say that the flags are adjacent *along an edge*, *across an edge*, and *through a corner*, if they share an edge of M^b with vertices labelled $\{2,1\}$, $\{1,0\}$, and $\{0,2\}$, respectively.

What happens if one interchanges the labels 0 and 2 across the barycentric subdivision M^b of M? This operation leads to the well known concept of the *dual map* M^* of M; the new labelling means that every vertex of

M^* corresponds to a face centre of M and vice versa, and for every edge e of M the dual edge e^* 'perpendicularly crosses' the original edge. One also sees that, up to corner labels, the maps M and M^* have the same flag set, as their barycentric subdivisions can be naturally identified.

Let us return to symmetries. Formally, we define an *automorphism* of a map M to be an arbitrary permutation of the set $\mathcal{F}$ of its flags with the property that whenever two flags are adjacent across an edge, along an edge, or through a corner, then their images have the same type of adjacency. The collection of all automorphisms of M under composition forms the *automorphism group* $\text{Aut}(M)$ of the map M. It follows that $\text{Aut}(M)$ is a subgroup of the symmetric group on $\mathcal{F}$.

If an automorphism of M fixes a flag incident with a vertex, then it must fix all flags incident with the vertex. But then the automorphism must fix all flags with sides contained in edges incident with the vertex, and also all flags contained in faces incident with the vertex. Extending this argument and invoking connectivity we conclude that if an automorphism fixes a flag, then the automorphism is the identity permutation of $\mathcal{F}$. This implies the following folklore fact.

Proposition 2.1 *The automorphism group of a map is a semiregular permutation group on the flag set of the map.*

Elements of every orbit of $\text{Aut}(M)$ on $\mathcal{F}$ are thus in one-to-one correspondence with elements of the group $\text{Aut}(M)$.

2.2 Introducing regular maps

What, then, are the 'most symmetric' maps? The answer is more complex than it may at first appear. At this point, however, it is natural to expect that the most symmetric maps are the ones having just one orbit on flags, or, equivalently, those with a transitive automorphism group on flags. Since permutation groups which are both transitive and semiregular are called regular, we call a map *regular* if its automorphism group is regular on its flag set. The regularity of a map implies that all faces are bounded by closed walks of the same length and all vertices of the underlying graph have the same degree. We will reserve the symbols ℓ and m for these quantities and call them the *face length* and the *degree* of a map, respectively, and speak about a regular map of *type* (ℓ, m). Note also that, by our characterisation of an automorphism as a permutation of flags that preserves corner labels and flag adjacency, it follows that if M is a regular map, then so is its dual M^* since the two maps have, up

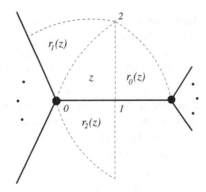

Figure 2: The flag z and its images $r_i(z)$.

to an interchange of the labels 0 and 2, the same flag set. Observe that M^* is a regular map of type (m, ℓ).

From here on, let M be a regular map. We continue by identifying three particular symmetries of M. Let z be a fixed flag of M. By regularity, for $i = 0, 1, 2$ there exists an automorphism r_i of M taking z onto the flag $r_i(z)$ adjacent to z along an edge, through a corner, and across an edge, respectively. Equivalently, the flags z and $r_i(z)$ for $i \in \{0, 1, 2\}$ share an edge of M^b labelled $i - 1$ and $i + 1$ (subscripts being read mod 3). The rationale for choosing this notation is that the only corner of z moved by r_i is the one whose label corresponds to an i-dimensional object of the map – a vertex, and edges, and a face for $i = 0, 1$, and 2, respectively. The situation is depicted in Fig. 2.

Observe that for every $i \in \{0, 1, 2\}$, the automorphism r_i is an involution. Indeed, by our definition of an automorphism of a map, for $i = 0, 1, 2$ the flags z and $r_i(z)$ are adjacent along an edge, through a corner, or across an edge, respectively, if and only if the flags $r_i(z)$ and $r_i^2(z)$ are adjacent in the same way. But for $i = 0, 1, 2$ the only flag adjacent to $r_i(z)$ along an edge, through a corner, or across an edge, respectively, is the flag z. It follows that $r_i^2(z) = z$ and by Proposition 2.1 we have $r_i^2 = id$ for every $i \in \{0, 1, 2\}$.

An important property of the three involutions r_i $(i = 0, 1, 2)$ is that they generate the automorphism group of our regular map M. To demonstrate this, we associate with every $g \in \mathrm{Aut}(M)$ a non-negative integer $k = k_g$ defined as the smallest k for which there is a sequence $z_0, z_1, \ldots, z_k$ of flags of M such that $z_0 = z$, $z_k = g(z)$, and for every $j \in \{1, 2, \ldots, k\}$ the flags z_{j-1} and z_j are adjacent. The existence of k_g for every automorphism

g of M follows from connectivity of the underlying surface of M. We now prove by induction on k_g that for every $g \in \mathrm{Aut}(M)$ we have $g \in \langle r_0, r_1, r_2 \rangle$. This is obvious for the automorphisms g satisfying $k_g \leq 1$, since in this case $g \in \{id, r_0, r_1, r_2\}$. Let now $g \in \mathrm{Aut}(M)$ be such that $k_g \geq 2$. The regularity of M implies the existence of an $h \in \mathrm{Aut}(M)$ such that $h(z) = z_{k-1}$. Obviously $k_h < k_g$ and by the inductive hypothesis we have $h \in \langle r_0, r_1, r_2 \rangle$. Applying the automorphism h^{-1} to the pair of adjacent flags $h(z) = z_{k-1}$ and $g(z) = z_k$ we conclude that the flags z and $h^{-1}g(z)$ are adjacent, and so $h^{-1}g(z) = r_i(z)$ for some $i \in \{0, 1, 2\}$. By semiregularity (Proposition 2.1) we have $h^{-1}g = r_i$ and hence $g = hr_i \in \langle r_0, r_1, r_2 \rangle$. In summary, we have:

Proposition 2.2 *Let M be a regular map and for $i = 0, 1, 2$ let r_i be the automorphism of M taking a flag z of M onto the flag adjacent to z along and edge, through a corner, and across an edge, respectively. Then, the automorphisms r_i are involutions and $\langle r_0, r_1, r_2 \rangle = \mathrm{Aut}(M)$.*

Observe that, for $i = 0, 1, 2$, the automorphism r_i acts locally on the map as a reflection in the side of z with corners labelled $i - 1$ and $i + 1$ (subscripts being read mod 3), taking z onto the flag $r_i(z)$ adjacent to z along that side.

How do compositions of the three involutions act? Let the flag z be associated with a vertex v, and edge e and a face F of our regular map M. Obviously, $r_0 r_1$ fixes the corner labelled 2, that is, the vertex w_F of M^b. The automorphism $r_0 r_1$ therefore acts locally as a rotation about w_F, moving the flag z to the flag $r_0 r_1(z)$ 'two steps away' from z. It follows that the order of $r_0 r_1$, equal to the face length of M, is also equal to a half of the number of flags in F. Similarly, $r_1 r_2$ fixes the corner of z labelled 0, that is, the vertex v, and so $r_1 r_2$ acts locally as a rotation about v and the order of $r_1 r_2$ is the degree of M. The same reasoning applied to $r_2 r_0$ shows that this automorphism is a rotation about the corner of z labelled 1, that is, the vertex w_e of M^b. The product $r_2 r_0$ has order 2 and r_0 commutes with r_2. The action of the three rotations is indicated in Fig. 3.

Letting ℓ and m denote the length of F and the degree of v, we have thus arrived at the following partial presentation of the automorphism group $G = \mathrm{Aut}(M)$ of the regular map M:

$$G = \langle r_0, r_1, r_2 \mid r_0^2, r_1^2, r_2^2, (r_0 r_1)^\ell, (r_1 r_2)^m, (r_2 r_0)^2, \ldots \rangle \qquad (2.1)$$

where the dots indicate a possible presence of additional relators.

In all group presentations in this article we will assume that the powers at generators and relators are their *true orders*. In principle, we could

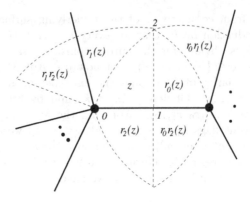

Figure 3: The images of the flag z under the rotations r_0r_1, r_1r_2 and r_0r_2.

have worked in a more general setting and permitted degeneracy regarding orders. For example, if the relator r_2r_0 was allowed to have order 1 while keeping the remaining orders in (2.1) unchanged, then we would have $r_2 = r_0$ and $\ell = m$, resulting in a *semistar map*, that is, an embedding of a semistar (m semiedges attached to a single vertex) on a sphere, with a dihedral automorphism group G of order 2ℓ. One could go even further and allow some generators to be of order 1 (i.e., equal to the identity), leading to maps on surfaces with boundary. As already stated, however, we will generally *not* consider such degeneracies. In very few exceptional cases we will allow semistar maps and the reader will be adequately alerted about this. Reiterating our general convention, our (regular) maps will have no semiedges unless explicitly stated, and the powers in our presentations will be true orders. We note that one or both of ℓ, m in (2.1) may be infinite, in which case we consider the corresponding relator(s) vacuous.

2.3   Introductory examples of regular maps

The ubiquitous five Platonic solids are perhaps the most quoted examples of regular maps when regarded as graphs embedded on the sphere. The five regular maps inherit the names of the solids – the tetrahedron, the cube, the octahedron, the dodecahedron and the icosahedron; see Fig. 4.

The five maps are not the only regular maps on the sphere. In some sense, however, they are the only non-trivial ones, as we shall see in Subsection 4.2 where the classification of spherical regular maps will be discussed. Note

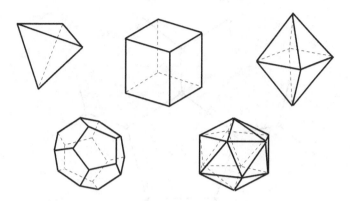

Figure 4: The five Platonic solids.

that the cube and the octahedron are a pair of mutually dual maps, and so are the dodecahedron and the icosahedron. The dual of the tetrahedron is a tetrahedron again; such 'self-dual' regular maps will be studied later in Subsection 5.2.

With the exception of the tetrahedron, the Platonic solids can be represented as maps on a sphere in $\mathbb{R}^3$ centered at the origin in such a way that all their automorphisms are spatial rotations about the origin or compositions of such rotations with the antipodal reflection that maps every point onto the point centrally symmetric with respect to the origin. The quotient space obtained from the sphere by identifying pairs of antipodal points into a single point (that is, by the action of the group of order 2 consisting of the the identity and the antipodal reflection) is the projective plane, the simplest non-orientable surface. This also means that the projective plane can be represented as the quotient space of a closed unit disc, with pairs of antipodal points on the boundary circle identified.

A similar process gives rise to quotient maps, discussed in greater generality and in more detail in Subsections 2.4 and 3.5. For now let us just observe that antipodal quotients of the cube and the octahedron, and also of the dodecahedron and the icosahedron, are pairs of mutually dual regular maps on the projective plane. The second pair of maps is in Fig. 5, where antipodal pairs on the dashed outer circle are assumed to be identified. The underlying graph of the map in thick and thin lines, respectively, is the Petersen graph and K_6, the complete graph on 6 vertices.

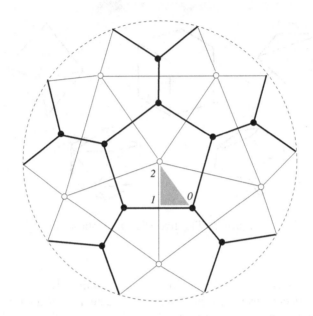

Figure 5: A pair of mutually dual regular maps on the projective plane.

Regularity of the two maps in Fig. 5 follows from the fact that the subgroup consisting of the antipodal reflection and the identity is normal in the automorphism group of the dodecahedron (and the icosahedron), as it will be clear from the material in Subsection 2.4. (In fact, the antipodal reflection of the sphere commutes with every isometry of $\mathbb{R}^3$ that fixes the origin.) In some cases, however, regularity may be simply decided by producing a slightly different drawing, such as the one in Fig. 6 for the quotient of the dodecahedron.

In Fig. 5 one sees that the reflections of the picture in the axes through the segments $1, 2$ and $0, 2$ can be identified with the involutions r_0 and r_1 applied to the shaded flag, while from Fig. 6 it is obvious that the reflection in the horizontal axis can be identified with the involution r_2 applied to the same flag. Hence, the map in Fig. 6 is regular, and by the remark at the end of Subsection 2.2 so is its dual in Fig. 5.

An example of a toroidal regular map is given in Fig. 7. The torus is obtained from the dashed rectangle in the diagram by identifying the pairs of sides labelled a into a single segment, followed by the identification of the two circles resulting from the pair of sides labelled b into a single circle, in both cases consistently with the indicated direction. The underlying graph

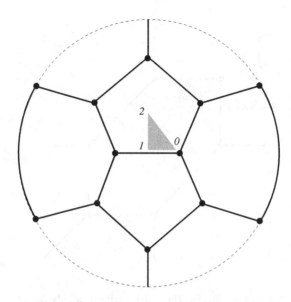

Figure 6: A different drawing of the Petersen graph on the projective plane.

of the map is $K_{3,3}$, the complete bipartite (in fact, equipartite) graph on 6 vertices.

The regularity of this map can again be seen from the picture, this time without re-drawing. Indeed, the reflection in the diagonal through the top right corner induces the involution r_1 applied to the flag F. With a little stretch of imagination the reader will realise that the same diagram can be drawn but with F appearing on the spot occupied by F' and F'', respectively, and then the reflection in the diagonal through the top left corner can be identified with the involutions r_1 and r_2. It follows that the map in Fig. 7 is regular. In Subsections 4.2 and 4.8 we will see generalisations of this situation in terms of toroidal regular maps and regular maps with underlying graphs $K_{n,n}$.

Regular maps on more complicated surfaces are harder to display. We will have a discussion of surfaces in Subsection 2.5. For now, as a taster, in Fig. 8 we give an example of a regular map on the double-torus, an orientable surface of genus 2. In our diagram, the surface is obtained by identifying pairs of sides designated by the same letter, consistently with the direction shown. This map was apparently discovered by Threlfall [108] and its regularity may again be proved by considering symmetries of

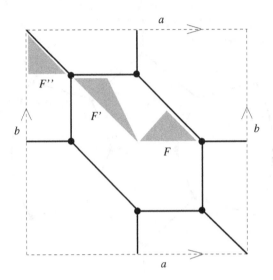

Figure 7: A regular map on the torus, with underlying graph $K_{3,3}$.

the diagram. It is a less trivial exercise to show that its automorphism group is isomorphic to $\mathrm{GL}(2,3)$, the group of invertible 2×2 matrices over the finite field of order 3.

2.4   Regular maps and groups

Let M be a regular map and let $G = \mathrm{Aut}(M)$ be its automorphism group. The regularity of G on the set of flags of M allows us to identify the map M with the group G acting on itself by left multiplication. To see this, choose a flag z of M and consider the bijection $g \mapsto g(z)$ for $g \in G$ that identifies flags with elements of G. This assignment also induces a bijection between left cosets of the dihedral subgroups $\langle r_0, r_1 \rangle$, $\langle r_1, r_2 \rangle$ and $\langle r_2, r_0 \rangle$ of order 2ℓ, $2m$ and 4, and the faces, vertices and edges of M, respectively.

Conversely, given a group G with presentation as in (2.1), one can reconstruct the corresponding regular map, which we denote $(G; r_0, r_1, r_2)$, by letting its flag set be equal to G and realising a flag $g \in G$ as a triangle labelled g with corners labelled 0, 1 and 2. For $i \in \{0, 1, 2\}$, two such flags $g, g' \in G$ will be adjacent, or, more specifically, i-*adjacent*, if $g' = gr_i$. This adjacency can be topologically realised by gluing the two flags along the sides joining the corners labelled $i - 1$ and $i + 1$ (mod 3) in both g

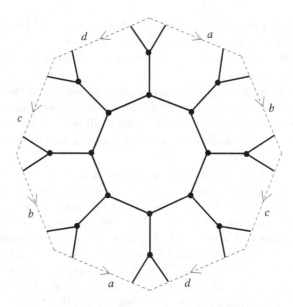

Figure 8: A regular map on the double-torus.

and g'. The group G then indeed acts on itself by left multiplication as the automorphism group of the map $(G; r_0, r_1, r_2)$, since for any $h \in G$ the flags $g, g' \in G$ are adjacent if and only if the flags hg and hg' are. This all leads to the following important conclusion, which could be stated precisely in terms of categories but we prefer to stay at a more informal level here:

The study of regular maps is equivalent to the study of presentations of type (2.1).

From this point on we will be using this principle and will identify regular maps with the quadruples $(G; r_0, r_1, r_2)$, where G is a group generated by r_0, r_1, r_2 and presented as in (2.1).

Let us point out that directed face boundary walks in the setting (2.1) are in a one-to-one correspondence with orbits of the *right* action of the cyclic group $\langle r_0 r_1 \rangle$ on G, since every left coset $g \langle r_0 r_1 \rangle$ representing a face is a union of the two orbits $g \langle r_0 r_1 \rangle$ and $g r_0 \langle r_0 r_1 \rangle$ corresponding to the two directions of tracing round the face boundary. In this connection it is useful to point out the distinction between the left and right actions of the group G on the map $(G; r_0, r_1, r_2)$. While left multiplication by elements of G corresponds to automorphisms, right multiplication by r_i for

$i = 0, 1, 2$ determines i-adjacency of flags. For instance, the three reflection automorphisms of such a map in the three sides of the flag represented by an element $g \in G$ are given by left multiplication by the elements $g r_i g^{-1} \in G$ for $i = 0, 1, 2$.

The identification of regular maps with presentations of the type (2.1) has a number of advantages in terms of the algebraic description of various issues related to regular maps. For now let us illustrate this with two examples: forming normal quotients and determining isomorphism between regular maps.

Let $M = (G; r_0, r_1, r_2)$ be a regular map of type (ℓ, m) and let K be a normal subgroup of G. Assuming that $r_0, r_1, r_2, r_0 r_2 \notin K$, the quadruple $(G/K, r_0 K, r_1 K, r_2 K)$ determines a regular map, called a *(normal) quotient* of M and denoted M/K. (The assumption is needed to avoid degeneracies discussed at the end of Subsection 2.2.) The quotient map M/K is of type (ℓ_K, m_K), where ℓ_K and m_K are the smallest positive integers κ and λ for which $(r_0 r_1)^\kappa \in K$ and $(r_1 r_2)^\lambda \in K$, respectively. Clearly, ℓ_K is a divisor of ℓ and m_K is a divisor of m. The relationship between the supporting surfaces of M and M/K will be explained in the next subsection.

As regards isomorphism of two regular maps, that is, deciding when there is an isomorphism of the corresponding underlying graph that extends to a homeomorphism between the supporting surfaces, a necessary and sufficient condition in the language of groups is as follows.

Proposition 2.3 *Two regular maps $(G; r_0, r_1, r_2)$ and $(G'; r_0', r_1', r_2')$ are isomorphic if and only if there is a group isomorphism $\theta : G \to G'$ such that $\theta(r_i) = r_i'$ for every $i \in \{0, 1, 2\}$.*

There is another way a regular map $M = (G; r_0, r_1, r_2)$ with flag set $\mathcal{F} = G$ on a supporting surface $\mathcal{S}$ can be algebraically represented – namely, by the Cayley graph $\Gamma = \text{Cay}(G, \{r_0, r_1, r_2\})$, embedded on $\mathcal{S}$. To describe the embedding, for each $g \in G$ we identify the *vertex* $g \in \Gamma$ with a midpoint of the *flag* $g \in \mathcal{F}$. Then, for $i = 0, 1, 2 \mod 3$, the edge of Γ joining the vertex g with the vertex $g r_i$ is realised by a segment joining the midpoints of the flags g and $g r_i$, intersecting the common side of the two flags. The embedded Cayley graph Γ on $\mathcal{S}$ has three types of faces, namely, those 'surrounding' the original face centres, vertices, and edge midpoints. These faces have length 2ℓ, $2m$, and 4, respectively (with the last entry replaced with 2 in the case of a semistar), corresponding to the relators $(r_0 r_1)^\ell$, $(r_1 r_2)^m$, and $(r_2 r_0)^2$; see Fig. 9. The Cayley graph embedded as described will be denoted $\text{Cay}(M)$ and called the *Cayley map associated with M*.

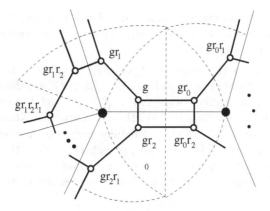

Figure 9: The associated Cayley map (in thick lines) of a regular map.

2.5 Compact, and simply connected supporting surfaces

We assume familiarity of the reader with the classification of compact surfaces, cf. [87]. Up to homeomorphism, any compact connected surface S is uniquely determined by its Euler characteristic $\chi = \chi(S)$ and orientability. The Euler characteristic χ can be computed by considering an arbitrary (not necessarily regular) map supported by S. If the map has n_0 vertices, n_1 edges and n_2 faces, then the Euler-Poincaré formula tells us that $\chi = n_0 - n_1 + n_2$, independently of the map. If S is orientable, it is also uniquely determined by its genus g, related to the Euler characteristic by $\chi = 2 - 2g$, and then S can be obtained from the sphere by 'attaching g handles'. If S is non-orientable, then it is uniquely determined by its non-orientable genus, or crosscap number, h, where $h = 2 - \chi$. In such a case, S arises from the sphere by 'attaching h crosscaps'.

Now, the supporting surface S of a regular map $M = (G; r_0, r_1, r_2)$ is compact if and only if the flag set $\mathcal{F}$ of M is finite, which is equivalent to finiteness of the group G (after all, we know that $\mathcal{F}$ can be identified with G). Suppose that S is compact, and hence G is finite. Then the number of vertices, edges, and faces of M is simply the number of flags divided by the order of the corresponding stabiliser. That is, if M is of type (ℓ, m) and not a semistar, then M has $|G|/(2m)$ vertices, $|G|/4$ edges, and $|G|/(2\ell)$ faces. Feeding these into the Euler-Poincaré formula leads to an equation giving the Euler characteristic χ of S in the form

$$\chi = \frac{|G|}{2m} - \frac{|G|}{4} + \frac{|G|}{2\ell} = -\mu(\ell, m) \cdot \frac{|G|}{2} \quad \text{where} \quad \mu(\ell, m) = \frac{1}{2} - \frac{1}{\ell} - \frac{1}{m}. \quad (2.2)$$

It follows that the underlying surface of M has positive, zero, and negative Euler characteristic, respectively, depending on whether $\mu(\ell, m)$ is negative, zero, and positive.

It is well known that $\chi(\mathcal{S}) \leq 2$ for every compact surface $\mathcal{S}$. The only compact surfaces of positive Euler characteristic are the sphere (with $\chi = 2$) and the projective plane (with $\chi = 1$), and the only compact surfaces of Euler characteristic 0 are the torus and the Klein bottle. We will give details of a classification of regular maps on these surfaces later in Subsection 4.2. At this stage we just say that while the sphere, the projective plane and the torus support an infinite number of (pairwise non-isomorphic) regular maps each, there is no regular map on the Klein bottle.

Compact surfaces of negative Euler characteristic (in other words, when $\mu(\ell, m) > 0$) are particularly interesting and important, since (2.2) implies that if $\chi < 0$, then $|G| = 2(-\chi)/\mu(\ell, m)$. It can be checked that the quantity $\mu(\ell, m)$ achieves its minimum positive value over all positive integers ℓ, m exactly when $\{\ell, m\} = \{3, 7\}$, the minimum being $1/42$. Combining this with the previous equality yields the following upper bound on $|G|$, named after H. Hurwitz who first proved its orientable variant for groups of conformal automorphisms acting on Riemann surfaces; see e.g. Tucker [110] for more details.

Proposition 2.4 *Let* $(G; r_0, r_1, r_2)$ *be a regular map on a compact supporting surface of negative Euler characteristic* χ. *Then,* $|G| \leq 84(-\chi)$. *In particular, a compact surface of negative Euler characteristic supports only a finite number of regular maps.*

We conclude this subsection by a remark on simply connected surfaces, assuming familiarity with rudiments of algebraic topology. A connected but not necessarily compact surface is simply connected if every loop (that is, a closed curve) on the surface is homotopic to zero, i.e., contractible to a point. It is known (cf. [87]) that, up to homeomorphism, the only two simply connected surfaces are the sphere and the plane.

Given a (finite or infinite) regular map $M = (G; r_0, r_1, r_2)$, it is important to be able to determine just from the presentation (2.1) of G if the supporting surface $\mathcal{S}$ of M is simply connected or not. A corresponding necessary and sufficient condition can be derived with the help of the associated Cayley map $\mathrm{Cay}(M)$ introduced in Subsection 2.4. We describe the essence of the process, leaving out numerous technicalities.

One first needs to realise that every loop on $\mathcal{S}$ is homotopic to a closed walk consisting of edges of the underlying graph of $\mathrm{Cay}(M)$, and every such

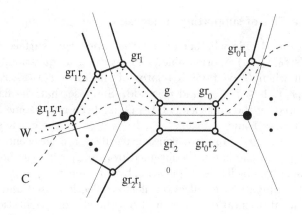

Figure 10: A loop C on $\mathcal{S}$ and a walk W homotopic to C in the associated Cayley map, corresponding to the relator $[W] = \ldots r_1 r_2 r_1 r_0 r_1 \ldots$.

closed walk represents a loop on $\mathcal{S}$. Suppose that a loop on $\mathcal{S}$ is homotopic to a closed walk $W = (g_0, g_1 \ldots, g_{k-1}, g_k)$, $g_k = g_0$, in the Cayley graph $\Gamma = \mathrm{Cay}(G, \{r_0, r_1, r_2\})$. Since $g_{i+1} = g_i r_{j_i}$ for $i = 0, 1, \ldots, k-1 \pmod{k}$ and $j_i \in \{0, 1, 2\}$, the closed walk W corresponds to the product $\Pi_{i=0}^{k-1} r_{j_i} = 1$, that is, to the relator $[W] = \Pi_{i=0}^{k-1} r_{j_i}$ of the group G; see Fig. 10.

If W' is a different walk in Γ homotopic to the same loop, then W' can be obtained from W by application of a finite number of transformations of the following type: If a closed walk has the form (ABC) where A, B, C are open walks and there is a face bounded by a closed walk of the form BD, then replace (ABC) by the closed walk $(AD^{-1}C)$, where D^{-1} is the reverse of D. But the face boundaries of $\mathrm{Cay}(M)$, which are obviously contractible, correspond only to the three relators $(r_0 r_1)^\ell$, $(r_1 r_2)^m$ and $(r_2 r_0)^2$. One then concludes that W and W' are homotopic if and only if the relator $[W']$ can be derived from $[W]$ using the relators $(r_0 r_1)^\ell$, $(r_1 r_2)^m$ and $(r_2 r_0)^2$. In particular, all walks of the associated Cayley map $\mathrm{Cay}(M)$ on the surface $\mathcal{S}$ are contractible if and only if all relators of G are consequences of the three relators $(r_0 r_1)^\ell$, $(r_1 r_2)^m$ and $(r_2 r_0)^2$. This way one arrives at the following result.

Proposition 2.5 *The supporting surface $\mathcal{S}$ of a regular map M is simply connected if and only if every closed walk in the Cayley map $\mathrm{Cay}(M)$ associated with M is a contractible curve on $\mathcal{S}$, which is equivalent to (2.1) being a full presentation of the group $\mathrm{Aut}(M)$.*

2.6   Orientability of supporting surfaces

How can one decide whether or not the supporting surface of a regular map $(G; r_0, r_1, r_2)$ is orientable? To answer this question, consider the situation when G contains a relator of the form $x_1 x_2 \ldots x_n$, where $x_j \in \{r_0, r_1, r_2\}$, $1 \leq j \leq n$, and n is odd. Suppose that the flag of the map represented by the unit element $1 \in G$ has been given some local orientation. Let $y_0 = 1$ and $y_j = x_1 \ldots x_j$, $1 \leq j \leq n$. In the cyclic sequence of flags $(y_0, y_1, \ldots, y_n)$, any two consecutive flags are adjacent. But for every j, $1 \leq j \leq n$, the flag y_j is obtained from y_{j-1} with the help of the automorphism given by left multiplication by the element $z_j = y_{j-1} x_j y_{j-1}^{-1}$. Since z_j is a conjugate of a reflection, it is a reflection itself and hence it reverses the local orientation. Starting from the chosen orientation of the flag 1 by successively applying the reflections z_j and recording how the reflections have changed the local orientations of the resulting flags one sees that the local orientation of the flag y_j agrees with the chosen orientation of y_0 if and only if j is even. The length n of our relator, however, was assumed to be odd. It follows that when one glues the flags in an analogous procedure to that used for regular maps, the part of the supporting surface corresponding to our cyclic sequence of flags contains a Möbius band.

On the other hand, if all relators in the presentation (2.1) of G in terms of generators $\{r_0, r_1, r_2\}$ have even length, then the supporting surface of the regular map $(G; r_0, r_1, r_2)$ is orientable. Indeed, in such a case, if $x_1 x_2 \ldots x_k = 1$ with $x_j \in \{r_0, r_1, r_2\}$ for $j = 1, 2, \ldots, k$, then k must be even. This is because $x_1 x_2 \ldots x_k = 1$ means that the 'word' $x_1 x_2 \ldots x_k$ can be re-written as a product of a finite number of conjugates of the relators in the presentation (2.1) of G, each such conjugate is a word of even length in the generators, and eventual cancellation of neighbouring terms does not change the parity of the length of a word. In this case the mapping $G \to \mathbb{Z}_2$ assigning to a word the parity of its length is a well defined group epimorphism and its kernel is the normal subgroup of G index 2 formed by all even-length products of elements in $\{r_0, r_1, r_2\}$. A global orientation to the supporting surface can then be given by choosing an orientation of the flag 1 and carrying it over to any flag by the action of the even-length products subgroup.

In any case, given a regular map $(G; r_0, r_1, r_2)$, we may define G^+, the *even subgroup* of G, to be the subgroup formed by all elements of G that can be written as a product of an even number of generators taken from the set $\{r_0, r_1, r_2\}$. Clearly, either $G^+ = G$, which (by the facts encountered above) necessarily means that G contains a relator in terms of $\{r_0, r_1, r_2\}$ of odd length and the map is non-orientable, or G^+ is a (normal) subgroup of index 2 in G and the map is orientable; in the latter case the even subgroup G^+ is the subgroup of orientation-preserving automorphism of the map. Summing up, we have:

Proposition 2.6 *The supporting surface of a regular map* $(G; r_0, r_1, r_2)$ *is orientable if and only if the even subgroup of G has index 2 in G; in such a case G^+ is the group of orientation-preserving automorphisms of the map.*

In Subsection 2.4 we saw that the automorphisms $r = r_0 r_1$ and $s = r_1 r_2$ act locally as rotations about midpoint of a fixed face and about a fixed vertex incident with the face. Proposition 2.6 can now be restated by saying that the supporting surface of a regular map $(G; r_0, r_1, r_2)$ is non-orientable if and only if $G = \langle r, s \rangle$. In such a case any of the original involutions r_i can be expressed as a word w in the generators r, s and their inverses, and then $w r_i$ is a relator of odd length. Note that since $(rs)^2 = 1$, the group G in this case admits a partial presentation of a different form, namely, $\langle r, s \mid r^\ell, s^m, (rs)^2, \ldots \rangle$. We emphasise that the last presentation relates to regular maps on *non-orientable* surfaces, although later in Section 3 we will encounter such a presentation in the context of maps on orientable surfaces as well.

2.7 Full triangle groups and tessellations

We continue with an inspection of the situation when there are no additional independent relators in the presentation (2.1). The groups

$$T_{\ell,m} = \langle R_0, R_1, R_2 \mid R_0^2, R_1^2, R_2^2, (R_0 R_1)^\ell, (R_1 R_2)^m, (R_2 R_0)^2 \rangle \qquad (2.3)$$

are known as *full triangle groups*. We recall that, by our convention, powers of generators and relators in (2.3) are true orders. The corresponding regular maps $U_{\ell,m} = (T_{\ell,m}; R_0, R_1, R_2)$ will be called *universal tessellations* for reasons that will become clear later. Proposition 2.5 shows that the supporting surface $\mathcal{U}$ of the map $U_{\ell,m}$ is simply connected, that is, a sphere or a plane. By (2.2), $\mathcal{U}$ is a sphere if and only if $1/\ell + 1/m > 1/2$, and the types (ℓ, m) satisfying this inequality are *spherical*. By the same token, $\mathcal{U}$ is a plane if and only if $1/\ell + 1/m \leq 1/2$; the types (ℓ, m) for which $1/\ell + 1/m = 1/2$ and $1/\ell + 1/m < 1/2$ are called *Euclidean* and *hyperbolic*, respectively.

Spherical types

The spherical types are $(3, 3)$, $(3, 4)$, $(4, 3)$, $(3, 5)$, $(5, 3)$, together with $(\ell, 2)$ for all $\ell \geq 1$ and $(2, m)$ for all $m \geq 1$. The universal spherical tessellations $U_{\ell,m}$ for the first five types have a unique realisation each in the form of maps coming from Platonic solids; see Fig. 4 in Subsection

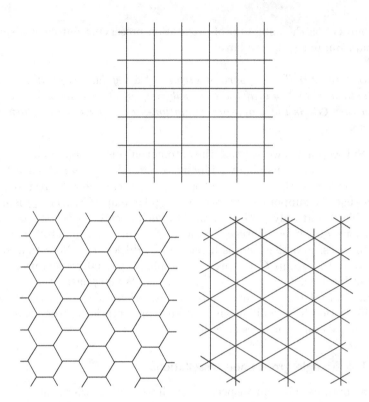

Figure 11: Fragments of the Euclidean tessellations $U_{4,4}$, $U_{6,3}$ and $U_{3,6}$.

2.3. The remaining universal spherical tessellations are realised by ℓ-cycles and their duals, m-dipoles; more details are given in Subsection 4.2. (Had we allowed semistar maps, these would have to be included in this list.) Each spherical regular map $U_{\ell,m}$ admits a geometric realisation on a unit sphere in $\mathbb{R}^3$ such that all flags are mutually congruent triangles with geodesic sides, and with angles of magnitude π/ℓ, π/m and $\pi/2$; here 'congruence' refers to the underlying Riemannian geometry of a sphere. The spherical triangle groups are finite, with isomorphism types $T_{3,3} \simeq S_4$, $T_{3,4} \simeq T_{4,3} \simeq S_4 \times Z_2$, $T_{3,5} \simeq T_{5,3} \simeq A_5 \times \mathbb{Z}_2$, and $T_{2,m} \simeq T_{m,2} \simeq D_m \times \mathbb{Z}_2$.

Euclidean types

There are only three Euclidean types: $(4,4)$, $(6,3)$ and $(3,6)$. The corresponding tessellations $U_{\ell,m}$ of a plane are an infinite square grid, a honeycomb, and its dual; see Fig. 11. They have a symmetric geometric realisation in a Euclidean plane, with all flags pairwise congruent (with respect to Euclidean isometries) and formed by Euclidean right triangles with the remaining angles π/ℓ and π/m. The Euclidean triangle groups are infinite and soluble; they are subgroups of $AGL(2, \mathbb{R}) \rtimes \mathbb{Z}_2$ and it can be shown that $T_{4,4} \simeq ((\mathbb{Z} \times \mathbb{Z}) \rtimes \mathbb{Z}_4) \rtimes \mathbb{Z}_2$ and $T_{3,6} \simeq T_{6,3} \simeq ((\mathbb{Z} \times \mathbb{Z}) \rtimes \mathbb{Z}_6) \rtimes \mathbb{Z}_2$.

Hyperbolic types

Every pair (ℓ, m) with $\ell \geq m$ such that $m \geq 5$, or $m \geq 4$ and $\ell \geq 5$, or else $m \geq 3$ and $\ell \geq 7$, is hyperbolic; of course, (ℓ, m) is hyperbolic if and only if (m, ℓ) is. If one tries to make a drawing of $U_{\ell,m}$ on a plane for a hyperbolic pair (ℓ, m), one would end up with a picture in which the ℓ-sided polygons get smaller and smaller as one moves away from the 'centre' of the drawing. Fortunately, even in this case there is a geometry of a plane in which all flags can be realised by mutually congruent triangles with angles π/ℓ, π/m and $\pi/2$ that sum to less than π – the *hyperbolic geometry*. It is highly non-trivial to prove that for every hyperbolic pair (ℓ, m) the map $U_{\ell,m}$ can be realised in a *hyperbolic plane* in such a way that its flags are triangles with sides being hyperbolic straight line segments and angles having magnitude π/ℓ, π/m and $\pi/2$, cf. [83]. Since a hyperbolic triangle is, up to congruence (i.e., hyperbolic isometry), uniquely determined by its angles, the flags just described will automatically be congruent. A fragment of the hyperbolic tessellation $U_{7,3}$ on the disc model of the hyperbolic plane, drawn with the help of the applet [73], is in Fig. 12.

All hyperbolic triangle groups are infinite and have, from the group-theoretic point of view, an extremely complicated structure. To illustrate this fact we just quote one consequence of the results from [63] which says that if $\ell, m \geq 6$, then the hyperbolic triangle group $T_{\ell,m}$ contains uncountably many pairwise non-isomorphic infinite simple quotients.

From now on we will *assume* that the regular map $U_{\ell,m}$ is a tessellation of a unit sphere, or a Euclidean plane, or a hyperbolic plane, by pairwise congruent regular ℓ-sided polygons, m of which meet at every vertex. Congruence, of course, refers here to spherical, Euclidean or hyperbolic geometry, depending on whether $1/\ell + 1/m$ is larger than, equal to, or smaller than $1/2$, and by a regular polygon we mean one whose sides are geodesic line segments of equal length (in the appropriate geometry) and have equal angles. Note also that the comparison of $1/\ell + 1/m$ with $1/2$ is equivalent to comparison of the sum of angles $\pi/\ell + \pi/m + \pi/2$ in the flag triangles described above with the straight angle, π, determining thus the underlying geometry.

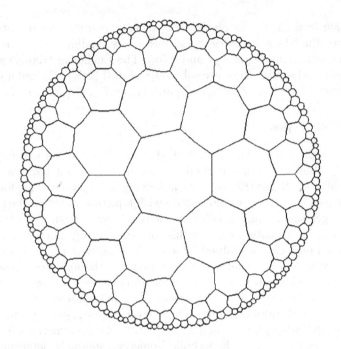

Figure 12: A fragment of the hyperbolic tessellation $U_{7,3}$.

2.8   Algebra and geometry coming together

Consider now a regular map $M = (G; r_0, r_1, r_2)$ of type (ℓ, m) with a presentation of G as in (2.1) on a supporting surface $\mathcal{S}$ (compact or not, orientable or not). Recall our non-degeneracy assumption, that is, the three generators and the product $r_2 r_0$ are assumed to be involutions. Let $U_{\ell,m} = (T_{\ell,m}; R_0, R_1, R_2)$ be the corresponding regular map of the same type on a simply connected surface $\mathcal{U}$, with the full triangle group presented as in (2.3). For brevity we will omit the subscript (ℓ, m) on U and T in the next few paragraphs, keeping the type fixed.

Since, up to notation, the presentation (2.1) is obtained from (2.3) by adding relators, there is an obvious group epimorphism $T \to G$ taking the ordered triple (R_0, R_1, R_2) onto the ordered triple (r_0, r_1, r_2). The kernel N of this epimorphism, a normal subgroup of T, will be called the *map subgroup* of M. By a non-trivial result on the full triangle groups (see [83] for a proof), if T is infinite, the only non-identity elements of finite order in T are elements conjugate to R_0, R_1, R_2, $R_0 R_1$, $R_1 R_2$, $R_2 R_0$ and their powers. Thus, if T is infinite, the map subgroup N is torsion-free, that is, containing no non-trivial elements of finite order, since U and M have the same type.

The map subgroup N allows us to identify the regular map $(G; r_0, r_1, r_2)$ with $(T/N; R_0 N, R_1 N, R_2 N)$ which we simply record by writing $M \simeq U/N$ as in Subsection 2.4. We may sum up these findings by saying that, at the level of automorphism groups, regular maps M of a given type are normal quotients of the regular maps U of the same type on simply connected surfaces; the quotients are torsion-free if the supporting surface of U is the plane. This is the reason why the maps $U = U_{\ell,m}$ are called universal tessellations, explaining also the usage of the letter U in their notation.

Note that we have $N = 1$ in the case of regular maps on a sphere, and $N \simeq \mathbb{Z}_2$ for regular maps of spherical types on a projective plane, the only non-trivial element of N being the antipodal involution on a sphere (discussed later in Subsection 3.5). This shows that one cannot claim that N is torsion-free if the supporting surface of U is the sphere. It is also useful to realise that the regular map M/N is finite if and only if the index $[T : N]$ is finite.

The isomorphism condition of Proposition 2.3 for regular maps implies that the associated epimorphisms from the corresponding full triangle group have the *same* kernels. This means that two regular maps of the same type are isomorphic if and only if they have *identical* map subgroups. Taking this into account and leaving spherical maps aside, our previous discussion can be summed up as follows.

Proposition 2.7 *The (isomorphism classes of) regular maps of a non-spherical type (ℓ, m) are in a one-to-one correspondence with torsion-free normal subgroups of the full triangle group $T_{\ell,m}$; the maps are finite if and only if the subgroups are of finite index.*

Since the universal tessellation U of a spherical, Euclidean or a hyperbolic type (ℓ, m) is now formed by congruent regular polygons in the appropriate geometry, the automorphism group T of the regular map U can be identified with a discrete subgroup of the group of all (orientation-preserving and also reversing) isometries of the sphere, the Euclidean plane, or the hyperbolic plane. This line of thought has a number of consequences of geometric and topological nature, of which we mention two.

The first consequence is that the three reflections R_0, R_1 and R_2 in the sides of a fixed flag can be expressed in terms of 2×2 or 3×3 matrices representing orientation-reversing isometries in the corresponding geometry. The matrices are defined over some finite extension $\mathbb{F}$ of $\mathbb{Q}$, the field of rational numbers; the extension $\mathbb{F}$ depends, of course, on the type (ℓ, m) of the tessellation. Since the matrices represent distance-preserving transformations, one arrives at a representation of $T_{\ell,m}$ in linear groups of the form $SL(3,\mathbb{F}) \rtimes \mathbb{Z}_2$ (for all types) or $PSL(2,\mathbb{F}) \rtimes \mathbb{Z}_2$ (for the hyperbolic types), where the extension by $\mathbb{Z}_2$ corresponds to the inclusion of orientation-reversing elements. As regards the orientation-preserving subgroups of the hyperbolic triangle groups $T_{\ell,m}$, which will be studied in Subsection 3.3, their representations in the linear groups $PSL(2,\mathbb{F})$ and $SL(3,\mathbb{F})$ are surveyed in [103]. We will build on this connection later in Subsection 3.4.

The second consequence is that the supporting surface $\mathcal{S}$ of *any* regular map M can be equipped with geometry in which the faces of M appear as congruent regular polygons. Indeed, if U is the corresponding universal tessellation of the same type as M on a simply connected surface $\mathcal{U}$ and if N is the map subgroup of M, the identification $M \simeq U/N$ means forming a quotient surface (and a quotient tessellation) from U in the standard way. That is, points of U/N are orbits of N, and orbits are equipped with quotient topology as usual, making the surface $\mathcal{U}/N$ homeomorphic to $\mathcal{S}$ and making the projection $\pi : \mathcal{U} \to \mathcal{S}$ given by $\pi(w) = N(w)$ continuous. We will return to this projection in a while in a different context, but for now let us stick with geometry. Recall that N acts semiregularly not only on $\mathcal{U}$ but also on the tessellation U. The cell structure of U therefore induces a cell structure on U/N, which is nothing but the map M. But since N is, at the same time, a group of isometries, we may use it to project down a metric structure onto the surface $\mathcal{S}$ in a natural way. Namely, the geodesic joining two distinct points u, v on $\mathcal{S}$ is simply the π-image of a shortest geodesic between a point in $\pi^{-1}(u)$ and a point in $\pi^{-1}(v)$. This makes all regular maps 'regular' also in a very strong geometric sense.

2.9   Regular maps and algebraic topology

A further consequence of the facts outlined in the previous subsection relates the theory of regular maps to algebraic topology. We will assume that the reader is at least passingly familiar with the concepts of a covering space, a covering projection, and the fundamental group of a surface. Let $M = (G; r_0, r_1, r_2)$ be a regular map of type (ℓ, m) on a supporting surface $\mathcal{S}$ and let $\mathcal{U}$ be the supporting (simply connected) surface of the corresponding universal map U. As before, let N be the map subgroup of M, that is, the kernel of the natural epimorphism $T \to G$; recall that N is regarded as a group of isometries of $\mathcal{U}$ and hence acts on the whole of $\mathcal{U}$.

The projection $\pi : \mathcal{U} \to \mathcal{S}$ given by $\pi(w) = N(w)$ is what is known in algebraic topology as a *smooth regular covering projection*. Smoothness means that the covering has no branch points (reflecting the fact that N is torsion-free) and the adjective 'regular' in this context is related to the fact that the group N acts regularly on each *fibre* $\pi^{-1}(v)$ for every point v of $\mathcal{S}$, implying that $\pi(w) = \pi(N(w))$ for every point $w \in \mathcal{U}$. The theory of covering spaces applied to surfaces also tells us that the group N is isomorphic to the *fundamental group* of the surface $\mathcal{S}$. This looks helpful at a first glance since fundamental groups of surfaces are well understood. The difficulty, however, is that copies of a particular fundamental group in a particular full triangle group appear to be hard to classify. This is notwithstanding the fact that, by Proposition 2.4, an isomorphic copy of a given fundamental group of a *compact* surface of negative Euler characteristic can appear as a normal subgroup of some full triangle group only for a finite number of map types.

If M is a *finite* regular map on a compact surface $\mathcal{S}$ (orientable or not) of Euler characteristic $\chi \le 0$, one can be more specific about the interplay between the corresponding universal tessellation U of a plane $\mathcal{U}$ (Euclidean if $\chi = 0$ and hyperbolic if $\chi < 0$), and the map subgroup N, viewed as a discrete group of isometries of U and also as a fundamental group of $\mathcal{S}$. To do so we will freely quote some well-known facts in algebraic topology. Since $[T : N]$ is assumed to be finite, the group N has a finite *fundamental polygon* $\mathcal{P}$ on $\mathcal{U}$, such that for every non-identity $g \in N$ the intersection $\mathcal{P} \cap g(\mathcal{P})$ is either empty, or a corner of $\mathcal{P}$, or else a side of $\mathcal{P}$. Moreover, $\mathcal{P}$ can be chosen to be a regular polygon (Euclidean if $\chi = 0$, and hyperbolic if $\chi < 0$) with $n = 4 - 2\chi$ sides, and the semiregular action of N on $\mathcal{U}$ applied to $\mathcal{P}$ generates a tessellation $N\mathcal{P} = \{g(\mathcal{P}); \; g \in N\}$ of $\mathcal{U}$ by n-sided polygons, the same number n of which meet at each corner. The tessellation $N\mathcal{P}$ is then congruent to $U_{n,n}$ in the appropriate geometry.

The elements of N for which $\mathcal{P}$ and $g(\mathcal{P})$ intersect in a side of the polygon $\mathcal{P}$ form a generating set X of N. Since N is torsion-free, generators in X come in pairs g, g^{-1} and the corresponding sides $s_g^+ = \mathcal{P} \cap g(\mathcal{P})$ and $s_g^- = \mathcal{P} \cap g^{-1}(\mathcal{P})$ determine a *side pairing* of $\mathcal{P}$. Geometrically, elements in N represent (Euclidean or hyperbolic) translations and glide reflections on the plane $\mathcal{U}$. The formation of the quotient surface $\mathcal{U}/N \simeq \mathcal{S}$ can then be realised by an appropriate identification of the pairs of sides s_g^+ and s_g^- of the fundamental polygon $\mathcal{P}$ for generators $g \in X$. But recall that we also have another tessellation of our plane $\mathcal{U}$, namely, the universal tessellation U we started with. The interplay between the tessellations U and $N\mathcal{P}$, geometrically realised on the same plane $\mathcal{U}$ as described, can further be revealed by looking at the 'part' of U contained inside and on

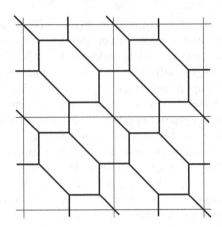

Figure 13: The tesselation $U_{6,3}$ in thick lines and the tessellation $N\mathcal{P}$ of fundamental polygons for the regular map from Fig. 7 of Subsection 2.3 in thin lines.

the boundary of the particular polygon $\mathcal{P}$. Namely, after identification of sides of $\mathcal{P}$ as indicated above, this 'part' of U turns exactly into the original map M on the quotient surface $\mathcal{S}$. An example of such a situation in the Euclidean geometry is in Fig. 13.

To explain Fig. 13 let us recall the regular map M of type $(6,3)$ in Fig. 7 of Subsection 2.3. The fundamental polygon $\mathcal{P}$ for this map may be identified, say, with the bottom left thin-line square. The map subgroup N is now isomorphic to $\mathbb{Z} \times \mathbb{Z}$ and generated by the obvious horizontal and vertical translations by one square to the right and up, respectively. In Fig. 13 the tessellation $N\mathcal{P}$ is drawn in thin lines and the universal tessellation $U_{6,3}$, which may be thought of as a 'continuation' of the drawing of the original map M, is drawn in thick lines.

Analogous examples in the hyperbolic plane are difficult to display. Namely, in the disc model of the hyperbolic plane the diagrams tend to be extremely dense when moving away from the centre (cf. Fig. 11 in Section 2.7); a similar feature is observed in other models (e.g. the complex upper half-plane). The thin-line 'central octagon' $\mathcal{P}$ in Fig. 14 represents a fundamental region for the regular map M of type $(8,3)$ which we have encountered in Fig. 8 of Subsection 2.3. The hyperbolic translations that take the sides labelled a, b, c, d in the top-right part of the octagon to the correspondingly labelled sides in the bottom-left part generate the map subgroup N of the map M. Fig. 14 displays a fragment of the resulting

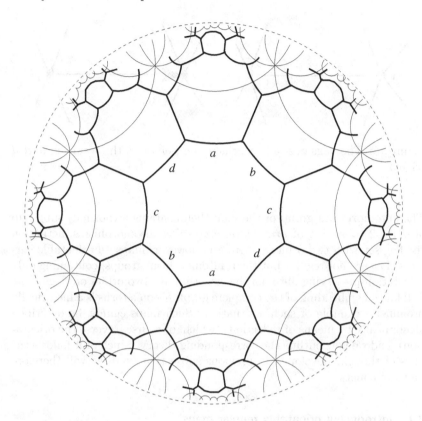

Figure 14: The tesselation $U_{8,3}$ in thick lines and the tessellation $N\mathcal{P}$ of fundamental polygons for the regular map from Fig. 8 of Subsection 2.3 in thin lines.

tessellation $N\mathcal{P}$ of type $(8,8)$ in thin lines, together with the tessellation $U_{8,3}$ in thick lines as a 'continuation' of the drawing of the map M from Fig. 8.

3 Orientably regular maps – algebra and geometry

Although our focus is on the highest 'level of symmetry', we will now discuss maps that are a little less rich in symmetries than the regular ones, but are important from the point of view of theory and arise quite naturally. As motivation, consider a regular map on an *orientable* surface.

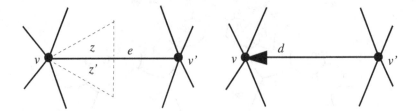

Figure 15: The flags z, z' adjacent across e (left) and the resulting dart d (right).

The automorphism group of the map then contains a (normal) subgroup of index 2 consisting of *orientation-preserving* automorphisms, acting on the flag set with two orbits. There are, however, maps on orientable surfaces that are *not* regular but their automorphism groups, containing only orientation-preserving automorphisms, still have two orbits on flags. The well known embedding of the complete graph K_7 on a torus is a historically prominent example of such a situation. Such maps cannot be described algebraically by means of reflections (which, of course, reverse the orientation), and we now outline the corresponding theory. Much of it follows the lines of thought developed in previous subsections and we will therefore be more concise.

3.1 Introducing orientably regular maps

Let M be a map on an *orientable* surface S and let $\text{Aut}^+(M)$ be the subgroup of $\text{Aut}(M)$ consisting of all automorphisms of M that preserve an orientation of S. Such automorphisms will be called *orientation-preserving* and $\text{Aut}^+(M)$ will be called the *orientation-preserving automorphism group* of M. For any flag z from the flag set $\mathcal{F}$ of M there is a unique flag z' of M with the property that z and z' share the side with corner labels 0 and 1. Let e be the (unique) edge associated with the two flags, so that the two flags are adjacent across e. In a local picture of the barycentric subdivision of M the edge e together with the flags z and z' remotely resemble an arrow with a 'thick tip'. Accordingly, a pair of flags z, z' as above will be called a *dart*. This dart is said to be *incident* with the (unique) vertex associated with both z, z', and also *incident* with (possibly two) faces associated with z or z'; see Fig. 15. Any edge of the underlying graph of M can then be thought as consisting of two oppositely directed darts; the *reverse* of a dart d will be denoted d^-. Let $\mathcal{D}$ be the dart set of M. Obviously, we have $|\mathcal{D}| = |\mathcal{F}|/2$ if the sets are finite.

If necessary or desirable, one may simply think of a dart as of an edge endowed with one of two possible directions, making no prior reference to an embedding. This is of advantage, for instance, in the identification of face boundary walks of a map on an orientable surface. Choosing one of the two possible orientations of the supporting surface makes a map *oriented*. Then, the *oriented* face boundary walk of a face is a cyclic sequence of darts as they appear when walking along the boundary of the face, such that every dart points in a direction consistent with the orientation of the face inherited from the chosen orientation of the supporting surface.

From the analysis towards the end of Subsection 2.1 it is immediate that the group $\text{Aut}^+(M)$ acts semiregularly on the set $\mathcal{D}$ of darts of M. At this point we may ask a slight modification of the philosophical question which we posed at the very beginning of Subsection 2.2: What are the 'most symmetric' maps when restricted to preservation of orientation? At the moment a plausible answer is to say 'those with a transitive (and hence regular) group of orientation-preserving automorphisms on darts'. We therefore define a map M on an orientable surface to be *orientably regular* if the group $\text{Aut}^+(M)$ acts regularly on the dart set of M.

It is again obvious that all faces of an orientably regular map must have the same length, say, ℓ, and all vertices must have the same degree, say, m. As before, we refer to ℓ and m as the *face length* and *degree* of the map, and say that the map is of *type* (ℓ, m). Note that an orientably regular map M may or may not be regular, depending on whether $\text{Aut}^+(M)$ is a (normal) subgroup of $\text{Aut}(M)$ of index 2 or $\text{Aut}^+(M) = \text{Aut}(M)$, and the first case happens if and only if M has some automorphism that reverses the orientation of the supporting surface (an *orientation-reversing* automorphism, for short).

We warn the reader that the terminology regarding orientably regular maps has not been unified. Orientably regular maps have also been called *rotary*, cf. [116]. Regular maps on orientable surfaces (that is, maps that are both regular and orientably regular) are often called *reflexible*, while orientably regular maps that are not regular (i.e., contain no orientation-reversing automorphism) are called *chiral*. Two examples of toroidal chiral maps are in Fig. 16; the underlying graphs are K_5 and K_7.

In Subsection 4.2 we will present a classification of toroidal chiral and regular maps. Interestingly, as we will see later in Subsection 4.3, the next genus of an orientable surface that supports a chiral map is 7.

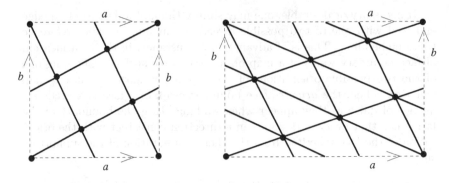

Figure 16: The toroidal chiral maps with underlying graphs K_5 and K_7.

3.2 Orientably regular maps and groups

Assume now that an orientably regular map M has been oriented. Choose a dart d of M; let v be the vertex incident with d (i.e. the vertex d points at) and let F be the face containing the reverse dart d^- on its oriented face boundary walk. By orientable regularity, there is an automorphism $s \in \mathrm{Aut}^+(M)$ taking d onto the next dart, $s(d)$, that is incident with v when moving around v consistently with the chosen orientation. Locally, s acts as a rotation about v and the sense of rotation agrees with the orientation of M. By the same token, there is an automorphism $r \in \mathrm{Aut}^+(M)$ taking the dart $s(d)$ onto d^-, and r acts locally as a rotation about the midpoint of F consistently with the orientation of M. Since $rs(d) = d^-$, the automorphism rs acts locally as a rotation of the dart d about the midpoint of the corresponding edge, and by regularity we know that $(rs)^2$ is the identity automorphism. The situation is depicted in Fig. 17.

A connectivity argument as in Subsection 2.2 shows that the group $\mathrm{Aut}^+(M)$ is generated by the two rotations r and s. If M is of type (ℓ, m), the order of r and s is obviously equal to ℓ and m, respectively. We have thus arrived at the following partial presentation of the orientation-preserving automorphism group of an orientably regular map M:

$$\mathrm{Aut}^+(M) = \langle r, s \mid r^\ell, s^m, (rs)^2, \ldots \rangle \tag{3.1}$$

where, again, powers are true orders and dots indicate a possible presence of additional relators independent from those displayed. Letting $t = rs$ we

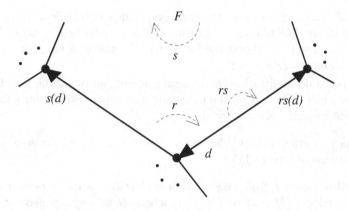

Figure 17: The action of s, r and rs on a dart d.

may rewrite (3.1) in the form

$$\text{Aut}^+(M) = \langle s, t \mid s^m, t^2, (ts^{-1})^\ell, \ldots \rangle \tag{3.2}$$

and we will later also use this type of presentation.

The regular action of the group $H = \text{Aut}^+(M)$ on the dart set $\mathcal{D}$ of an orientably regular map M of type (ℓ, m) enables one to identify the set $\mathcal{D}$ with the group H, and hence the map M with the left action of the group H on itself. Similar to what we saw in Subsection 2.4, faces, vertices and edges of M correspond to left cosets of the cyclic groups $\langle r \rangle$ of order ℓ, $\langle s \rangle$ of order m and $\langle t \rangle$ of order 2, respectively, and orientation-preserving automorphisms are simply given by left multiplication by elements of H.

In the converse direction, for any group H generated by two elements r, s or s, t with a presentation as in (3.1) or (3.2), one can construct the corresponding orientably regular map, denoted $(H; r, s)$ or $(H; s, t)$, respectively, in essentially the same way as in Subsection 2.4, except replacing individual flags with pairs of flags adjacent across edges. One may, however, use a different approach serving the same purpose. We will illustrate it on maps $(H; r, s)$ and to avoid technical issues we will assume that both ℓ and m are finite.

We will think of edges of the map $(H; r, s)$, that is, left cosets of $\langle rs \rangle$ of the form $\{h, hrs\}$ for $h \in H$, as arising from pairs of oppositely directed darts h and hrs. The directed closed walks bounding the faces to be constructed are formed by left cosets $h\langle r \rangle$ for $h \in H$ listed in the cyclic order $(h, hr, hr^2, \ldots, hr^{\ell-1})$. The set of all such closed walks partitions the set $\mathcal{D}$ and has the property that every edge is contained in exactly two oppositely directed walks. The map M is now formed by attaching a 2-cell to each walk as above; the formation of vertices as left cosets of $\langle s \rangle$ will be automatic.

We now informally state the general principle for orientable regularity of maps, which follows from the preceding discussion (see [69] for a formal statement in terms of categories):

The study of orientably regular maps is equivalent to the study of group presentations of type (3.1) or (3.2).

As in Subsection 2.4, from now on we will identify orientably regular maps with the triples $(H; r, s)$ or $(H; s, t)$, where H is a group generated in terms of generators r, s or s, t as in (3.1) or (3.2). This way, properties of orientably regular maps and relations between them can be conveniently expressed. We will state the results for presentations of type (3.1) and leave the translation to the type (3.2) to the reader. For instance, we have an obvious analogue of Proposition 2.3 in the orientable case when one wants to preserve orientations of the supporting surfaces as well.

Proposition 3.1 *Orientably regular maps $(H; r, s)$ and $H'; r', s')$ are isomorphic if and only if there is a group isomorphism from H onto H' taking r onto r' and s onto s'.*

An analogue of the equation (2.2) can be derived by noting that, in a *finite* orientably regular map $M = (H; r, s)$, the number of vertices, edges, and faces is equal to the number of *darts* divided by the order of the corresponding stabiliser, which gives $|H|/m$, $|H|/2$, and $|H|/\ell$, respectively. Assuming that the supporting surface has genus g, tied to Euler characteristic χ by $\chi = 2 - 2g$, the Euler-Poincaré formula leads to

$$\chi = \frac{|H|}{m} - \frac{|H|}{2} + \frac{|H|}{\ell} = -\mu(\ell, m) \cdot |H| \qquad (3.3)$$

where $\mu(\ell, m)$ is the same as in (2.2).

Suppose that an orientably regular map $M = (H; r, s)$ is also regular. This is equivalent to saying that the group $H = \mathrm{Aut}^+(M)$ is a subgroup of the group $G = \mathrm{Aut}(M)$ of index 2. It follows that G is a split extension of H, since G obviously contains an involution outside H. Applying conjugation if necessary, we may – and will – assume that the presentations (2.1) and (3.1) of the groups G and H of such a map are related by $r = r_0 r_1$ and $s = r_1 r_2$. We then have $r_1 r r_1 = r^{-1}$ and $r_1 s r_1 = s^{-1}$, giving the extension $G \simeq H \rtimes \mathbb{Z}_2$ an explicit form. The converse is easily seen to hold as well, leading to a useful result:

Proposition 3.2 *An orientably regular map $(H; r, s)$ is reflexible if and only if the group H admits an involutory automorphism that inverts both r and s.*

This fact can also be interpreted by saying that the map $(H; r, s)$ is regular if and only if it is isomorphic to the map $(H; r^{-1}, s^{-1})$. Observe that, in such a case, H is just the even subgroup G^+ introduced in Subsection 2.5 when discussing orientability of supporting surfaces of regular maps.

From now on we will be using the following convention. Whenever we identify an orientably regular map with a triple $M = (H; r, s)$, we will automatically assume that the map has also been *oriented*; the orientation is defined to be consistent with the rotation r. The maps M and $M^{-1} = (H; r^{-1}, s^{-1})$, which may or may not be isomorphic, then induce opposite orientations of the supporting surface.

We conclude by a note on forming quotients. If $M = (H; r, s)$ is an orientably regular map and K is a normal subgroup of H such that $r, s, rs \notin K$, then the orientably regular map $M/K = (H/K; rK, sK)$ is a *(normal) quotient* of M. The type of a quotient map is related to the type of the original map as stated in Subsection 2.4, and the condition $r, s, rs \notin K$ is only needed to avoid degeneracy. In the orientable case, however, we will allow one type of degeneracy in Subsections 3.6 and 3.7 – namely, when $K = H$, in which case the quotient map consists just of one semiedge embedded on a sphere.

3.3 Triangle groups, quotients, and chirality

In Subsection 2.7 we examined the regular maps on simply connected surfaces. Uniqueness of universal tessellation of a given type together with our discussion leading to Proposition 3.2 quickly implies that *every orientably regular map on a simply connected surface is reflexible*. Thus, the 'universal' objects for orientably regular maps of type (ℓ, m) are again the tessellations $U_{\ell,m}$, but equipped just with the orientation-preserving subgroups $T^+_{\ell,m}$ of the full triangle groups $T_{\ell,m}$. The groups $T^+_{\ell,m}$ with presentation

$$T^+_{\ell,m} = \langle R, S \mid R^\ell = S^m = (RS)^2 = 1 \rangle \tag{3.4}$$

are known as *triangle groups*, with ℓ, m and 2 being true orders as before. The relation of (3.4) to the presentation (2.3) of the corresponding full triangle groups is given by $R = R_0 R_1$ and $S = R_1 R_2$, copying the connection between the presentations (3.1) and (2.1). By the same token we have $T_{\ell,m} \cong T^+_{\ell,m} \rtimes \mathbb{Z}_2$, since the full triangle group $T_{\ell,m}$ is a split extension of the triangle group $T_{\ell,m}^+$ by adjoining the involution R_1 that inverts both R and S.

Further bits of theory of orientably regular maps can be developed along the lines of Subsections 2.8 and 2.9 and we will again be more concise in our presentation. Let $M = (H; r, s)$ be an orientably regular map (finite or not) of type (ℓ, m), with the group H presented as in (3.1). Comparing this presentation with the one in (3.4) shows that H is an image of $T_{\ell,m}^+$ under the obvious epimorphism taking R onto r and S onto s. The kernel N of this epimorphism is the *oriented map subgroup* of M. The triangle group $T_{\ell,m}^+$, as a subgroup of the full triangle group $T_{\ell,m}$, can, and will, be identified with the group of orientation-preserving (spherical, Euclidean, or hyperbolic) isometries which, in addition, act on the regular universal tessellation $U = U_{\ell,m}$. The map M is therefore a quotient, in both the geometric and topological sense, of U by the action of N, which we formally record by writing $M = U/N$. The remaining facts to do with 'projecting down' the geometry from U onto M and identifying N with the fundamental group of the supporting surface of M carry over, *mutatis mutandis*, to orientably regular maps. This leads to an analogue of Proposition 2.7 for orientable regularity.

Proposition 3.3 *The (isomorphism classes of) oriented regular maps of type (ℓ, m) are on a one-to-one correspondence with torsion-free normal subgroups of the oriented triangle group $T_{\ell,m}^+$; the maps are finite if and only if the subgroups are of finite index.*

By Proposition 3.1, isomorphic orientably regular maps yield the *same* kernels of epimorphisms from the corresponding triangle group. It follows that two oriented regular maps of the same type are isomorphic if and only if they have *identical* oriented map subgroups.

We have already noted that there exist orientably regular maps that are chiral. How is this feature reflected in terms of map subgroups? If the oriented map subgroup $N \lhd T_{\ell,m}^+$ of an orientably regular map $M = (H; r, s)$ is also normal in the corresponding *full* triangle group $T_{\ell,m}$, then M is obviously regular and hence not chiral. Conversely, if the supporting surface of a regular map $M \simeq T_{\ell,m}/K$ for some normal subgroup K of the full triangle group $T_{\ell,m}$ is orientable, then (by Proposition 2.6) the group K cannot contain an element that is a product of an odd number of generating involutions from the set $\{R_0, R_1, R_2\}$ and hence is a (normal) subgroup of $T_{\ell,m}^+$.

It follows that an orientably regular map $(H; r, s)$ of type (ℓ, m) is chiral if and only if its oriented map subgroup $N \lhd T_{\ell,m}^+$ is *not* a normal subgroup of $T_{\ell,m}$. If this is the case, then there are two conjugacy classes of N in $T_{\ell,m}$, namely, $N_1 = N$ and $N_2 = R_1 N R_1$. These are both normal in $T_{\ell,m}^+$ and the corresponding quotient maps $U_{\ell,m}/N_1 \simeq (H; r, s)$ and $U_{\ell,m}/N_2 \simeq (H; r^{-1}, s^{-1})$ are *not* isomorphic, which agrees with our note after Proposition 3.2. The fact that the two maps are related by replacing r and s by their inverses has a geometric counterpart: namely, the two maps appear on the same oriented surface as non-isomorphic 'mirror images' of each other. We thus have

Proposition 3.4 *If a normal subgroup N of $T_{\ell,m}^+$ is torsion-free, it is the map subgroup of a regular map if and only if N is also normal in the full triangle group $T_{\ell,m}$. Otherwise, N induces two congruence classes in $T_{\ell,m}$ and the oriented map subgroups of the corresponding pair of chiral maps are N and $R_1 N R_1$, where R_1 is as in (2.3).*

The phenomenon of chirality was studied in greater detail by Breda, Jones, Nedela and Škoviera [12] and we direct the reader to this article for more information on the subject.

3.4 Residual finiteness and orientably regular maps of a given type

A group is *residually finite* if for any finite set of its non-identity elements the group contains a normal subgroup of finite index avoiding the set. Since by Propositions 2.7 and 3.3 regular and orientably regular maps can be identified with normal subgroups of the (full) triangle groups, the question of residual finiteness of triangle groups is a very natural one. As mentioned in Subsection 2.8, the (full) triangle groups are subgroups of isometries (in an appropriate geometry) and hence they are representable as matrix groups. And, indeed, the triangle and full triangle groups of any given type (ℓ, m), including the ones where one or both entries are equal to infinity, are residually finite by a very general theorem of Mal'cev [86] on matrix groups.

A long time ago, Grünbaum [52] asked if there is a finite orientably regular map of any given hyperbolic type (ℓ, m). The question can be answered rather quickly with the help of residual finiteness of triangle groups. Given a hyperbolic type (ℓ, m), in the triangle group $T_{\ell,m}^+$ presented as in (3.4) it is sufficient to select the finite set $X = \{R^i, 1 \le i < \ell; S^j, 1 \le j < m; RS\}$. By residual finiteness, $T_{\ell,m}^+$ contains a normal subgroup N of finite index that avoids X. Combining this with the known fact [83] that the only non-trivial elements in $T_{\ell,m}^+$ of finite order are those conjugate to powers of R, S and RS it follows that N is torsion-free, and hence $(G/N; RN, SN)$ is a finite orientably regular map of type (ℓ, m). Extending X to any finite superset implies, in the same way, the existence of an *infinite number* of finite orientably regular maps of any given non-spherical type (ℓ, m).

Proposition 3.5 *For every non-spherical type (ℓ, m) there exist infinitely many finite orientably regular maps of that type.*

We note that the above argument is essentially the one given by Vince [112]. Gray and Wilson [48] gave a different proof of Proposition 3.5 in terms of permutations.

Another interesting property of regular maps related to the residual finiteness is 'local planarity' which we now introduce. Let M be a (not necessarily regular) map on a compact surface S different from the sphere. We say that M has *planar width at least ρ* if every non-contractible loop on S intersects the map in at least ρ points. One clearly may assume that in this definition it is sufficient to consider only the non-contractible loops that intersect M at vertices and at no other points. Intuitively, if M has planar width at least ρ, then, for each face F of M, the collection of faces of distance (in the natural sense) less than $\rho/2$ from F lies on a subset of S homeomorphic to a disc. This is why planar width can be considered to be a measure of 'local planarity' of a map.

With the help of coverings, Nedela and Škoviera [92] proved that for any hyperbolic type (ℓ, m) and any positive integer ρ there exists a finite orientably regular map of type (ℓ, m) of planar width at least ρ. In [103, 104] the author showed that the findings of [92] are, in fact (and perhaps surprisingly at a first glance), equivalent to the residual finiteness of the hyperbolic triangle groups $T_{\ell,m}^+$ and also derived bounds on the order of the automorphism group of the corresponding maps.

The technique of [104] can be described in terms of representing non-contractible loops on the supporting compact orientable surface of a regular map M of a hyperbolic type by closed walks in the associated Cayley map $\mathrm{Cay}(M)$, and hence by relators in the presentation of $\mathrm{Aut}(M)$, as explained in Subsection 2.5. The property of M having planar width at least ρ then turns out to be equivalent to the following fact: Except for the obvious relators $r_0^2, r_1^2, r_2^2, (r_0r_1)^\ell, (r_1r_2)^m, (r_2r_0)^2$ and their products and conjugates (representing contractible loops), no other relator of length smaller than $f(\rho)$ for some function f can appear in the presentation (2.1), as such relators would correspond to contractible loops.

Now, let Y be an extension of the set X by all non-identity elements of $T_{\ell,m}^+$ which, expressed in terms of generators R_0, R_1, R_2, have length smaller than $f(\rho)$. Applying residual finiteness of $T_{\ell,m}$ to the (finite) set Y one obtains the existence of a normal subgroup N in $T_{\ell,m}$ of finite index, which avoids Y. The intersection $N \cap T_{\ell,m}^+$ is a normal subgroup of finite index in both $T_{\ell,m}^+$ and $T_{\ell,m}$ and contains no 'short' non-identity words in terms of generators. Consequently, $N \cap T_{\ell,m}^+$ is the oriented map subgroup of a finite regular map of 'large' planar width:

Proposition 3.6 *For any non-spherical type (ℓ, m) and any ρ there exist infinitely many finite regular maps on orientable surfaces of type (ℓ, m) and with planar width at least ρ.*

Lower and upper bounds in terms of ℓ, m and ρ on the order of the automorphism group of orientably regular maps of a given hyperbolic type (ℓ, m) with planar width at least ρ were proved by Mačaj, Ipolyiová and the author in [81].

Chiral and non-orientable analogues of the above results appear to be much harder to obtain. A deeper study of Jones, Mačaj and the author [64] implies the existence of a finite regular map of any given hyperbolic type on some non-orientable surface. The fact that for every non-spherical type there exists a finite chiral regular map of that type was proved only recently by Hucíková, Nedela and the author [53]. We will encounter more results proved with the help of residual finiteness later in Section 5.

3.5 Double covers

In this subsection we make a digression to non-orientable regular maps and link them to some natural regular maps on orientable surfaces. In the theory of 2-dimensional surfaces one learns that every non-orientable surface $\mathcal{S}$ has a *double cover* $\tilde{\mathcal{S}}$ with a natural projection $\tilde{\mathcal{S}} \to \mathcal{S}$, a smooth two-sheeted covering that reverses the orientation of $\tilde{\mathcal{S}}$. For the Euler characteristics of the two surfaces we then have $\chi(\tilde{\mathcal{S}}) = 2\chi(\mathcal{S})$. In the context of regular maps one can say a lot more. Assume that our non-orientable surface $\mathcal{S}$ supports a regular map $M = (G; r_0, r_1, r_2)$. We first show a way to explicitly construct the double cover of $\mathcal{S}$ by knowing M.

Let $\tilde{G} = G \times \mathbb{Z}_2$ with $\mathbb{Z}_2 = \{1, -1\}$ given as a multiplicative group, and for $i = 0, 1, 2$ let $\tilde{r}_i = (r_i, -1)$. We claim that $\{\tilde{r}_i; \ i = 0, 1, 2\}$ is a generating set for $\tilde{G}$. Indeed, since M is non-orientable, Proposition 2.6 tells us that G contains a relator $\prod_j r_{i_j}$ that is a product of an *odd* number of involutions from the set $\{r_0, r_1, r_2\}$. But we then have $\prod_j \tilde{r}_{i_j} = (id, -1)$. It follows that the subgroup $\langle \tilde{r}_0, \tilde{r}_1, \tilde{r}_2 \rangle \leq \tilde{G}$ is equal to $\tilde{G}$, as claimed. The quadruple $\tilde{M} = (\tilde{G}; \tilde{r}_0, \tilde{r}_1, \tilde{r}_2)$ thus defines a regular map.

The supporting surface of $\tilde{M}$, which we denote by $\tilde{\mathcal{S}}$, is orientable. This follows from Proposition 2.6, because the even subgroup $\tilde{G}^+$ is isomorphic to G and clearly has index 2 in $\tilde{G}$. The two-to-one natural projection $\tilde{G} \to G$ given by $(g, j) \mapsto g$ for any $g \in G$ and $j \in \mathbb{Z}_2$ with kernel $K \simeq \mathbb{Z}_2$ extends to a mapping $\pi : \tilde{S} \to S$ and, of course, to the corresponding projection $\tilde{M} \to M$, still denoted by π. It follows that M coincides with the normal quotient $\tilde{M}/K$. Recalling the concept of smoothness (discussed very briefly in Subsection 2.9), it can be checked that π is a *regular smooth double covering*, that is, every point on S has exactly two pre-images and $\mathbb{Z}_2$ acts regularly on every fibre. We will say that the just constructed regular map $\tilde{M}$ is the *canonical double cover* of M.

We know that left multiplication by elements of $\tilde{G}$ represent automorphisms of the regular map $\tilde{M}$. The important observation to be made is that the left multiplication by the (only) non-trivial element $(id, -1)$ of the kernel K represents an automorphism of $\tilde{M}$ of order two that *reverses* the orientation of the supporting surface $\tilde{S}$. Moreover, this automorphism is fixed-point-free and commutes with every other automorphism of $\tilde{M}$ since $(id, -1)$ lies in the centre of $\tilde{G}$. For short, a fixed-point-free and orientation-reversing map automorphism of order 2 of a regular map on an orientable surface that commutes with every automorphism of the map is said to be an *antipodal reflection*.

We have seen how a non-orientable regular map gives rise to a (canonical) orientable double cover. One may now reverse the approach and ask: Given a regular map M' on an orientable surface, when does there exist a regular map M on a non-orientable surface such that its canonical double cover $\tilde{M}$ is isomorphic to M'? In a somewhat looser form, this is to ask when a regular map on an orientable surface smoothly double-covers a non-orientable regular map. Expanding the above arguments one can show that this is the case if and only if the map M' has an antipodal reflection, say, u, and then $M \simeq M'/\langle u \rangle$. We have seen an example of this situation on the sphere and the projective plane in Subsection 2.3. Such an antipodal reflection need not be unique, as was observed e.g. in [42].

The significance of antipodal reflections is that, in the classification of regular maps, knowing a presentation of the automorphism group of M' with an antipodal reflection u automatically yields a presentation of $M \simeq M'/\langle u \rangle$ by adding u as a relator. This means that knowing all regular maps on an orientable surface of (even) Euler characteristic χ in terms of group presentations, would yield a corresponding classification of regular maps on a non-orientable surface of Euler characteristic $\chi/2$ simply by looking for antipodal reflections. The reader will encounter this in Subsections 4.2 and 4.3.

We conclude by mentioning a related issue. In a regular map $M = (G; r_0, r_1, r_2)$, with $r = r_0 r_1$ and $s = r_1 r_2$ being the 'rotations' about a fixed face and about a fixed vertex incident to the face, conjugation by r_1 inverts both r and s. In addition, if M is non-orientable, an automorphism f of G is said to be a *reflector* of M if f inverts both r and s, and an *inner reflector* if the automorphism f is inner. In this terminology, conjugation by r_1 as above is an inner reflector, in particular, induced by an involution. The fact that there exist non-orientable regular maps admitting inner reflectors that are not induced by an involution was proved in [31].

3.6 Orientably regular maps and Riemann surfaces

Let $M = (H; r, s)$ be an orientably regular map of type (ℓ, m) with an underlying graph Γ on an (orientable) supporting surface $\mathcal{S}$. Further, let $U = U_{\ell,m}$ be the corresponding universal tessellation of a simply connected surface $\mathcal{U}$ with orientation-preserving automorphism group $T^+ = T_{\ell,m}$ and let $N \triangleleft T^+$ be the oriented map subgroup of M. We have seen that there is a smooth regular covering $\mathcal{U} \to \mathcal{S}$ that carries over to a smooth regular covering $U \to M$ of maps; on the algebraic side the covering is simply induced by the natural epimorphism $T^+ \to T^+/N \simeq H$. This covering can be used to 'project down' onto M the geometry of $\mathcal{U}$ (hyperbolic, Euclidean, or spherical). Accordingly, identification of $\mathcal{U}$ with the complex upper half-plane as a model of the hyperbolic plane, the complex (Euclidean) plane, or the complex Riemann sphere, enables one to 'project down' onto $\mathcal{S}$ a complex structure as well, turning $\mathcal{S}$ into a Riemann surface. This way, in addition to the properties mentioned earlier, automorphisms of M can be considered to be conformal mappings of the underlying Riemann surface $\mathcal{S}$.

One may, however, consider *lifting* the map structure of $M = (H; r, s)$ from its smallest normal quotient $M_0 = M/H$. This is a degenerate map, which we now exceptionally consider: it is formed by a semiedge on a sphere $\mathcal{S}_0$. The associated covering $\beta : \mathcal{S} \to \mathcal{S}_0$ has at most three branch points in $\mathcal{S}_0$ of order ℓ, m and 2, occurring, respectively, in the centre of the face, at the vertex, and at the non-vertex end (the 'free' end) of the semiedge of M_0. Identifying the sphere $\mathcal{S}_0$ with the complex projective line $\mathrm{P}^1(\mathbb{C})$, the key point is that β can be identified with a meromorphic function realising the covering $\mathcal{S} \to \mathcal{S}_0$ of *Riemann surfaces*, ramified at no more than 3 points.

Conversely, assume that we are given a meromorphic function β from a compact Riemann surface $\mathcal{S}$ onto a Riemann sphere $\mathcal{S}_0$ with at most three critical values, one of order at most 2. Let us embed a semiedge on $\mathcal{S}_0$ in such a way that the free end is identified with the branch point of order at most 2 and the remaining (at most) two branch points are identified with the vertex and the face centre. In other words, we endow $\mathcal{S}_0$ with the structure of a map – the most trivial map one can think of, consisting just of one dart. Borrowing the above notation we will denote this map M_0 and say that it is *compatible with* β. But then, the inverse image $\beta^{-1}(M_0)$ is a map on the surface $\mathcal{S}$, which we denote M_β. Moreover, with the help of β^{-1} the complex structure on sphere $\mathcal{S}_0$ lifts onto a complex structure on $\mathcal{S}$; we may speak of a complex structure on $\mathcal{S}$ *induced* by the map M_β.

An orientably regular map M with automorphism group H and oriented map subgroup N can therefore be 'looked at' from the 'top' and from the 'bottom'. Looking at M from the top means considering the normal quotient $U/N \simeq M$ of a universal tessellation U of the same type as M, which enables to 'project down' onto M the geometric and complex structure of U. On the other hand, M can be regarded to be a lift of the trivial spherical one-dart map M_0 by a covering with at most three branch points, 'dragging up' a complex structure onto the supporting surface of M and thus turning it into a Riemann surface. For worked-out examples of this situation we refer e.g. to Jones [59]. The existence of such a covering has remarkable implications in the theory of Riemann surfaces and since these are closely related to maps, we will now make a digression into Riemann surfaces.

3.7   Riemann surfaces, maps, and Galois theory

We begin with a fundamental fact discovered by Riemann around 1860: Given a non-constant function $f(x,y)$ of two complex variables satisfying some natural smoothness restrictions the details of which will not be discussed here, the 'solution space', or *locus*, of all $(x,y) \in \mathbb{C} \times \mathbb{C}$ such that $f(x,y) = 0$, is homeomorphic to an orientable surface, now called the *Riemann surface* defined by $f(x,y)$. By a major theorem of Riemann, such a surface defined by $f(x,y)$ is *compact* if and only if $f(x,y)$ is birationally equivalent to some *polynomial* $p(x,y)$ over $\mathbb{C}$.

Sticking with *compact* orientable surfaces, one may be interested in 'efficient' ways to define a surface using a polynomial $p(x,y)$. It would be plausible to hope for some 'tameness' if the (complex) coefficients of $p(x,y)$ were contained in the field $\overline{\mathbb{Q}}$ of *algebraic numbers* at the very least. We will thus say that a compact orientable surface is *algebraic* if there is a polynomial $p(x,y)$ with algebraic coefficients which defines the surface. Of course, a polynomial over $\overline{\mathbb{Q}}$ contains just a finite number of coefficients and so is actually defined over a *number field*, that is, over a finite-degree extension of the field $\mathbb{Q}$ of rational numbers.

By an important result of Belyĭ (cf. [59]), a compact Riemann surface $\mathcal{S}$ is algebraic if and only if there is a meromorphic function β from $\mathcal{S}$ onto the Riemann sphere $\mathcal{S}_0 = \mathrm{P}^1(\mathbb{C})$ with at most three critical values; one can require in addition that one of the values has multiplicity at most two. Such functions β are called *Belyĭ functions*.

The fact that the existence of a Belyĭ function on a compact Riemann surface implies that the surface is algebraic has been known as the Weil rigidity theorem; see [59] again. The ingenious contribution of Belyĭ was to show the converse by taking a defining polynomial $p(x, y) \in \overline{\mathbb{Q}}(x, y)$ of a compact Riemann surface, observing that the composition of $p(x, y)$ with the projection onto the x coordinate is a meromorphic function from the Riemann surface onto the Riemann sphere with a finite number of critical values, and devising a procedure to reduce the number of critical values to at most three. In addition, this reduction procedure renders β in terms of a rational function with all coefficients algebraic, and one of the critical values can be made of degree at most two.

As we have seen in Subsection 3.6, given a Belyĭ function $\beta : \mathcal{S} \rightarrow \mathcal{S}_0$ and a trivial map M_0 on $\mathcal{S}_0$ compatible with β, the inverse image $M_\beta = \beta^{-1}(M_0)$ is a map on $\mathcal{S}$ inducing a complex structure on $\mathcal{S}$. Belyĭ's theorem may therefore be informally understood by saying that a compact Riemann surface $\mathcal{S}$ is defined over a number field if and only if its complex structure is induced by a map.

What has this all to do with Galois theory? As Grothendieck outlined in his influential 'Esquisse d'un programme', the Belyĭ-Weil theorem enables one to study the absolute Galois group by means of its action on maps on orientable surfaces, and also on orientably regular maps. The absolute Galois group, which is the automorphism group of the field $\overline{\mathbb{Q}}$ of algebraic numbers, is one of the most mysterious groups in mathematics and plays a substantial role in a number of profound mathematical theories, for example, in the proof of the Fermat's last theorem. The way one can study its action on maps on compact orientable surfaces is as follows.

Let $\mathcal{S}$ be an algebraic surface. By Belyĭ's theorem there is a meromorphic function $\beta : \mathcal{S} \rightarrow \mathcal{S}_0$ with at most three critical values, one of degree at most two, such that all coefficients appearing in the defining equation(s) for β are algebraic. Letting M_0 be a trivial map compatible with β, then, as we know, the pre-image $\beta^{-1}(M_0)$ can be identified with a map $M = M_\beta$ on $\mathcal{S}$. (In the degenerate case M may end up being a semistar map, but this causes no problems here.)

Let us now take an element σ of the absolute Galois group, that is an automorphism of the field $\overline{\mathbb{Q}}$ of algebraic numbers. Let $\mathcal{S}$ be a compact (orientable) Riemann surface and let $\beta : \mathcal{S} \to \mathcal{S}_0$ be a Belyĭ function. Consider the map $M_\beta = \beta^{-1}(M_0)$. Since all coefficients appearing in the defining equation(s) for β are, by Belyĭ's theorem, algebraic, the action of σ on these coefficients produce, in general, a new Belyĭ function β_σ. But repeating the construction of the previous paragraph gives a map $M_{\beta_\sigma} = \beta_\sigma^{-1}(M_0)$ on the same supporting surface as M_β. The action of the absolute Galois group on maps as inverse images of the trivial map under Belyĭ functions is then given by assigning, to every $\sigma \in \overline{\mathbb{Q}}$, the mapping $M_\beta \mapsto M_{\beta_\sigma}$.

It is known that this action is *faithful* even if restricted to trees embedded on a sphere. Some of the properties of maps, such as the number of vertices, edges, faces, and the automorphism group, are preserved under the described action of the absolute Galois group. Thus, in the above description of the action of $\overline{\mathbb{Q}}$ on maps, if M_β is an orientably regular map, then so is M_{β_σ} for every $\sigma \in \overline{\mathbb{Q}}$. Relating the defining polynomials and the Belyĭ functions to *orientable regularity* of maps, however, is a completely open question.

Our overview regarding the action of the absolute Galois groups on maps is somewhat simplified; a detailed version would involve discussing the so-called Belyĭ pairs (M, β), where M is a map as above and β is a Belyĭ function from the supporting surface of M onto $\mathcal{S}_0$. A large number of details regarding the action of the absolute Galois group and explicit examples can be found in the monographs by Lando and Zvonkin [78] and Girondo and González-Diez [46]. In the context of regular maps, worked-out examples of the action of $\overline{\mathbb{Q}}$ on orientably regular embeddings of complete bipartite graphs are included in the article by Jones, Wolfart and Streit [71]. We do not reproduce any of the examples due to their length and refer the interested reader to the three sources.

4   The classification of regular maps

Regular and orientably regular maps have been classified mostly by their supporting surface, automorphism group, or underlying graph. In this section we will survey results in all three directions and select the ones that are, in our opinion, most important and influential.

To save space we will be displaying presentations of automorphism groups of regular maps in the abbreviated form $\langle (r_i), \ldots \rangle$, where the symbol (r_i) will stand for the expression '$r_0, r_1, r_2 |\ r_0^2, r_1^2, r_2^2, (r_2 r_0)^2$' that repeatedly appears in our presentations, that is, for the three generators and the first four 'obvious' relators. We will also be using throughout the notation $r = r_0 r_1$ and $s = r_1 r_2$ that we have used before.

4.1   Structural results on automorphism groups of regular maps

We begin with a selection of important observations and results of a group-theoretic character which have proved useful in deriving structural results on groups with presentation (2.1) and (3.1).

Restrictions on the structure of automorphism groups of regular maps on a given surface can sometimes be extracted from the Euler-Poincaré formula by simple arithmetic. An example of this situation is an observation by Conder, Potočnik and the author [27].

Proposition 4.1 *Let G be the automorphism group of a finite regular map on a surface with Euler characteristic $\chi \neq 0$ and let p be a prime divisor of $|G|$ relatively prime with χ. Then, the Sylow p-subgroups of G are cyclic if p is odd, and dihedral if $p = 2$. In particular, the Sylow 2-subgroups of G are automatically dihedral if χ is odd. Moreover, if $-\chi$ is equal to a second power of an odd prime greater than 3, then every odd-order Sylow subgroup of G is cyclic.*

It is thus of interest to study regular maps with automorphism groups having dihedral Sylow 2-subgroups and cyclic odd-order Sylow subgroups. This is very close to some classical results in group theory; namely, finite groups in which all Sylow subgroups are cyclic were considered by Frobenius (see [16]) and completely determined by Burnside [16].

Allowing Sylow 2-subgroups to be dihedral required the development of a more substantial machinery. A little more generally, let us say that a group is *almost Sylow-cyclic* if all its odd-order Sylow subgroups are cyclic and its Sylow 2-subgroups contain a cyclic subgroup of index 2.

The *solvable* almost Sylow-cyclic groups were explicitly determined by Zassenhaus [122]; this result (too long to be reproduced here) is nicely presented in Wolf's monograph [119]. A characterisation of non-solvable almost Sylow-cyclic groups follows from the work of Suzuki [107] and Wong [120]:

Theorem 4.2 *Let G be a non-solvable almost Sylow-cyclic group. Then G contains a normal subgroup G_0 of index at most 2 such that $G_0 \simeq (\mathbb{Z}_m \times \mathbb{Z}_n) \rtimes L$, where $L \simeq PSL(2, p)$ or $L \simeq SL(2, p)$ for some prime $p \geq 5$ and some odd m, n relatively prime to the order of L. Moreover if the Sylow 2-subgroup of G is dihedral, then $L \simeq PSL(2, p)$.*

If the automorphism group of a regular map satisfies the hypotheses of Proposition 4.1, then its Sylow 2-subgroups are dihedral. In terms of general groups, relaxation of the Sylow subgroup condition to requiring only that the Sylow 2-subgroups be dihedral leads to a much harder classification problem. The ultimate result on such groups is that of Gorenstein and Walter [47] obtained as part of the project towards the classification of finite simple groups. For a group G let $O(G)$ denote the *odd part* of G, which is the (unique) maximal normal subgroup of G of odd order.

Theorem 4.3 [47] *If G is a group with a dihedral Sylow 2-subgroup, then $G/O(G)$ is isomorphic to either a Sylow 2-subgroup of G, or to the alternating group A_7, or to a subgroup of $\mathrm{Aut}(PSL(2,q))$ containing $PSL(2,q)$, where q is an odd prime power.*

The shortest known proof of this result, due to Bender and Glauberman [6, 7], still depends on the Feit-Thompson odd-order theorem.

Another class of structural results that have appeared in the literature relate automorphism groups of regular maps with surfaces of Euler characteristic of particular type. For example, suppose that G is the automorphism group of a regular map on a surface of an odd negative prime Euler characteristic χ, and let p be a prime distinct from $-\chi$. Applying Proposition 4.1 to G one sees that G also fulfils the assumptions of Theorem 4.3. Examining the possible types of regular maps that can arise this way one can derive the following corollary, implicitly proved in [14]:

Corollary 4.4 *Let G be the automorphism group of a regular map on a surface of Euler characteristic $-p$ for a prime $p \geq 29$. Then, $G/O(G)$ is isomorphic to S_4, A_5, or to a Sylow 2-subgroup of G.*

Remarkable results on the structure of the orientation-preserving automorphism groups of orientably regular maps on surfaces with arithmetically restricted Euler characteristic have been recently obtained by Gill [45] and we present here a sample of two. The first is about a surprising implication of the number of prime divisors of the Euler characteristic on the structure of the group.

Theorem 4.5 *Let k be a positive integer. There exists a finite set S_k of finite simple groups such that the following holds: If H is the orientation-preserving automorphism group of a finite orientably regular map on a surface with Euler characteristic χ, if K is a non-abelian composition factor of H, and if χ is divisible by exactly k distinct primes, then either K is a simple group of Lie type of rank at most k, or $K \in S_k$.*

The proof uses some simple ideas connected to the structure of Sylow subgroups (similar to those leading to Proposition 4.1), but the important ingredient is knowledge of some structural information of the so-called prime graph of simple groups obtained by Vasiljev and Vdovin [111].

The second result of Gill [45] we present here is an interesting counterpart (although with a restriction on the Euler characteristic) to Theorem 4.3 in that one has to factor a group H by its 2-*part*, that is, the largest normal 2-subgroup of H.

Theorem 4.6 *Let H be the orientation-preserving automorphism group of an orientably regular map on a surface of Euler characteristic -2^a for some positive integer a, and let $J = H/O_2(H)$. Then, J has a normal subgroup K such that J/K is isomorphic to a subgroup of $\mathbb{Z}_2 \times \mathbb{Z}_2$, and K is a direct product of a cyclic group and a group L of coprime order, where L is trivial, or $L \simeq SL(2,q)$ where $q \geq 8$ is a power of 2, or else $L \simeq PSL(2,p)$ for some odd prime p of the form $p = 2^a \pm 1$.*

To conclude our selection we mention a remarkable classification result relating the genus with group structure in a still different way. Namely, in [29] Conder, Tucker and the author showed that if H is the automorphism group of an orientably regular map on an orientable surface of (necessarily even) Euler characteristic $\chi \leq -2$ such that the order of H is relatively prime to $\chi/2$, then H is one of the groups listed in Table 2 of Theorem 8.4 of [29]. We do not reproduce the classification here due to length and invite the reader to consult [29] for details.

4.2 Regular maps on surfaces of non-negative Euler characteristic

There are just four compact surfaces, up to homeomorphism, with non-negative Euler characteristic: the sphere, the projective plane, the torus and the Klein bottle. In all four cases a classification of regular maps has been known for a long time.

The sphere

Leaving degenerate maps aside, the only spherical types (ℓ, m) are those with $\ell = 2$ or $m = 2$ or else the five types coming from Platonic solids. We therefore just give a list of the corresponding regular maps $(G; r_0, r_1, r_2)$

by giving complete presentations of the groups G which, of course, are full triangle groups since the sphere is simply connected:

embedded ℓ−cycles (duals: ℓ−dipoles)	$\langle (r_i), r^\ell, s^2 \rangle \cong D_\ell \times \mathbb{Z}_2$
tetrahedron (self − dual)	$\langle (r_i), r^3, s^3 \rangle \cong S_4$
octahedron (dual: cube)	$\langle (r_i), r^3, s^4 \rangle \cong S_4 \times \mathbb{Z}_2$
icosahedron (dual: dodecahedron)	$\langle (r_i), r^3, s^5 \rangle \cong A_5 \times \mathbb{Z}_2$

The projective plane

The projective plane, a non-orientable surface of Euler characteristic 1, may be identified as a quotient of a unit sphere embedded in $\mathbb{R}^3$ and centered at the origin by the reflection given by multiplication of the negative of the identity matrix of dimension 3. Since this reflection commutes with all isometries of the unit sphere and is fixed-point-free, it is an antipodal reflection as introduced in Subsection 3.5. By the theory outlined there it is sufficient to look for such antipodal reflections in the above presentations for spherical regular maps. One finds that the groups of odd cycles, odd dipoles, and the tetrahedron have no antipodal reflection. In the remaining cases, the (unique) antipodal reflection u is given by $u = r_0 s r^k$ for ℓ-cycles with $\ell = 2k$, $u = rs^2 r^{-1} sr_1$ for the octahedron, and $u = rs^{-2} rs^{-1} rs^2 r_1$ for the icosahedron; the corresponding elements for the duals of these maps can be worked out in an obvious way. Since the antipodal reflection is central, we have $G = \langle r_0, r_1, r_2 \rangle \cong \langle r, s \rangle \times \langle u \rangle \cong G^+ \times \mathbb{Z}_2$ and the quotients arise by dividing out by the normal subgroup $\langle u \rangle$ of G. Since antipodal quotients of the above maps are the projective-planar embeddings of k-cycles and their duals (bouquets of k loops), K_4 and its dual (denoted by $K_3^{(2)}$, which is K_3 with doubled edges), and K_6 and its dual (the Petersen graph); a complete list of presentations of groups G of projective-planar regular maps $(G; r_0, r_1, r_2)$ is:

embedded k−cycles (duals: k−bouquets)	$\langle (r_i), r^{2k}, s^2, r_2 r^k \rangle \cong D_{2k}$
embedded $K_3^{(2)}$ (dual: embedded K_4)	$\langle (r_i), r^3, s^4, rs^2 r^{-1} sr_1 \rangle \cong S_4$
embedded K_6 (dual: embedded P)	$\langle (r_i), r^3, s^5, rs^{-2} rs^{-1} rs^2 r_1 \rangle$ $\cong A_5$

The torus

Classification of toroidal regular and chiral maps on the torus started with the work of Brahana [11] and was completed later by Coxeter [32]. There are only three types of such maps, namely, $(4, 4)$ and the dual pairs

$(3,6)$ and $(6,3)$. By Subsections 2.8 and 3.3 and by duality, classification of toroidal regular and chiral maps amounts to finding torsion-free normal subgroups of finite index of the groups $T_{4,4}$, $T_{4,4}^+$, $T_{6,3}$ and $T_{6,3}^+$. Using linear representations of these groups of Euclidean isometries (either over $\mathbb{R}$ or $\mathbb{C}$) one finds that in each of the four cases, all finite-index torsion-free normal subgroups are abelian and generated by what one may call a 'linear combination' of a pair of translations, specific to each type, with integer coefficients, say, b, c. Finally, one concludes that for any integers b, c such that $b \geq 0$ and $bc(b - c) \neq 0$, presentations of groups H of all toroidal *chiral* regular maps $(H; r, s)$ are, up to duality, given by

$$\text{Type } (4,4): \quad H = \langle r, s \mid r^4, s^4, (rs)^2, (rs^{-1})^b (r^{-1}s)^c \rangle$$
$$\text{Type } (3,6): \quad H = \langle r, s \mid r^6, s^3, (rs)^2, (rs^{-1}r)^b (s^{-1}r^2)^c \rangle$$

If $bc(b - c) = 0$ one arrives at *regular*, that is, reflexible toroidal maps $(G; r_0, r_1, r_2)$, with groups G having presentations

$$\text{Type } (4,4): \quad \langle (r_i), r^4, s^4, (rs^{-1})^b (r^{-1}s)^c \rangle$$
$$\text{Type } (3,6): \quad \langle (r_i), r^6, s^3, (r^2 s^{-1})^{b+c} (r^{-2}s)^c \rangle$$

The Klein bottle

Toroidal regular maps do not have antipodal reflections, which implies that there are no regular maps on the Klein bottle, a non-orientable surface of Euler characteristic 0 double-covered by a torus. This also follows from the fact that every automorphism of the Klein bottle acting locally as a non-trivial rotation about a point is an involution.

This completes our classification of regular maps on surfaces of non-negative Euler characteristic. Although the sphere, the projective plane and the torus each support infinitely many regular maps, only a finite number of these in the first two cases can be considered non-trivial.

4.3 Regular maps on surfaces of negative Euler characteristic: Before 2005

The Hurwitz bound from Proposition 2.4 implies that, in contrast with the sphere, projective plane and the torus, the number of regular maps on any compact surface of negative Euler characteristic is finite. More explicitly, the order of the automorphism group of a regular map on a surface with Euler characteristic $\chi < 0$ is bounded above by -84χ. By the same token, if the underlying surface is orientable, of genus $g \geq 2$ which is related to χ by $\chi = 2 - 2g$, then the order of the orientation-preserving automorphism group of an orientably regular map on such a surface is at most $42(g - 1)$.

Table 1: Authors and years of classification of orientably regular maps of genus g, $2 \leq g \leq 7$ and non-orientable regular maps of Euler characteristic χ, $-6 \leq \chi \leq -2$.

	Orientable surfaces			Non-orientable surfaces	
g	Author(s)	Year	χ	Author(s)	Year
2	Threlfall	1932	-2	Grek	1963
3	Sherk	1959	-3	Grek	1966
4	Garbe	1969	-4	Grek	1966
5, 6	Bergau, Garbe	1989	-5	Scherwa	1985
7	Garbe	1978	-6	Bergau, Garbe	1989

By the principles outlined in Subsections 2.8 and 3.3, the classification of all regular and orientably regular maps on a given compact surface of Euler characteristic $\chi < 0$ amounts to finding normal torsion-free subgroups N of the full triangle groups $T_{\ell,m}$ or their even subgroups $T_{\ell,m}^+$ such that the orders of the factor groups $G = T_{\ell,m}/N$ and $H = T_{\ell,m}^+/N$, the entries ℓ, m and the Euler characteristic χ, are all tied together by the Euler-Poincaré formula (2.2) and (3.3). This, however, is very hard to do for hyperbolic types (ℓ, m) since our knowledge of the normal quotients of the corresponding triangle groups is very limited. The findings regarding classification of orientably regular and regular maps by the late 1980's are listed in Table 1.

We now comment on the entries in Table 1. In the early years of classification, results were generated by *relation-chasing*, which consisted in trying to make educated guesses about the extra relations to be added to a presentation of a (say, full) triangle group to obtain a quotient map on a given surface. With some group theory input this led to the classification of orientably regular maps on surfaces of genus 2 and 3. The case of genus 2 was begun by Erréra [41] and was completed by Threlfall [108] (cf. Coxeter and Moser [33]), and the case of genus 3 was done by Sherk [102]; none of these maps are chiral. Since regular maps of genus 2 turn out to have no antipodal reflections, the theory developed in Subsection 3.5 shows that there are no regular maps on non-orientable surfaces of Euler characteristic -1.

Developing such methods further, in a series of papers Grek [49, 50, 51] derived a classification of regular maps on non-orientable surfaces of Euler characteristic -2, -3 and -4. Somewhat later, Garbe [42] proposed a more systematic method, introduced in the course of classifying regular maps on the orientable surface of genus 4. His technique was based on working with suitable low-degree permutation representations of full triangle groups and their subgroups. This led to success in a by-hand classification of regular and chiral maps on orientable surfaces of genus 5 and 6 by Bergau and Garbe [8], and of genus 7 by Garbe in [43]. The non-orientable classification counterpart was accomplished by Scherwa [101] for Euler characteristic -5 and by Bergau and Garbe [8] for Euler characteristic -6. Independently, Wilson [114] has outlined a similar classification approach using a geometric language.

By the late 1980's, regular maps were classified on all non-orientable surfaces of Euler characteristic $\chi \geq -6$ and orientably regular maps were classified up to genus 7 of the supporting surface. An interesting by-product was the observation that there are no chiral maps on orientable surfaces of genus between 2 and 6.

About a decade later and marking a new era, Conder and Dobcsányi [22] published a computer-generated classification of all regular and chiral maps on orientable surfaces up to genus 15, and regular maps on non-orientable surfaces of Euler characteristic $\chi \geq -28$. The method was to systematically generate all normal subgroups N of the full triangle group $T(\infty, \infty)$, isomorphic to the free product of $\mathbb{Z}_2 \times \mathbb{Z}_2$ with $\mathbb{Z}_2$, of a manageable (finite) index, so that the elements $(r_0 r_1)^\ell$ and $(r_1 r_2)^m$ were in N for some finite ℓ and m; the largest index to be considered was calculated with the help of the Euler-Poincaré formula (2.2) to guarantee that the procedure yielded all regular maps on surfaces of Euler characteristic $\chi \geq -28$. For this task the authors of [22] used their own adaptation of the then available low-index subgroup algorithm.

4.4 Regular maps on surfaces of negative Euler characteristic: After 2005

At the same time as Conder and Dobcsányi were publishing their computer census, progress was being made in a different, equally significant, direction. Namely, in 2005 Breda, Nedela and the author [14] classified regular maps on *all* surfaces of negative prime Euler characteristic – the first classification result that applied to an *infinite* number of surfaces. To present the result, for any prime $p \equiv -1 \mod 4$ we let $\nu(p)$ denote the number of ordered pairs (j, k) such that $j > k \geq 3$, both j and k are

odd, coprime, and $(j-1)(k-1) = p+1$. For all primes p we let $n(p)$ be
the number of regular maps, up to isomorphism and duality, on a surface
with Euler characteristic $-p$. In view of the earlier census we just need to
consider $p \geq 29$.

Theorem 4.7 *Let p be an odd prime, $p \geq 29$. Then, writing $n(p)$ for the
number of regular maps on a surface with Euler characteristic $-p$ and G
for the corresponding automorphism groups of these maps, we have:*

$n(p) = 0$ *if* $p \equiv 1 \bmod 12$;
$n(p) = 1$ *if* $p \equiv 5 \bmod 12$, *with* $G \cong G_p = \langle (r_i), r^{p+4}, s^4, sr^3s^{-1}r^3 \rangle$;
$n(p) = \nu(p)$ *if* $p \equiv 7 \bmod 12$, *with* $G \cong G_{j,k}$ *where*

$$G_{j,k} = \langle (r_i), r^{2j}, s^{2k}, (rs^{-1})^2, r^j s^k r_1 \rangle;$$

$n(p) = \nu(p) + 1$ *if* $p \equiv -1 \bmod 12$; *the groups G are G_p and the $\nu(p)$
groups $G_{j,k}$.*

We note that the involution r_1 can be uniquely determined from r and s in
the above presentations (which is not true in general). It can be checked
that the group G_p is an extension of $\mathbb{Z}_{(p+4)/3}$ by S_4, and $G_{j,k} \simeq D_j \times
D_k$. The proof of Theorem 4.7 given in [14] relies on the deep result of
Gorenstein and Walter [47], stated as Theorem 4.3 in Subsection 4.1. A
different proof of Theorem 4.7 was later obtained by Conder, Tucker and
the author [29] but it still depends on the Gorenstein-Walter result.

On the computational front, Conder [19] was able to vastly extend
the census of regular and orientably regular maps to surfaces of Euler
characteristic $\chi \geq -200$. This was made possible by a new algorithm for
generating low index *normal* subgroups which appears to behave well when
applied to triangle groups. The new census contains a wealth of additional
information about regular maps and has proved useful in making working
conjectures about regular maps by looking for patterns in the data.

This takes us back to infinite sets of surfaces. With the help of structural results on almost Sylow-cyclic groups mentioned in Subsection 4.1,
Conder, Potočnik and the author [27] derived a classification of regular
maps on surfaces with Euler characteristic equal to $-p^2$ for a prime p. If
$p = 2$, the list of such maps can be extracted from the early census of
Conder and Dobcsányi [22]. For odd primes we give, with the help of the
extended census of Conder [19] a complete list, which, quite surprisingly,
is very short:

Theorem 4.8 *If p is an odd prime such that a surface of Euler characteristic $-p^2$ supports a regular map, then $p = 3$ or $p = 7$. Up to isomorphism*

and duality, there are only three such regular maps and the corresponding automorphism groups can be presented as follows:

$p = 3:$ $\quad \langle (r_i), r^4, s^6, s^{-1}rs^{-1}r^2sr^{-1}r_1 \rangle$ *and* $\langle (r_i), r^6, s^6, rs^{-1}r^{-2}s^{-2}r_1 \rangle;$

$p = 7:$ $\quad \langle (r_i), r^3, s^{13}, s^{-1}rs^{-2}rs^{-1}rs^2r^{-1}sr_1, s^5rs^{-5}r^{-1}s^4r^{-1}s^{-1}r_1 \rangle.$

Very recently, the above classifications have been extended to surfaces of Euler characteristic $-3p$ for p prime. In view of the extended census of [19] we just state the part of the classification that refers to sufficiently large p, borrowing the notation $G_{j,k}$ and the corresponding presentation from Theorem 4.7.

Theorem 4.9 *Let* $(G; r_0, r_1, r_2)$ *be a regular map on a surface of Euler characteristic* $\chi = -3p$ *where* $p \geq 13$ *is an odd prime. Then up to isomorphism and duality, one of these cases occurs:*

(i) $G = G_{j,k} \simeq D_j \times D_k$ *for odd* j *and* l *satisfying* $(j-1)(l-1) = 3p+1$ *where* $j \geq l \geq 3$ *and* $\gcd(j,l) \leq 3$ *but* $j \equiv l \not\equiv 1$ mod 3, *presented as in Theorem 4.7.*

(ii) $G \simeq (\mathbb{Z}_3 \times \mathbb{Z}_{3j}) \rtimes D_4$ *for some odd* j *satisfying* $4j - 3 = p$, *of order* $72j$, *with presentation* $\langle (r_i), r^6, s^{4j}, r^2s^2r^2s^{-2}, r^2s^{-1}r^2s^{2j+1}r^{-1}r_1 \rangle.$

(iii) $G \simeq \mathbb{Z}_n \rtimes PGL(2,7)$, *where* $p = 21n - 8 \equiv -8$ mod 21 *but* $p \not\equiv -8$ mod 49 *(so* $n = (p+8)/21$ *is coprime to 14), of order* $336n = 16(p+8)$, *with presentation either* $\langle (r_i), r^{7n}, s^8, s^{-1}r^7sr^7, s^2r^3s^3r^cr_1 \rangle,$ *or* $\langle (r_i), r^{7n}, s^8, s^{-1}r^7sr^7, sr^2s^2r^dr_1 \rangle,$ *where* c *and* d *are uniquely determined modulo* $7n$ *by* $c \equiv 1$ mod 7, $c \equiv 3$ mod n, *and* $d \equiv 1$ mod 7, $d \equiv -2$ mod n.

This was the state-of-the-art regarding classification of regular maps on infinite sets of non-orientable surfaces at the time of writing this article. An analogous result for non-orientable surfaces of Euler characteristic $-2p$ for prime p is still not available. There is, however, a classification result for regular and orientably regular maps on *orientable* surfaces of Euler characteristic $-2p$, that is, of genus $p+1$, for prime p. It was initiated by Belolipetsky and Jones [5] for automorphisms groups of order larger than $12p$ in the regular case and larger than $6p$ in the chiral case, and a complete classification was given by Conder, Tucker and the author in [29]. Since the corresponding statements get more and more complex and the same applies to the corresponding group presentations, we state the result in an abbreviated form, referring to [29] and [19] for details.

Theorem 4.10 *Let* M *be an orientably regular map on a surface of genus* $p+1$, *where* $p > 13$ *is prime. Then, up to isomorphism and duality,* M *is* *one of the 6 chiral maps described in Theorem* 3.1 *of* [29], *or M is one of the 7 reflexible maps described in Theorem* 8.4 *of* [29].

The proof of Theorem 4.10 is based on deeper results on the classification of orientably regular maps $(H; r, s)$ on surfaces of genus $p + 1$ with p prime and such that p divides the order of H (Theorem 3.1 of [29]) and on a similar (but much harder) classification result on orientably regular maps $(H; r, s)$ on a surface of genus g with $g - 1$ relatively prime to $|H|$ (Theorem 8.4 of [29]). We note that all these classifications are very explicit and given in terms of complete presentations of automorphism groups.

4.5 Surfaces supporting no regular map

Let us continue by discussing surfaces admitting no regular maps at all, or no regular maps with some extra properties. Every orientable surface supports a regular map. We have seen this for surfaces of low genus, but for every $g \geq 1$ an orientable surface of genus g admits a regular map formed by single-face embedding of a bouquet of $2g$ loops. Such maps, however, may be seen to be 'less worthy' and we return to this issue soon. For now let us point out that there is no analogue of this situation on non-orientable surfaces.

We will informally say that a value of the Euler characteristic χ of a non-orientable surface is a *gap* if this surface supports no regular map at all. The fact that there are gaps at $\chi = 0$ (the Klein bottle) and $\chi = -1$ is folklore, and a number of other gaps can be identified by examining the census [19]; all the gaps in the range $-50 \leq \chi \leq 0$ were also determined by Wilson and Breda [118]. In a much stronger form this follows from Theorems 4.7, 4.8 and 4.9, by which there are infinitely many gaps, and we state the result separately.

Corollary 4.11 *There are no regular maps on surfaces of Euler characteristic* $\chi = -p$ *where p is a prime congruent to* 1 mod 12 *and* $p \neq 13$, $\chi = -p^2$ *for any prime* $p \geq 11$, *and* $\chi = -3p$ *for any prime p such that* $p \geq 13$, $p \equiv 3$ mod 4, *and* $p \not\equiv 55$ mod 84.

On the positive side, however, Conder and Everitt [23] showed that, asymptotically, more than three quarters of compact non-orientable surfaces do support a regular map.

We conclude by commenting on gaps in a slightly different context. As noted above, there are no gaps in the genera of orientable surfaces, but we have filled these by regular maps containing loops. Such regular maps, and perhaps also those containing multiple edges, may be considered somewhat 'less worthy' by pure graph theorists. It thus may be of interest to look for 'gaps' among genera of *orientable* surfaces by trying to identify those supporting no regular map with a *simple* underlying graph. In a different direction one may also look for 'orientable gaps' with respect to supporting no chiral maps. The infinitude of 'orientable gaps' of both kinds again follows from the structural analysis due to Conder, Tucker and the author in [29], which we illustrate by a sample result:

Theorem 4.12 *For each prime p such that $p-1$ is not divisible by 3, 5 or 8, there is no chiral map on an orientable surface of genus $p+1$. If p satisfies the additional condition $p \equiv 1$ mod 6, then there is no regular map with a simple underlying graph on an orientable surface of genus $p+1$.*

4.6 Regular maps with a given automorphism group

Regular and orientably regular maps with cyclic and dihedral automorphism groups (or, more generally, with automorphism groups isomorphic to the split extension of an abelian group by a group of order two) are easy to describe and we leave this as an exercise for the reader. The underlying graphs of such maps are, up to isomorphism and duality, dipoles and bouquets of loops (and semistars if one wishes to include them).

We proceed with considering the class of nilpotent groups. Orientably regular maps M with nilpotent automorphism groups were studied by Ban et al. [4]. The characterization of nilpotent groups as direct products of their Sylow subgroups implies that each such map M is isomorphic to a direct product (discussed later in Subsection 5.1) of two maps M_1 and M_2, where $\mathrm{Aut}^+(M_1)$ is a 2-group and M_2 is a semistar with an odd number of edges, and hence $\mathrm{Aut}^+(M_2)$ is cyclic of odd order. The main results of [4] are a classification of orientably regular maps admitting an orientation-preserving automorphism group of nilpotency class 2 or 3, and a proof of the fact that for every positive integer n there are only a finite number of orientably regular maps M with a simple underlying graph such that $\mathrm{Aut}^+(M)$ has nilpotency class n.

Another natural step regarding group structure would be to look at metacyclic groups, or, more generally, at (general) products of two abelian groups – that is, groups H of the form $H = AB$ where A and B are abelian subgroups of H. By a result of Ito [55], a product of two abelian groups has an abelian commutator subgroup. Such groups are also known as *metabelian*, lying 'between' nilpotent and solvable. But even products of two cyclic groups appear to be too difficult to handle. Progress in the classification of orientably regular maps M has been made in the special case when the group $\mathrm{Aut}^+(M)$ contains a subgroup acting regularly on the vertices of M. Such maps are known as *Cayley maps* and were investigated in great detail by Richter et al. [99]. (A special case of such maps, namely, the Cayley maps *associated* with a regular map, were briefly mentioned at the end of Section 2.3.)

In Section 3.2 we introduced an alternative notation $(H; s, t)$ for orientably regular maps by means of the rotation s about a fixed vertex and the involutory rotation t about the midpoint of an incident edge. In these terms it is not hard to show [99, 24] that $(H; s, t)$ is an orientably regular Cayley map if and only if H contains a subgroup K such that $K\langle s \rangle = H$ and $K \cap \langle s \rangle$ is the trivial group. Important results towards a classification of the abelian groups K that *admit* such orientably regular Cayley maps together with a wealth of related results can be found in the papers of Conder, Jajcay and Tucker [24, 25]. As a taster we present just the following very interesting result of [24]:

Theorem 4.13 *The only finite cyclic groups failing to admit a nontrivial orientably regular Cayley map are those of order a product of distinct Fermat primes one of which is* 3.

A complete classification of all finite cyclic groups K admitting an orientably regular map was finally obtained by Conder and Tucker [30], including the enumeration of all such maps for a cyclic group of given order, the calculation of the type of each map, and the derivation of simple number-theoretic conditions for reflexibility and other interesting properties.

In view of the difficulties with handling products of an abelian and a cyclic group it appears to be rather hopeless to attempt to arrive at substantial results within even broader families of groups, for example, solvable groups. Nevertheless, interesting classification results have been proved for certain restricted classes of (solvable as well as non-solvable) groups. For example, in Subsection 4.1 we have encountered the almost Sylow-cyclic groups in connection with regular maps on surfaces with negative prime Euler characteristic. With the help of Theorems 4.2 and 4.3 Conder, Potočnik and the author [27] classified *all regular maps with almost Sylow-cyclic automorphism groups*. Since details of the classification are too involved to be reproduced here, we just give a brief summary.

If such a group G is solvable, the maps fall into eight classes and the group is either dihedral, or a split extension of a cyclic group by the Klein four-group, or a split extension of the Klein four-group by a dihedral group with certain congruence restrictions on the orders. (It is perhaps worth mentioning that the classification of [27] is independent of

Zassenhaus' description [122, 119] of the class of all solvable almost Sylow-cyclic groups.) If G is a non-solvable almost Sylow-cyclic automorphism group of a regular map, then there exists a prime p such that either $G \cong PGL(2,p)$, or G is an extension of a cyclic group of order relatively prime to $p(p^2 - 1)$ by the group $PGL(2,p)$.

4.7 Regular maps with simple and almost simple automorphism groups

What can one say about groups that are structurally 'very far from being abelian', namely, the non-abelian finite simple and almost simple groups? A complete classification of orientably regular maps M with $\mathrm{Aut}^+(M)$ isomorphic to a non-abelian (almost) simple group is available only for the two-dimensional projective linear groups $PSL(2,q)$ and $PGL(2,q)$ for prime powers q, which follows from the work of Macbeath [82] and Sah [100]. These results have been reproved (adding considerably more detail) and extended to a classification of *regular maps* having such an automorphism group by Conder, Potočnik and the author in [28].

The remaining classes of non-abelian finite simple groups appear to be much less tractable, as similar (and less complete) results exist just for two other classes of finite simple non-abelian groups – the Ree and Suzuki groups. Jones [58] gave a classification of all orientably regular maps M of type (ℓ, m) with $\mathrm{Aut}^+(M)$ isomorphic to a Ree group $^2G_2(3^n)$ for odd $n > 1$ for the values of $\ell = 3$ and $m = 7$, $m = 9$, and for all primes m such that $m \equiv 11 \bmod 12$; all these maps turn out to be chiral. Regarding the second class, results of Jones and Silver [68] imply a classification of all orientably regular maps M of type $(4,5)$ with $\mathrm{Aut}^+(M)$ isomorphic to a Suzuki group $Sz(2^n)$ for odd $n > 1$, and all these maps are again chiral.

By a result of a deep study done by Malle, Saxl and Weigel, [84], every finite non-abelian simple group can be generated by an involution and a non-involution. Equivalently, in our terms:

Theorem 4.14 *Every finite non-abelian simple group is the orientation-preserving automorphism group of an orientably regular map.*

This is, of course, very far from any classification, but it at least shows that there is a wealth of orientably regular maps whose automorphism groups are structurally very complicated. It is rather unrealistic to expect more specific information about possible types of orientably regular maps arising from simple groups. The current state of knowledge appears to be summed up in the work of Lübeck and Malle [80]:

Theorem 4.15 *Let G be a finite non-abelian simple group other than $Sp_4(2^n)$, $Sp_4(3^n)$ and $^2B_2(2^{2n+1})$. Then, up to a finite number of possible exceptions, G is the automorphism group of an orientably regular map of type $(3,m)$.*

A simple group can only be the automorphism group of a *regular* map on a *non-orientable* surface, because the automorphism group of a regular map on an orientable surface has a proper subgroup of index two. Regarding regular maps with simple automorphism groups, the following theorem (arising from a communication with M. Conder [21]) summarises results of Nuzhin [94, 95, 96, 97] on infinite classes of finite simple groups, Timofeenko [109] on sporadic simple groups, Ashaev (unpublished) on the Baby Monster group and Norton (unpublished) on the Fischer-Griess Monster group.

Theorem 4.16 *A finite simple group is the automorphism group of a regular map if and only if the group is not isomorphic to any of the following: $PSL(3,q)$ and $PSU(3,q)$ for q a prime power, $PSL(4,q)$ and $PSU(4,q)$ for q a power of 2, A_6, A_7, the sporadic Mathieu groups M_{11}, M_{22}, M_{23}, and the McLaughlin group McL.*

Note that the above list includes also the groups $PSL(2,7) \simeq PSL(3,2)$ and $A_8 \simeq PSL(4,2)$.

To conclude we mention a result on almost simple groups H, i.e., groups H such that $G \leq H \leq \text{Aut}(G)$ for some simple group G. The following result of Gill [45] could as well be included in Subsection 4.1 because it makes reference to a surface of a given Euler characteristic, but we preferred to view it as more strongly related to groups with a preassigned structure – namely, to almost simple groups.

Theorem 4.17 *Let H be an almost simple automorphism group of a finite orientably regular map on a surface of Euler characteristic $\chi = -2^a p^b$ for some integers $a \geq 1$, $b \geq 0$ and some odd prime p. If H is simple, let $J = H$; otherwise let J be the unique non-trivial normal subgroup of H. Then, up to isomorphism and duality, we have the following:*

(i) $b = 0$, the map has type $(3,5)$ for $\chi = 2$, $(3,7)$ for $\chi = -4$, $(5,6)$ for $\chi = -16$, and $(6,7)$ for $\chi = -64$, and H is isomorphic, respectively, to $PSL(2,5)$, $PSL(2,7)$, $PGL(2,5)$, and $PGL(2,7)$, with a unique map in each case;

(ii) $b \geq 1$, $J \simeq PSL(2,q)$ for some prime power $q \geq 5$, and either $H = J$, or $H \simeq PGL(2,q)$, or else H is one of three exceptional groups listed in Table 1 of [45];

(iii) $b \geq 1$, H *is one of the 22 groups listed in Table 2 of* [45] *of the form* $H \simeq J.\mathbb{Z}_n$ *for* $n \in \{1, 2, 3\}$, *where J is a finite simple group distinct from* $PSL(2, q)$.

Although this result is an important step towards a classification of orientably regular maps with almost simple automorphism groups on surfaces with Euler characteristic as above, a complete classification is still not available.

4.8 Regular maps with a given underlying graph

Which graphs can underlie regular or orientably regular maps? Clearly, such a graph must be regular itself, that is, all vertices of the graph must have the same degree. A characterization of graphs underlying regular maps was given by Gardiner, Nedela, Škoviera and the author in [44]:

Theorem 4.18 *Let Γ be a connected regular graph of degree at least three. Then:*

(a) Γ *is the underlying graph of a regular map if and only if its automorphism group contains a subgroup G acting transitively on the darts of Γ, such that the stabiliser G_e of every edge e is dihedral of order 4 and the stabiliser G_v of each vertex v is a dihedral group the cyclic subgroup of index 2 of which is regular on the darts incident with v.*

(b) Γ *is the underlying graph of an orientably regular map if and only if its automorphism group contains a subgroup H acting regularly on the darts of Γ with cyclic vertex stabilisers.*

This result may be interpreted by saying that a graph underlies an (orientably) regular map if and only if the automorphism group of the graph contains 'an appropriate subgroup acting on the graph in the right way'. It is therefore not a surprise that investigations in this direction have focused on families of 'highly symmetric' graphs, such as complete graphs and, more generally, complete equipartite graphs, hypercubes over finite fields, and Johnson graphs. We will discuss these families separately. To simplify terminology, instead of saying that a graph Γ is the underlying graph of an (orientably) regular map we will often say that Γ *has an (orientably) regular embedding*.

Complete graphs

Classification of orientably regular embeddings of complete graphs was initiated by N. L. Biggs [9] and completed by L. A. James and G. A. Jones [57]. It turns out that all such embeddings are obtained as follows.

Let q be a prime power and let $GF^+(q)$ and $GF^*(q)$ be the additive and the multiplicative group of the Galois field $GF(q)$ of order q. Consider the group $H = AGL(1, q) \simeq GF^+(q) \rtimes GF^*(q)$ with multiplication given by $(x, a)(y, b) = (x + ay, ab)$. Choose a primitive element ξ of $GF(q)$ and define $s = s_\xi = (0, \xi)$ and $t = (1, -1)$. It can be checked that $\{s, t\}$ is a generating set for H and since t is an involution, the triple $M = M_\xi = (H; s, t)$ is an orientably regular map. Its underlying graph is K_q, the complete graph on q vertices, which we prove next.

We know that vertices of M are left cosets of $\langle s \rangle$, and so M has $|G|/|\langle s \rangle| = q$ vertices. Since M is regular, it is sufficient to check the situation at the vertex $\langle s \rangle$. Edges emanating from this vertex are one-to-one correspondence with left cosets of $\langle t \rangle$ that intersect with $\langle s \rangle$ non-trivially. There are exactly $q - 1$ such cosets, namely, those of the form $s^i \langle t \rangle$, $0 \le i \le q - 2$, which means that the underlying graph is regular of degree $q - 1$. It remains to show that there are no parallel edges. If there was a pair of parallel edges, then, by regularity of the map, we may assume that the two edges would be $\langle t \rangle$ and $s^i \langle t \rangle$ for some i such that $1 < i \le q - 2$. If the two edges had another common vertex $w\langle s \rangle \ne \langle s \rangle$ for some $w \in G$, then we would have $t, s^i t \in w\langle s \rangle$, and hence $s^i t = ts^j$ for some j. But with s and t as above this can happen only if $i = j = 0$, a contradiction. Thus, all the $q - 1$ edges emanating from (any) vertex end at distinct vertices, implying that the underlying graph of M is K_q, as claimed.

It is not difficult to check that for distinct primitive elements ξ, ζ of $GF(q)$ the maps M_ξ and M_ζ are isomorphic if and only if ξ and ζ are conjugate under a Galois automorphism of $GF(q)$. If $q = p^e$ for some prime p and a positive integer e, it follows that there are $\phi(q-1)/e$ pairwise non-isomorphic orientably regular embeddings of K_q, where ϕ is the Euler totient function; the maps are reflexible if and only if $q = 2, 3, 4$. The maps are of type $((q - 1)/2, q - 1)$ and genus $(q^2 - 7q + 4)/4$ if $q \equiv 3$ mod 4, and of type $(q-1, q-1)$ and genus $(q-1)(q-4)/4$ otherwise. By results of James [56] and Wilson [116], a complete graph K_n has a regular embedding on a non-orientable surface if and only if $n = 2, 3, 4, 6$.

Complete bipartite graphs

The classification of regular and orientably regular embeddings of complete bipartite graphs $K_{n,n}$ turned out to be a much harder problem. For every n there exists an orientably regular embedding of $K_{n,n}$, now called a *standard embedding*, which was introduced by Biggs and White [10]. In the notation of this article, the standard embedding can be presented in the form $(H; s, t)$, where $H = K \rtimes L$, $K = \mathbb{Z}_n \times \mathbb{Z}_n$, $L = \mathbb{Z}_2$, the non-trivial element of L transposes coordinates of elements of K, and $s = (1, 0, 0)$,

$t = (0, 0, 1)$. However, other regular embeddings exist in general, as noted e.g. by Kwak and Kwon [74].

A complete solution to the classification problem for orientably regular embeddings of $K_{n,n}$ was given by Jones [61], preceded by solutions for special cases of n: for n equal to a prime by Nedela, Škoviera and Zlatoš [93], for n a number relatively prime to $\phi(n)$ by Jones, Nedela and Škoviera [66] (these are the only cases when the standard embedding is the unique orientably regular embedding of $K_{n,n}$), for n an odd prime power again by Jones, Nedela and Škoviera [65], and for n a power of 2 by Du, Jones, Kwak, Nedela and Škoviera [34, 35]. The maps in the complete classification (which is too complicated to be restated here) arise from a combination of product and covering constructions from embeddings of $K_{q,q}$ for prime powers q appearing in the prime factorisation of n. For completeness, regular embeddings of $K_{n,n}$ on non-orientable surfaces were classified by Kwak and Kwon [75].

Complete multipartite graphs

A study of orientably regular embeddings of complete multipartite graphs was initiated by Du, Kwak and Nedela [38] who proved that, for any prime p, the graph $K_{p,p,\ldots,p}$ has such an embedding if and only if the number of parts is 2, 3 or p; in the same paper the authors also prove more general results on orientably regular embeddings of certain highly symmetric graphs arising as lexicographic products. The classification of orientably regular embeddings of $K_{n,n\ldots,n}$ for general n was taken much further by Zhang and Du [123] and completed by Du and Zhang [40].

Cubes

Catalano, Conder, Du, Kwon, Nedela and Wilson [17] classified the orientably regular embeddings of n-dimensional (binary) cubes Q_n. The first result in this direction, due to Nedela and Škoviera [91], was a covering construction of $\phi(n)$ pairwise non-isomorphic orientably regular embeddings of Q_n for every $n \geq 3$. Later, Du, Kwak and Nedela [39] proved that these are the only such embeddings if n is odd. Kwon [76] then constructed new orientably regular embeddings of Q_n for even n, and Xu [121] proved that Kwon's embeddings are the only extra ones if $n \equiv 2 \bmod 4$. Still new orientably regular embeddings of Q_n were constructed by Catalano and Nedela [18], and the already mentioned work of Catalano et al. [17] sealed the process. The nonexistence of non-orientable regular embeddings of n-cubes for $n \geq 3$ was proved by Kwon and Nedela [77].

Other graphs

We comment on other classes of graphs only briefly. Orientably regular embeddings of Hamming graphs and merged Johnson graphs were classified by Jones [62, 60]. The regular embeddings of graphs of order (i.e., number of vertices) equal to a prime can easily be classified using folklore facts about the structure of permutation groups of prime degree. Orientably regular embeddings of graphs of order a product of two (not necessarily distinct) primes were classified by Du, Kwak and Nedela [37] and corresponding results for non-orientable embeddings can be found in Wang and Du [113] and Du and Kwak [36]. Finally, a method of lifting the classification of orientably regular embeddings of graphs to their canonical double covers was devised by Nedela and Škoviera [89].

5 Regular maps: Constructions, operations, and external symmetries

We have encountered two fundamental and equivalent ways of representing regular maps of a given type (ℓ, m), namely, the correspondence between regular maps and groups with presentations (2.1), or, equivalently, normal, torsion-free subgroups of the full triangle groups $T_{\ell,m}$ summed up in Proposition 2.7; their orientable analogues are (3.1), (3.2) and Proposition 3.3. This language turns out to be convenient to describe constructions and operations with regular and orientably regular maps. We present here a limited selection of those that we find most important.

5.1 Covers, intersection and join

In Subsection 2.4 we saw that factorisation of a regular map by a normal subgroup (avoiding degeneracy) leads to a regular covering of maps, which can be viewed as producing a 'smaller' regular map from a 'larger' one. For the generation of new regular maps it is of great interest to be able to reverse this line of thought and construct regular maps that cover a given regular map, possibly with branch points at vertices and face centres. More precisely, given a regular map M and a group L, one would like to produce all regular maps $\tilde{M}$ with the property that $\tilde{M}/L \simeq M$. Descriptions of this process in terms of the so-called *voltage assignments* on flags, corners, or darts of M were given e.g. by Archdeacon, Richter, Škoviera and the author [3], Archdeacon, Gvozdjak and the author [2], Nedela and Škoviera [89] and Malnič, Nedela and Škoviera [85]. Since details are quite involved we invite the reader to consult these sources.

Let $M = (G; r_0, r_1, r_2)$ be a regular map of type (ℓ, m) and let $N \triangleleft T = T_{\ell,m}$ be its map subgroup, that is, $G \simeq T/N$. In Subsection 2.4 we saw that if K is a normal subgroup of G not containing any of $r_0, r_1, r_2, r_0 r_2$, then the quotient map $M/K = (G/K; r_0 K, r_1 K, r_2 K)$ is a regular map and the natural projection $M \to M/K$ is a regular covering, possibly with branch points at vertices and face centres. If (ℓ_K, m_K) is the type of M/K, then ℓ_K and m_K are divisors of ℓ and m; in fact ℓ_K and m_K are the smallest positive powers of $r_0 r_1$ and $r_1 r_2$ such that both $(r_0 r_1)^{\ell_K}$ and $(r_1 r_2)^{m_K}$ are in K. By the correspondence theorem in group theory, there is a normal subgroup J of T with $N \lhd J \lhd T$ such that $K \simeq J/N$.

Recalling that $G \simeq T/N$ and invoking further basic facts from group theory, we have $G/K \simeq (T/N)/(J/N) \simeq T/J$, so that J plays the role of a 'map subgroup' of the quotient map M/K. However, one has to be careful because if $(\ell_K, m_K) \neq (\ell, m)$, then J is not torsion-free. In order to incorporate this feature in the theory we will identify a regular map M of type (ℓ, m) as above also with a normal (not necessarily torsion-free) subgroup J of any full triangle group $T = T_{\ell', m'}$ with $\ell \mid \ell'$ and $m \mid m'$ such that $\mathcal{U}_{\ell', m'}/J \simeq M$; we will even allow one or both ℓ' and m' to be infinity if necessary or desirable. In such a situation we will call J the map $subgroup$ $relative$ to $T_{\ell', m'}$. Replacing the full triangle groups with their orientation-preserving subgroups $T^+ = T^+_{\ell', m'}$, orientably regular maps can be identified with $oriented$ map subgroups relative to T^+ in a similar way.

With this in mind we explain the notions of $intersection$ and $join$ for orientably regular maps. Let M and M' be orientably regular maps (not necessarily finite or of the same type) with oriented map subgroups N and N' relative to the triangle group $T^+ = T^+(\infty, \infty)$. The $join$ $M \vee M'$ and the $intersection$ $M \wedge M'$ are the orientably regular maps uniquely defined by their oriented map subgroups $N \cap N'$ and NN' relative to T^+, respectively. Such binary operations were studied by Wilson [117] and later in more detail and generality by Breda and Nedela [13]. Note that by the third isomorphism theorem for groups we have $N/N \cap N' \simeq NN'/N'$; let us denote this group by L. Then, the regular coverings $M \vee M' \to M$ and $M' \to M \wedge M'$ induce the same quotient L. A similar statement is valid when one interchanges the dashed and undashed symbols. From the map covering perspective, $M \vee M'$ is the smallest orientably regular map that regularly covers both M and M', and $M \wedge M'$ is the largest orientably regular map regularly covered by both M and M'.

For a full development of the theory one would have to allow also degenerate maps, since it may well happen that $NN' = T^+$ and in such a case, as we know, the quotient map T^+/NN' would be a semiedge on a sphere. Letting $H = T^+/N$ and $H' = T^+/N'$, it can be shown [12, 13] that this happens if and only if the orientation-preserving automorphism group $T^+/N \cap N'$ of the map $M \vee M'$ is isomorphic to the direct product

$H \times H'$. More explicitly, if $M = (H; r, s)$ and $M' = (H'; r', s')$ are (say) both finite and have types (ℓ, m) and (ℓ', m'), then $T^+ = NN'$ if and only if ℓ is coprime to ℓ' and m is coprime to m', in which case $M \vee M'$ is the *direct product* of orientably regular maps and $M \vee M' = (H \times H'; (r, r'), (s, s'))$. Such product constructions have been used; see e.g. [91] and earlier references in this subsection.

Interesting results can be derived if M and M' are related, for instance, if they are dual of each other or if they are a pair of chiral maps, or a *chiral pair*, for short; cf. the last part of Proposition 3.4. Taking on the latter example, if M and M' form a chiral pair with oriented map subgroups N and $N' = R_1 N R_1$ relative to $T^+ = T^+(\infty, \infty)$, then both NN' and $N \cap N'$ are invariant under conjugation by R_1 and hence the corresponding maps are reflexible. In this situation, $M \vee M'$ is the smallest *reflexible* map that regularly covers both M and M', and $M \wedge M'$ is the largest *reflexible* map regularly covered by both M and M'. An extreme occurs if $NN' = T^+$ and then the map M is called *totally chiral*, meaning that the only regular map covered by both M and M' is the trivial map. For much more detail on total chirality we refer to Breda et al. [12].

5.2   Dualities of regular maps

As indicated in Subsection 2.1, the dual of a regular map is formed by taking the topological realisation of the regular map $(G; r_0, r_1, r_2)$ by gluing flags in the usual way but then interchanging the labels 0 and 2 throughout. But swapping the labels is equivalent with interchanging the roles of the involutions r_0 and r_2. The corresponding *dual* regular map can therefore be identified with the quadruple $(G; s_0, s_1, s_2)$ with $s_0 = r_2$, $s_2 = r_0$ and $s_1 = r_1$. The *duality operator* on regular maps is the mapping D that assigns to a regular map $M = (G; r_0, r_1, r_2)$ its dual (regular) map $D(M) = (G; r_2, r_1, r_0)$. A regular map M is *self-dual* if $M \simeq D(M)$, i.e., if M is isomorphic to its dual – that is, if M is a fixed point of D. Combining this with Proposition 2.3 it follows that a regular map $(G; r_0, r_1, r_2)$ is self-dual if and only if there is an automorphism, still denoted D, of the group G, which fixes r_1 and interchanges r_0 with r_2.

If $(G; r_0, r_1, r_2)$ is orientably regular, then so is its dual. If we now take the even subgroup G^+ of G and represent M in the form (G^+, r, s) with $r = r_0 r_1$ and $s = r_1 r_2$, the interchange of r_0 with r_2 corresponds to the interchange of r with s^{-1} and s with r^{-1}. This induces not only duality but also a change in the orientation of the supporting surface; to retain the orientation one may use conjugation by r_1 to invert both r and s. This feature will show up in a distinction in the treatment of

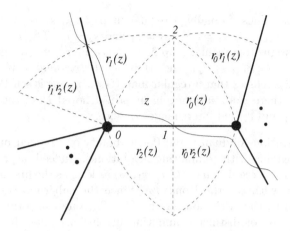

Figure 18: A part of a Petrie walk, illustrated by the thin line.

duality in the orientable case. Namely, if $(H; r, s)$ is an orientably regular map, not necessarily reflexible, the orientably regular maps $(H; r', s')$ and $(H; r'', s'')$ with $r' = s$, $s' = r$ and $r'' = s^{-1}$, $s'' = r^{-1}$ are the *positive* and the *negative dual* of $(H; r, s)$. A criterion for positive and negative *self-duality* follows by applying Proposition 3.1 in an obvious way and we leave the details for the reader.

Our second and slightly more involved example is an algebraic way to describe the Petrie dual of a regular map. This arises from $(G; r_0, r_1, r_2)$ by bending the flag-gluing rule for $i = 0$ by stipulating that two flags $g, g' \in G$ be 0-adjacent in the new map if $g' = gr_0r_2$, leaving the i-adjacency rule intact for $i = 1, 2$. The new face boundary walks are then (pairs of) orbits of the cyclic group $\langle (r_0r_2)r_1 \rangle$. The topological effect of replacing r_0 with r_0r_2 is that face boundary walks of the *new* map are obtained by walking along edges on the sides of the supporting surface and following the map corners of the *original* map, except that whenever reaching the midpoint of an edge one has to switch the side. The new face boundary walks thus look on the original map as 'zigzag' or 'left-right' walks, known as *Petrie walks*; see Fig. 18.

The operation of taking a Petrie dual preserves the underlying graph but *not* the supporting surfaces in general. In fact, taking a Petrie dual of a map on a non-orientable surface may result in a regular map on an orientable surface, and, of course, vice versa. A formal definition of a Petrie dual is obtained by expressing the construction outlined above, that is, the interchange of r_0 with r_0r_2, in our language of encoding of maps by quadruples. Thus, formally, a regular map $(G; s_0, s_1, s_2)$ is the *Petrie dual* of $(G; r_0, r_1, r_2)$ if $s_0 = r_0 r_2$, $s_1 = r_1$ and $s_2 = r_2$. The *Petrie duality operator* P assigns to a regular map $M = (G; r_0, r_1, r_2)$ its (regular) Petrie dual $P(M) = (G; r_0 r_2, r_1, r_2)$, and M is *self-Petrie-dual* if $M \simeq P(M)$. Proposition 2.3 tells us that a regular map $(G; r_0, r_1, r_2)$ is self-Petrie-dual if and only if there is an automorphism, still denoted P, of the group G, which fixes r_1 and r_2 and interchanges r_0 with $r_0 r_2$.

The descriptions of duality and Petrie duality of a regular map have a flavour of 'juggling with commuting involutions'. Indeed, a presentation (2.1) determines a regular map $(G; r_0, r_1, r_2)$ as long as the first involution, r_0, commutes with the third one, r_2. Since the subgroup $\langle r_0, r_2 \rangle$ of G is isomorphic to the Klein four-group $\mathbb{Z}_2 \times \mathbb{Z}_2$, one can form a total of six ordered pairs of distinct commuting involutions taken from the set $\{r_0, r_2, r_0 r_2\}$ and these ordered pairs correspond to the action of the group $\langle D, P \rangle \simeq \mathrm{Aut}\langle r_0, r_2 \rangle \simeq S_3$ on this set; see Jones and Thornton [72] for details. This way a regular map $M = (G; r_0, r_1, r_2)$ gives rise to an orbit of the group $\langle D, P \rangle$ containing 1, 2, 3 or 6 non-isomorphic maps of the form $(G; s_0, r_1, s_2)$ for the six ordered pairs (s_0, s_2) of distinct elements from $\{r_0, r_2, r_0 r_2\}$. In the case when the orbit has length 1, i.e., if both D and P are automorphisms of the group G, the regular map $(G; r_0, r_1, r_2)$ is said to be *completely self-dual* (called also a regular map with *trinity symmetry* in [1]).

5.3 Exponents, or hole operators

Hole operators were introduced a long time ago by Wilson [115] and interpreted by Jones and Thornton [72] in terms of outer automorphisms of the triangle groups $T^+(\ell, \infty)$. They were later investigated by Nedela and Škoviera [90] under the name 'exponents' in the context of classification of orientably regular maps with a given underlying graph.

To give a brief explanation, we will use the alternative notation (3.2) for orientably regular maps. Thus, suppose that one has an orientably regular map $(H; s, t)$ of type (ℓ, m) and that one thinks of changing the map but keeping the underlying graph and the orientation-preserving automorphism group unchanged, by which we mean 'physically the same', that is, we are not allowed to replace the graph and the group by isomorphic copies. Keeping the graph intact means leaving the left cosets of the cyclic groups $\langle s \rangle$ and $\langle t \rangle$ unchanged, as they correspond to vertices and edges, with incidence given by non-empty intersection. But we are allowed to change the left cosets of $\langle r \rangle$ because we have no condition on the resulting surface and, in particular, on the new faces.

This means that to define a new map of the form $(H; s', t')$ with these restrictions, up to conjugation (representing a selection of a fixed dart) we have just one choice, namely, to let s' be some other generator of $\langle s \rangle$ and let $t' = t$ as this element has to be fixed. If s^j is some other generator of $\langle s \rangle$, we define the j-th power operator E_j on orientably regular maps of degree m as the mapping assigning to a regular map $M = (H, s, t)$ as above the (regular) map $E_j(M) = (H; s^j, t)$. We note that E_j has also been known as the j-th hole operator, or Wilson operator.

Without loss of generality we will identify the set of generators of $\langle s \rangle \cong \mathbb{Z}_m$ with $\mathbb{Z}_m^*$, the group of units of the ring $\mathbb{Z}_m$. We note that although M and $E_j(M)$ have the same underlying graphs and the same orientation-preserving automorphism group, their supporting surfaces may differ. If $j = -1$, however (which is a generator for all m and is the only choice distinct from 1 if m is infinite), the supporting surfaces of the new and the original map are the same, and the two maps are non-isomorphic if and only if they form a chiral pair.

Our motivation was to identify the situation when the maps M and $E_j(M)$ are isomorphic. Formally, given an orientably regular map $M = (H; s, t)$ of type (ℓ, m) with m finite, then a $j \in \mathbb{Z}_m^*$ is an *exponent* of the map if $E_j(M) \simeq M$. By an appropriate modification of Proposition 3.1 for generating sets of the form $\{s, t\}$, an element $j \in \mathbb{Z}_m^*$ is an exponent of an orientably regular map $(H; s, t)$ of degree m if and only if there is a group automorphism of H taking s onto s^j and fixing t.

We will need to extend the concept of an exponent to non-orientable surfaces also. The motivation is the same, that is, to find a replacement of the three generating involutions in the description of a regular map $M = (G; r_0, r_1, r_2)$ in such a way that the new map has 'physically' the same underlying graph and the same automorphism group. Preservation of edges means no change of left cosets of $\langle r_2 \rangle = \{1, r_2\}$ and no change of left multiples of the coset $r_0 \langle r_2 \rangle = \{r_0, r_0 r_2\}$, representing darts and their reverses. This can only be achieved by fixing r_2 and either fixing r_0 or interchanging r_0 with $r_0 r_2$. Since the latter is nothing but the Petrie operation discussed in Subsection 5.2 we begin with focusing on the first possibility.

If r_0 and r_2 are kept unchanged, the only freedom we have to keep left cosets of $\langle r_1, r_2 \rangle = \langle r_1 r_2, r_2 \rangle$ representing vertices intact is to replace the element $r_1 r_2$ by another generator of the cyclic group $\langle r_1 r_2 \rangle$. Thus, under the above assumption and again up to conjugation, the only way to change the regular map $(G; r_0, r_1, r_2)$ to a regular map $(G; r_0, r_1', r_2)$

is to let $r_1'r_2 = (r_1r_2)^j$ for some j such that $(r_1r_2)^j$ generates the cyclic part of $\langle r_1, r_2 \rangle$. It follows that $r_1' = (r_1r_2)^jr_2$, which, of course, is an involution. This allows us to extend the definition of powers to regular maps in an obvious way. If $M = (G; r_0, r_1, r_2)$ is a regular map of degree m and $j \in \mathbb{Z}_m^*$, the j-th *power operator* E_j takes M onto the regular map $E_j(M) = (G; r_0, (r_1r_2)^jr_2, r_2)$. By Proposition 2.3 we have $M \simeq E_j(M)$ if and only if there is a group automorphism G fixing r_0 and r_2 while taking r_1 onto $(r_1r_2)^jr_2$; such an automorphism will also be denoted E_j without risking any confusion.

We now return to the possibility of preserving edges of a regular map $M = (G; r_0, r_1, r_2)$ of degree m by fixing r_2 and interchanging r_0 with r_0r_2. Since this is just the Petrie automorphism P, we obtain 'powers composed with the Petrie duality', that is, E_jP for $j \in \mathbb{Z}_m^*$. It is easy to check that the two operators commute, that is, $E_jP = PE_j$ for every $j \in \mathbb{Z}_m^*$. We will call E_jP the j-th *Petrie power operator*, and any of E_j and E_jP will be collectively called a *generalised power operator*. Clearly, for any fixed m the collection of all generalised power operators forms a group isomorphic to $\mathbb{Z}_m^* \times \mathbb{Z}_2$.

The above shows that it makes sense to extend the concept of an exponent from orientably regular maps to regular maps. We therefore say that, given a regular map $M = (G; r_0, r_1, r_2)$ of degree m, a pair (j, ε) for $j \in \mathbb{Z}_m^*$ and $\varepsilon \in \{0, 1\}$ is a *generalised exponent* of M if $E_jP^\varepsilon(M) \simeq M$. We note that this issue was first formally treated in [54] in a broader context of arbitrary maps. With the help of Proposition 2.3 one may easily translate the condition $E_jP^\varepsilon(M) \simeq M$ into the existence of an appropriate automorphism of the group G.

5.4   Kaleidoscopic and completely self-dual regular maps

Orientably regular maps M of type (ℓ, m) with m finite and with the property that $E_j(M) \simeq M$ for every $j \in \mathbb{Z}_m^*$ are called *kaleidoscopic*. This term was introduced by Archdeacon, Conder and the author [1], but orientably regular maps with this property have been investigated before. In their algebraic interpretation of Wilson's hole operators Jones and Thornton [72] observed that, in the current terminology, for any finite m the universal map $\mathcal{U}_{\infty,m}$ is kaleidoscopic. This is because for every $j \in \mathbb{Z}_m^*$ the assignment $S \mapsto S^j$ and $T \mapsto T$ extends to an automorphism of the triangle group $T_{\infty,m}^+ = \langle S, T | S^m, T^2 \rangle$, isomorphic to a free product $\mathbb{Z}_m * \mathbb{Z}_2$.

Combining this observation with the residual finiteness of the groups $T_{\infty,m}^+$, Wang and the author [106] proved that for any finite $m \geq 3$ there exist infinitely many finite, kaleidoscopic, orientably regular maps of degree m. Residual finiteness, however, does not offer control on the face length of resulting maps. In fact, for $\ell = 3$, that is, for orientably regular triangulations, it was shown in [106] that such a map of degree $m \equiv \pm 1$ mod 6 cannot have more than $\phi(m)/2$ exponents, and if $m \equiv -1$ mod 8 is a prime such that $(m-1)/2$ is also a prime, then such a triangulation cannot have exponents other than ± 1.

As regards the other end of the spectrum, namely, maps with fewest exponents, Archdeacon, Gvozdjak an the author [2] presented, for any finite m, a covering construction of an infinite number of finite orientably regular maps of degree m that admit only the trivial exponent, 1. For reflexible maps, that is, those admitting -1 as an exponent, Staneková, Olejár and the author [105] proved that for any finite non-spherical type (ℓ, m) there exist infinitely many finite, orientably regular and reflexible maps of type (ℓ, m) with no exponents except ± 1.

Residual finiteness can also be used to produce completely self-dual regular maps. Observe that the full triangle group

$$T_{\infty,\infty} = \langle R_0, R_1, R_2 |\ R_0^2, R_1^2, R_2^2, (R_0 R_2)^2 \rangle,$$

isomorphic to the free product $(\mathbb{Z}_2 \times \mathbb{Z}_2) * \mathbb{Z}_2$, admits both the duality and the Petrie duality automorphism. In equivalent terms, the universal map $\mathcal{U}_{\infty,\infty}$ is invariant under duality D and Petrie duality P. This observation together with the residual finiteness of $T_{\infty,\infty}$ led Richter, Wang and the author [98] to prove that for an infinite sequence of degrees m there exists a finite completely self-dual regular map of type (m, m). Unfortunately, this type of argument does not allow one to say anything specific about the sequence of such m's. By an interesting related result, proved by Jones and Poulton [67], there is an infinite sequence of degrees m for which there is a finite, orientably regular map of degree m invariant under the operator DP of order 3 but admitting no duality.

Can one have an 'absolute level of external symmetry' of regular maps, that is, regular maps that are kaleidoscopic and completely self-dual? It is not clear how such objects could be constructed with the help of residual finiteness. Namely, the universal tessellations $\mathcal{U}_{\infty,m}$ for constructions of kaleidoscopic maps are not completely self-dual, while the universal tessellations $\mathcal{U}_{\infty,\infty}$ for constructions of completely self-dual maps are not kaleidoscopic. The only known constructions of completely self-dual kaleidoscopic regular maps were given by Archdeacon, Conder and the author [1] and enable one to produce, from any given such map, an infinite number of maps with the same properties that branch-cover the initial map. Since Conder's census [19] yields 14 orientable and 3 non-orientable completely self-dual kaleidoscopic regular maps, the method of [1] gives an infinite number of such maps of various degrees.

The last issue we want to discuss in the context of our 'super-symmetric maps' is their group of 'external symmetries'. If $M = (G; r_0, r_1, r_2)$ is a finite completely self-dual kaleidoscopic regular map of degree m, we define its *external symmetry group* $\mathrm{Ext}(M)$ to be the group generated by the duality automorphism D, the Petrie automorphism P, and the exponent automorphisms E_j, $j \in \mathbb{Z}_m^*$, of the group G. While the subgroups $\langle E_j; \ j \in \mathbb{Z}_m^* \rangle$, $\langle P, E_j; \ j \in \mathbb{Z}_m^* \rangle$, and $\langle D, P \rangle$ have orders $\phi(m)$, $2\phi(m)$ and 6, it is not at all clear what the order of the group $\mathrm{Ext}(M)$ might be in general, or what the structure of this group might be. Hopes for a simple answer were shattered by a recent result of Conder, Kwon and the author [26], who show that there is an infinite sequence of finite completely self-dual kaleidoscopic regular maps M_n of constant degree 8 with $|\mathrm{Ext}(M_n)| \to \infty$. Thus, the behaviour of external symmetry groups appears to be mysterious even for the first non-trivial even degree, which is 8.

6 Conclusion

As we have indicated in the introduction, this survey is by no means exhaustive. For instance, we have not touched upon two important generalisations: hypermaps and higher-dimensional polytopes. Most of the theory of maps generalises naturally to hypermaps. An excellent reference in this direction, explaining also the connections between hypermaps, Riemann surfaces and Galois groups, is Jones and Singerman [70]; for regular hypermaps see e.g. Wilson and Breda [118]. The generalisations of regular maps to regular higher-dimensional polytopes are treated in detail in the monograph by McMullen and Schulte [88].

Even regarding results within the theory of regular maps we have restricted ourselves to the most important classification results, in most cases without including details of the methods used. We have thus skipped a number of topics, such as the theory and classification of Hurwitz maps (see Conder [20] and references therein), regular maps with multiple edges (see Nedela and Škoviera [90], Li and the author [79], and Conder, Tucker and the author [29]), and regular Cayley maps (see e.g. Conder, Jajcay and Tucker [24] and references therein), to name just a few.

It is not our aim to end with a list of open problems. The collection of results on the classification of regular and orientably regular maps by supporting surfaces, automorphism groups and underlying graphs given in Section 4 and on external symmetries in Section 5 represents the state of our knowledge at the time of writing this article. Any findings beyond these would be new and very likely interesting.

We conclude by returning to the question in the title, which can be rephrased by asking what the external symmetries of (regular and orientably regular) maps are. As we saw, dualities and exponents (hole operators) correspond to certain automorphisms of the automorphism group $Aut(M)$ of a regular map M. Filtering out inner automorphisms one can thus say that every outer automorphism of $Aut(M)$ could be elevated to the status of an external symmetry of M. This elevation would, however, allow 'external symmetries' that are hard to justify in terms of 'natural' operations on maps. Our question will therefore remain unanswered, at least for the time being.

Acknowledgements

The author thanks the anonymous referee for carefully reading the original manuscript and raising a number of helpful suggestions which led to a considerable improvement of the presentation.

The author also gratefully acknowledges support from the VEGA Research Grant 1/0781/11, the APVV Research Grant 0223-10, and the APVV support within the framework of the EUROCORES Programme EUROGIGA, project GREGAS, ESF-EC-0009-10, financed by the European Science Foundation.

References

[1] D. Archdeacon, M. Conder and J. Širáň, Kaleidoscopic regular maps with trinity symmetry, *Preprint* (2012), submitted.

[2] D. Archdeacon, P. Gvozdjak and J. Širáň, Constructing and forbidding automorphisms in lifted maps, *Mathematica Slovaca* **47** (1997) No. 2, 113–129.

[3] D. Archdeacon, B. Richter, J. Širáň and M. Škoviera, Branched coverings of maps and lifts of map homomorphisms, *Australas. J. Combin.* **9** (1994), 109–121.

[4] Y. F. Ban, S. F. Du, Y. Liu, A. Malnič, R. Nedela and M. Škoviera, Regular maps with nilpotent automorphism groups, Preprint, 2012.

[5] M. Belolipetsky and G. A. Jones, Automorphism groups of Riemann surfaces of genus $p + 1$, where p is a prime, *Glasgow Math. J.* **47** (2005), 379–393.

[6] H. Bender, Finite groups with dihedral Sylow 2-subgroups, *J. Algebra* **70** (1981), 216–228.

[7] H. Bender and G. Glauberman, Characters of finite groups with dihedral Sylow 2-subgroups, *J. Algebra* **70** (1981), 200–215.

[8] P. Bergau and D. Garbe, Non-orientable and orientable regular maps, in *Proceedings of "Groups-Korea 1988"*, *Lect. Notes Math. 1398*, Springer (1989), 29–42.

[9] N. L. Biggs, Automorphisms of imbedded graphs, *J. Combinat. Theory Ser. B* **11** (1971), 132–138.

[10] N. L. Biggs and A. T. White, Permutation groups and combinatorial structures, *London Mathematical Society Lecture Note Series 33* (Cambridge University Press, Cambridge), 1979.

[11] H. R. Brahana, Regular maps on an anchor ring, *Amer. J. Math.* **48** (1926), 225–240.

[12] A. Breda d'Azevedo, G. A. Jones, R. Nedela and M. Škoviera, Chirality groups of maps and hypermaps, *J. Algebraic Combin.* **29** (2009), 337–355.

[13] A. Breda d'Azevedo and R. Nedela, Join and intersection of hypermaps, *Acta Univ. M. Belii Math.* **9** (2001), 13–28.

[14] A. Breda d'Azevedo, R. Nedela and J. Širáň, Classification of regular maps of negative prime Euler characteristic, *Trans. Amer. Math. Soc.* **357** (2005) No. 10, 4175–4190.

[15] R. P. Bryant and D. Singerman, Foundations of the theory of maps on surfaces with boundary, *Quart. J. Math. Oxford Ser. (2)* **36** (1985) no. 141, 17–41.

[16] W. Burnside, Theory of Groups of Finite Order, *Cambridge Univ. Press*, 1911.

[17] D. Catalano, M. Conder, S. F. Du, Y. S. Kwon, R. Nedela and S. Wilson, Classification of regular embeddings of n-dimensional cubes, *J. Algebraic Combin.* **33** (2011) no. 2, 215–238.

[18] D. Catalano and R. Nedela, A characterization of regular embeddings of n-dimensional cubes, *Discrete Math.* **310** (2010) no. 17–18, 2364–2371.

[19] M. Conder, Regular maps and hypermaps of Euler characteristic −1 to −200, *J. Combin. Theory Ser. B* **99** (2009), 455–459.

[20] M. Conder, An update on Hurwitz groups, *Groups Complex. Cryptol.* **2** (2010) no. 1, 35–49.

[21] M. Conder, Personal communication, 2012.

[22] M. Conder and P. Dobcsányi, Determination of all regular maps of small genus, *J. Combinat. Theory Ser. B* **81** (2001), 224–242.

[23] M. Conder and B. Everitt, Regular maps on nonorientable surfaces, *Geom. Dedicata* **56** (1995), 209–219.

[24] M. Conder, R. Jajcay and T. Tucker, Regular Cayley maps for finite abelian groups, *J. Algebraic Combin.* **25** (2007), 259–283.

[25] M. Conder, R. Jajcay and T. Tucker, Regular *t*-balanced Cayley maps, *J. Combin. Theory Ser. B* **97** (2007), 453–473.

[26] M. Conder, Y. S. Kwon and J. Širáň, On external symmetry groups of regular maps, to appear in *Proc. Fields Institute Conf. on Symmetries of Maps*.

[27] M. Conder, P. Potočnik and J. Širáň, Regular maps with almost Sylow-cyclic automorphism groups, and classification of regular maps with Euler characteristic $-p^2$, *J. Algebra* **324** (2010), 2620–2635.

[28] M. Conder, P. Potočnik and J. Širáň, Regular hypermaps over projective linear groups, *J. Australian Math. Soc.* **85** (2008), 155–175.

[29] M. Conder, J. Širáň and T. Tucker, The genera, reflexibility and simplicity of regular maps, *J. Europ. Math. Soc.* **12** (2010), 343–364.

[30] M. Conder and T. Tucker, Regular Cayley maps for cyclic groups, *Preprint* (2011), submitted.

[31] M. Conder, S. Wilson, Inner reflectors and nonorientable regular maps, *Discrete Math.* **307** (2007), 367–372.

[32] H. S. M. Coxeter, Configurations and maps, *Rep. Math. Colloq (2)* **8** (1948), 18–38.

[33] H. S. M. Coxeter and W. O. J. Moser, Generators and Relations for Discrete Groups, 4th Ed., Springer-Verlag, Berlin, 1984.

[34] S. F. Du, G. A. Jones, J. H. Kwak, R. Nedela and M. Škoviera, Regular embeddings of $K_{n,n}$ where n is a power of 2, I: Metacyclic case, *European J. Combin.* **28** (2007) no. 6, 1595–1609.

[35] S. F. Du, G. A. Jones, J. H. Kwak, R. Nedela and M. Škoviera, Regular embeddings of $K_{n,n}$ where n is a power of 2, II: Non-metacyclic case, *European J. Combin.* **31** (2010) no. 7, 1946–1956.

[36] S. F. Du and J. H. Kwak, Nonorientable regular embeddings of graphs of order p^2, *Discrete Mathematics* **310** (2010), 1743–1751.

[37] S. F. Du, J. H. Kwak and R. Nedela, A classification of regular embeddings of graphs of order a product of two primes, *J. Algebraic Combin.* **19** (2004), 123–141.

[38] S. F. Du, J. H. Kwak and R. Nedela, Regular embeddings of complete multipartite graphs, *European J. Combin.* **26** (2005) no. 3–4, 505–519.

[39] S. F. Du, J. H. Kwak and R. Nedela, Classification of regular embeddings of hypercubes of odd dimension, *Discrete Math.* **307** (2007), 119–124.

[40] S. F. Du and J. Y. Zhang, A classification of orientably-regular embeddings of complete multipartite graphs, *arXiv:1202.1974v2*, 2012.

[41] A. Erréra, Sur les polyèdres réguliers de l'Analysis Situs, *Acad. Roy. Belg. Cl. Sci. Mem. Coll. in 8° (2)*, **7** (1922), 1–17.

[42] D. Garbe, Über die regulären Zerlegungen geschlossener orientierbarer Flächen, *J. Reine Angew. Math.* **237** (1969), 39–55.

[43] D. Garbe, A remark on non-symmetric Riemann surfaces, *Arch. Math.* **30** (1978), 435–437.

[44] A. Gardiner, R. Nedela, J. Širáň and M. Škoviera, Characterization of graphs which underlie regular maps on closed surfaces, *J. London Math. Soc. (2)* **59** (1999) No. 1, 100–108.

[45] N. Gill, Orientable regular maps with Euler characteristic divisible by few primes, *arXiv:1203.0138v2*, 2002.

[46] E. Girondo and G. González-Diez, Introduction to Compact Riemann Surfaces and Dessins d'Enfants, *London Mathematical Society Student Texts* No. 79, The London Mathematical Society, 2011.

[47] D. Gorenstein and J. H. Walter, The characterization of finite groups with dihedral Sylow 2-subgroups, I, II, III, *J. of Algebra* **2** (1965), 85–151, 218–270, 334–393.

[48] A. Gray and S. Wilson, A more elementary proof of Grünbaum's conjecture, *Congr. Numer.* **72** (1990), 25–32.

[49] A. S. Grek, Reqular polyhedra of simplest hyperbolic types (Russian), *Ivanov. Gos. Ped. Inst. Učen. Zap.* **34** (1963), 27–30.

[50] A. S. Grek, Regular polyhedra on a closed surface with the Euler characteristic $\chi = -3$ (Russian), *Izv. Vysš. Učebn. Zaved. Matematika* **55** (1966) no. 6, 50–53.

[51] A. S. Grek, Regular polyhedrons on surfaces with Euler characteristic $\chi = -4$ (Russian), *Soobšč. Akad. Nauk Gruzin. SSR* **42** (1966), 11–15.

[52] B. Grünbaum, Regularity of graphs, complexes and designs, in: *Problèmes Combinatoires et Théorie des Graphes*, Colloques Internationaux, Orsay, CNRS, Vol. 260 (1976), 191–197.

[53] V. Hucíková, R. Nedela and J. Širáň, Chiral maps of any given type, *Preprint* (2012), submitted.

[54] M. Hužvar, Exponents of maps, PhD Thesis, *Comenius University*, Bratislava, 2005.

[55] N. Ito, Über das Produkt von zwei abelschen Gruppen, *Math Z.* **62** (1955), 400–401.

[56] L. D. James, Imbeddings of the complete graphs, *Ars Combinat.* **16-B** (1983), 57–72.

[57] L. D. James and G. A. Jones, Regular orientable imbeddings of complete graphs, *J. Combinat. Theory Ser. B* **39** (1985), 353–367.

[58] G. A. Jones, Ree groups and Riemann surfaces. *J. Algebra* **165** (1994), 41–62.

[59] G. A. Jones, Maps on surfaces and Galois groups, *Math. Slovaca* **47** (1997), 1–33.

[60] G. A. Jones, Automorphisms and regular embeddings of merged Johnson graphs, *European J. Combin.* **26** (2005) no. 3-4, 417–435.

[61] G. A. Jones, Regular embeddings of complete bipartite graphs: classification and enumeration, *Proc. Lond. Math. Soc. (3)* **101** (2010) no. 2, 427–453.

[62] G. A. Jones, Classification and Galois conjugacy of Hamming maps, *Ars Math. Contemp.* **4** (2011) no. 2, 313–328.

[63] G. A. Jones and M. Jones, Infinite quotients of Fuchsian groups, *J. Group Theory* **3** (2000), 199–212.

[64] G. A. Jones, M. Mačaj and J. Širáň, Nonorientable regular maps over linear fractional groups, *Ars Math. Contemp.*, **6** (2013) no. 1, 25–35.

[65] G. A. Jones, R. Nedela and M. Škoviera, Regular embeddings of $K_{n,n}$ where n is odd prime power, *European J. Combin.* **28** (2007), 1863–1875.

[66] G. A. Jones, R. Nedela and M. Škoviera, Complete bipartite graphs with a unique regular embedding, *J. Combin. Theory Ser. B* **98** (2008), 241–248.

[67] G. A. Jones and A. Poulton, Maps admitting trialities but not dualities, *European J. Combin.* **31** (2010) no. 7, 1805–1818.

[68] G. A. Jones and S. A. Silver, Suzuki groups and surfaces, *J. London Math. Soc. (2)* **48** (1993), 117–125.

[69] G. A. Jones and D. Singerman, Theory of maps on orientable surfaces, *Proc. London Math. Soc. (3)* **37** (1978), 273–307.

[70] G. A. Jones and D. Singerman, Belyĭ functions, hypermaps, and Galois groups, *Bull. London Math. Soc.* **28** (1996), 561–590.

[71] G. A. Jones, M. Streit, J. Wolfart, Wilson's map operations on regular dessins and cyclotomic fields of definition, *Proc. Lond. Math. Soc. (3)* **100** (2010) no. 2, 510–532.

[72] G. A. Jones and J. S. Thornton, Operations on maps, and outer automorphisms, *J. Combin. Theory Ser. B* **35** (1983) no. 2, 93–103.

[73] D. E. Joyce, Java applet for drawing hyperbolic tessellations, available at http://aleph0.clarku.edu/ djoyce/poincare/ .

[74] J. H. Kwak and Y. S. Kwon, Regular orientable embeddings of complete bipartite graphs, *J. Graph Theory* **50** (2005) no. 2, 105–122.

[75] J. H. Kwak and Y. S. Kwon, Classification of nonorientable regular embeddings of complete bipartite graphs, *J. Combin. Theory Ser. B* **101** (2011), 191–205.

[76] Y. S. Kwon, New regular embeddings of n-cubes Q_n, *J. Graph Theory* **46** (2004), 297–312.

[77] Y. S. Kwon and R. Nedela, Non-existence of nonorientable regular embeddings of n-dimensional cubes, *Discrete Math.* **307** (2007), 511–516.

[78] S. Lando and A. Zvonkin, Graphs on Surfaces and Their Applications, *Springer*, 2004.

[79] C. H. Li and J. Širáň, Regular maps whose groups do not act faithfully on vertices, edges, or faces, *Europ. J. Combin.* **26** (2005), 521–541.

[80] F. Lübeck and G. Malle, $(2,3)$-generation of exceptional groups, *J. London Math. Soc. (2)* **59** (1999), no. 1, 109–122.

[81] M. Mačaj, J. Širáň and M. Iolyiová, Injective radius of representations of triangle groups and planar width of regular hypermaps, *Ars Math. Contemp.* **1** (2008), 223–241.

[82] A. M. Macbeath, Generators of the linear fractional groups, in: 1969 Number Theory (Proc. Sympos. Pure Math., Vol. XII, Houston, Tex.), *Amer. Math. Soc., Providence, R.I.*, 1967, 14–32.

[83] W. Magnus, Non-Euclidean Tessellations and Their Groups, *Acad, Press*, 1974.

[84] G. Malle, J. Saxl and T. Weigel, Generation of classical groups, *Geom. Dedicata* **49** (1994) no. 1, 85–116.

[85] A. Malnič, R. Nedela, and M. Škoviera, Regular homomorphisms and regular maps, *European J. Combin.* **23** (2002) no. 4, 449–461.

[86] I. Malcev, On the faithful representation of infinite groups by matrices, *Mat. Sb.* **8** (1940), 405–422 (Russian); *Amer. Math. Soc. Transl. (2)* **45** (1965), 1–18 (English).

[87] W. S. Massey, Algebraic Topology: An Introduction, *Harcourt, Brace and World*, New York, 1967.

[88] P. McMullen and E. Schulte, Abstract Regular Polytopes, *Cambridge Univ. Press*, 2003.

[89] R. Nedela and M. Škoviera, Regular maps of canonical double coverings of graphs, *J. Combin. Theory Ser. B* **67** (1996), 249–277.

[90] R. Nedela and M. Škoviera, Exponents of orientable maps, *Proc. London Math. Soc. (3)* **75** (1997) no. 1, 1–31.

[91] R. Nedela and M. Škoviera, Regular maps from voltage assignments and exponent groups, *Eur. J. Comb.* **18** (1997), 807–823.

[92] R. Nedela and M. Škoviera, Regular maps on surfaces with large planar width, *European J. Combin.* **22** (2001) no. 2, 243–261.

[93] R. Nedela, M. Škoviera and A. Zlatoš, Regular embeddings of complete bipartite graphs, *Discrete Math.* **258** (2002) no. 1-3, 379–381

[94] Ya. N. Nuzhin, Generating triples of involutions for alternating groups (Russian), *Mat. Zametki* **51** (1992) no. 4, 91–95; 142; English translation in: *Math. Notes* **51** (1992) no. 3-4, 389–392.

[95] Ya. N. Nuzhin, Generating triples of involutions of Chevalley groups over a finite field of characteristic 2 (Russian), *Algebra i Logika* **29** (1990), 192–206, 261; English translation in: *Algebra and Logic 29* (1990) no. 2, 134–143.

[96] Ya. N. Nuzhin, Generating triples of involutions of Lie-type groups over a finite field of odd characteristic I (Russian), *Algebra i Logika* **36** (1997), 77–96, 118; English translation in: *Algebra and Logic* **36** (1997) no. 1, 46–59.

[97] Ya. N. Nuzhin, Generating triples of involutions of Lie-type groups over a finite field of odd characteristic II (Russian), *Algebra i Logika* **36** (1997), 422–440, 479; English translation in: *Algebra and Logic* **36** (1997) no. 4, 245–256.

[98] B. Richter, J. Širáň and Y. Wang, Self-dual and self-Petrie-dual regular maps, J. Graph Theory, in press.

[99] B. Richter, J. Širáň, R. Jajcay, T. Tucker and M. E. Watkins, Cayley maps. *J. Combin. Theory Ser. B* **95** (2005), 189–245.

[100] C. H. Sah, Groups related to compact Riemann surfaces, *Acta Math.* **123** (1969), 13–42.

[101] J. Scherwa, Regulaere Karten geschlossener nichtorientierbarer Flaechen, Diploma Thesis, Bielefeld, 1985.

[102] F. A. Sherk, The regular maps on a surface of genus three, *Canad. J. Math.* **11** (1959) 452–480.

[103] J. Širáň, Triangle group representations and their applications to graphs and maps, *Discrete Math.* **209** (2001), 341–358.

[104] J. Širáň, Triangle group representations and constructions of regular maps, *Proc. London Math. Soc.* **82** (2001) no. 3, 513–532.

[105] J. Širáň, Ľ. Staneková and M. Olejár, Reflexible regular maps with no non-trivial exponents from residual finiteness, *Glasgow Math. J.* **53** (2011), 437–441.

[106] J. Širáň and Y. Wang, Maps with highest level of symmetry that are even more symmetric than other such maps: Regular maps with largest exponent groups, *Contemporary Mathematics* **531** (2010), 95–102.

[107] M. Suzuki, On finite groups with cyclic Sylow subgroups for all odd primes, *Amer. J. Math.* **77** (1955), 657–691.

[108] W. Threlfall, Gruppenbilder, *Abh. sächs. Akad. Wiss. Math.-Phys. Kl.* **41** (1932), 1–59.

[109] A. V. Timofeenko, On generating triples of involutions of large sporadic groups (Russian), *Diskret. Mat.* **15** (2003) no. 2, 103–112; English translation in: *Discrete Math. Appl.* **13** (2003) no. 3, 291–300.

[110] T. W. Tucker, Finite groups acting on surfaces and the genus of a group, *J. Combinat. Theory Ser. B* **34** (1983) No. 1, 82–98.

[111] A. V. Vasiľev and E. P. Vdovin, An adjacency criterion in the prime graph of a finite simple group, *Algebra Logika* **44** (2005), 682–725, 764.

[112] A Vince, Regular combinatorial maps, *J. Combin. Theory Ser. B* **35** (1983), 256–277.

[113] F. Wang, S. F. Du, Nonorientable regular embeddings of graphs of order pq, *Sci. China Math.* **54** (2011) no. 2, 351–363.

[114] S. Wilson, Riemann surfaces over regular maps, *Canad. J. Math.* **30** (1978), 763–782.

[115] S. Wilson, Operators over regular maps, *Pacific J. Math.* **81** (1979), 559–568.

[116] S. Wilson, Cantankerous maps and rotary embeddings of K_n, *J. Combin. Theory Ser. B* **47** (1989), 262–273.

[117] S. Wilson, Parallel products in graphs and maps, *J. Algebra* **167** (1994), 539–546.

[118] S. Wilson and A. Breda D'Azevedo, Surfaces having no regular hypermaps, *Discrete Math.* **277** (2004), 241–274.

[119] J .A. Wolf, Spaces of Constant Curvature, *McGraw-Hill*, New York, 1967.

[120] W. J. Wong, On finite groups with semidihedral Sylow 2-subgroups, *J. Algebra* **4** (1966), 52–63.

[121] J. Xu, A classification of regular embeddings of hypercubes Q_{2m} with m odd, *Sci. China Ser. A, Math.* **50** (2007), 1673–1679.

[122] H. Zassenhaus, Über endliche Fastkörper, *Abh. Math. Sem. Univ. Hamburg* **11** (1936), 187–220.

[123] J. Y. Zhang and S. F. Du, On the orientable regular embeddings of complete multipartite graphs, *European J. Combin.* **33** (2012), 1303–1312.

Department of Mathematics and Statistics
Open University, Walton Hall
Milton Keynes, MK7 6AA, UK
j.siran@open.ac.uk

Some open problems on permutation patterns

Einar Steingrímsson[1]

Abstract

This is a brief survey of some open problems on permutation patterns, with an emphasis on subjects not covered in the recent book by Kitaev, *Patterns in Permutations and words*. I first survey recent developments on the enumeration and asymptotics of the pattern 1324, the final pattern of length 4 whose asymptotic growth is unknown, and related issues such as upper bounds for the number of avoiders of any pattern of length k for any given k. Other subjects treated are the Möbius function, topological properties and other algebraic aspects of the poset of permutations, ordered by containment, and also the study of growth rates of permutation classes, which are containment closed subsets of this poset.

1   Introduction

The notion of permutation patterns is implicit in the literature a long way back, which is no surprise given that permutations are a natural object in many branches of mathematics, and because patterns of various sorts are ubiquitous in any study of discrete objects. In recent decades the study of permutation patterns has become a discipline in its own right, with hundreds of published papers. This rapid development has not only led to myriad new results, but also, and more interestingly, spawned several different research directions in the last few years. Also, many connections have been discovered between permutation patterns and other research areas, both inside and outside of combinatorics, showcasing the fundamental nature of patterns in permutations and other kinds of words.

Recently, Sergey Kitaev published a comprehensive reference work entitled *Patterns in permutations and words* [31]. In the present paper I highlight some aspects of some of the most recent developments and some areas that have hardly been touched, but which I think deserve more attention. This is speculative, of course, and strongly coloured by my own preferences and by the limits of my own knowledge in the field. The topics dealt with here are mostly left out in [31] (due to their very recent appearance) so there is little overlap here with that book.

In Section 2 I briefly describe the different kinds of patterns that are prominent in the field. In Section 3 I treat one case of pattern avoidance,

[1]Supported by grant no. 090038013 from the Icelandic Research Fund.

that of the pattern 1324, which is the shortest classical pattern whose avoidance is not known. Not even the asymptotics of the number of 1324-avoiders is known, even though significant effort has been put into this, resulting in ever better bounds. In fact, after a hiatus of a few years there were several successively improved results in 2012, which also seem likely to be applicable in a wider context.

Section 4 describes conjectures about which patterns are easiest to avoid, that is, are avoided by more permutations than other patterns of the same length. All the evidence points in the same direction, namely, that for patterns of any given length k, it is a layered pattern that is avoided by most permutations. It is still unclear, however, what form the most avoided layered pattern of length k has, although there are conjectures for particular families of values of k that seem reasonable, while others (in an abundance of conjectures, published and unpublished) have been shown false.

In Section 5 I discuss the poset (partially ordered set) $\mathcal{P}$ consisting of all permutations, ordered by pattern containment. This poset is the underlying object of all studies of pattern avoidance and containment. I mention the results so far on the Möbius function of $\mathcal{P}$, perhaps the most important combinatorial invariant of a poset, and some topological aspects of the order complexes of intervals in $\mathcal{P}$. Hardly anything is known so far on the topology of these intervals, but there are indications that large classes of them have a nice topological structure, the understanding of which might shed light on various pattern problems.

In Section 6 I mention some algebraic properties of the set of mesh patterns, a generalisation encompassing all the kinds of patterns considered here, and also the ring of functions counting occurrences of vincular patterns in permutations. Neither of these aspects has been studied substantially, but there are reasons to believe that they might be interesting.

Finally, in Section 7 I treat permutation classes, which are classes of permutations avoiding a set, finite or infinite, of classical patterns. The emphasis here is on the growth rates of these classes, that is, the growth of the number of permutations of length n in a given class. Substantial progress has been made here in recent years, although this only begins to scratch the surface of what promises to be an interesting study of fundamental properties of pattern containment and avoidance. In particular, the study of permutation classes and their growth rates is intimately related to the types of generating functions enumerating these classes, and generating functions are among the most important tools in the study of permutation patterns, as in all of enumerative combinatorics.

2   Kinds of patterns

We write permutations in one-line notation, as $a_1 a_2 \ldots a_n$, where the a_i are precisely the integers in $[n] = \{1, 2, \ldots, n\}$. For terminology not defined here see [31].

An occurrence of a classical pattern $p = p_1 p_2 \ldots p_k$ in a permutation $\pi = a_1 a_2 \ldots a_n$ is a subsequence $a_{i_1} a_{i_2} \ldots a_{i_k}$ of π whose letters appear in the same relative order of size as those in p. For example[2], 1-3-2 appears in 31542 as 142, 152, 154 and 354. Here, 154 and 354 are also occurrences of the *vincular* pattern 1-32, because the 5 and 4 are adjacent in 31542, which is required by the absence of a dash between 3 and 2 in 1-32. Also, 142 is an occurrence of the *bivincular* pattern $\bar{1}$-32, because 4 and 2 are adjacent (in position) and 1 and 2 are adjacent in value, as required by the bar over the 1 in $\bar{1}$-32.

These three kinds of patterns are illustrated in Figure 1, where the shaded column between the second and third black dots of the diagram for 3-24-1 indicates that in an occurrence of 3-24-1 in a permutation π no letter of π is allowed to lie between the letters corresponding to the 2 and the 4 or, equivalently, that those letters have to be adjacent in π. Patterns thus represented by diagrams with entire columns shaded are called vincular patterns, but were called generalised patterns when they were introduced in [6]. Similarly, the shaded row in the diagram for $p = $ 3-24-$\bar{1}$ indicates that in an occurrence of p in a permutation π, there must be no letters in π whose values lie between those corresponding to the 1 and the 2; that is, the letters corresponding to the 1 and 2 must be adjacent in value, because of the bar over the 1 in p. Patterns with some rows and columns shaded are called bivincular, and were introduced in [14].

Mesh patterns, introduced in [15] are now defined by extending the above prohibitions determined by shaded columns and rows to a shading of an arbitrary subset of squares in the diagram. Thus, in an occurrence, in a permutation π, of the pattern $(3241, R)$ in Figure 1, there must, for example, be no letter in π that precedes all letters in the occurrence and lies between the values of those corresponding to the 2 and the 3. This is required by the shaded square in the leftmost column. For example, in the permutation 415362, 5362 is not an occurrence of $(3241, R)$, since 4 precedes 5 and lies between 5 and 3 in value, whereas the subsequence 4362 is an occurrence of this mesh pattern.

---

[2]In this section I write classical patterns with dashes between all pairs of adjacent letters, to distinguish them from other vincular and bivincular patterns. In later sections, I will write classical patterns in the classical way, without any dashes, to keep the notation less cumbersome.

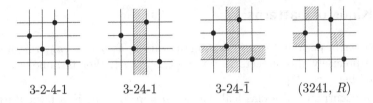

Figure 1: The patterns 3-2-4-1 (classical), 3-24-1 (vincular), 3-24-$\bar{1}$ (bivincular) and the mesh pattern $(3241, R)$, where $R = \{(0,2), (1,4), (3,2)\}$.

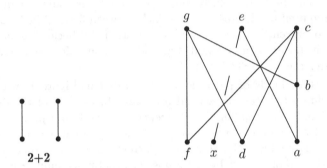

Figure 2: The poset $\mathbf{2+2}$, consisting of two disjoint chains, and a poset containing $\mathbf{2+2}$ in the subposet induced by b, c, e and x. The poset on the right is $(\mathbf{2+2})$-free if the vertex x is removed.

The bivincular patterns were introduced by Bousquet-Mélou, Claesson, Dukes and Kitaev in [14]. The original motivation behind their definition was to increase the symmetries of the vincular patterns, whose set is invariant under taking complement and reverse (corresponding to reflecting their diagrams horizontally and vertically), but not with respect to taking inverse, which corresponds to reflecting along the SW-NE diagonal. The study of the bivincular pattern 2-3$\bar{1}$ was the catalyst of the paper [14], where permutations avoiding this pattern were shown to be in bijective correspondence with two other families of combinatorial objects, the $(\mathbf{2+2})$-free posets (see Figure 2) and the *ascent sequences*. An ascent sequence is a sequence $a_1 a_2 \ldots a_n$ of nonnegative integers where $a_1 = 0$ and, for all i with $1 < i \leq n$,

$$a_i \leq \text{asc}(a_1 a_2 \ldots a_{i-1}) + 1,$$

where $\text{asc}(a_1 a_2 \ldots a_k)$ is the number of *ascents* in the sequence $a_1 a_2 \ldots a_k$, that is, the number of places $j \geq 1$ such that $a_j < a_{j+1}$. An example of such a sequence is 0101312052, whereas 0012143 is not, because the 4 is greater than $\mathrm{asc}(00121) + 1 = 3$.

This connection between $(2 + 2)$-free posets (also known as *interval orders*) and ascent sequences led to the determination of the elegant generating function for these families of objects, given in [14, Theorem 13] as

$$\sum_{n \geq 0} \prod_{i=1}^{n} \left(1 - (1 - t)^i\right).$$

This generating function, equivalent to an exact enumeration of the $(2+2)$-free posets, had eluded researchers for a long time. It became tractable because the previously little studied ascent sequences are more easily amenable to an effective recursive decomposition than the posets. The paper [14] has been followed by a great number of papers on these and bijectively related combinatorial objects, with no end in sight. The study of pattern avoidance by the ascent sequences in their own right was recently initiated [24] and has been furthered in [20, 35, 51], each of which proves some of the conjectures made in [24].

We say that two patterns are *Wilf equivalent*, and belong to the same *Wilf class*, if, for each n, the same number of permutations of length n avoids each. Of course, the symmetries mentioned above, corresponding to reflections of the diagram of a pattern, are (rather trivial) examples of Wilf equivalence, but there are many more, and they are often hard to prove. The smallest example of non-trivial Wilf equivalence is that for the classical patterns of length 3: the patterns 1-2-3 and 3-2-1 are trivially Wilf equivalent, and the same is true of the remaining four patterns of length 3, namely 1-3-2, 2-1-3, 2-3-1, and 3-1-2. All six of these patterns are Wilf equivalent, which is easy but non-trivial to prove; each is avoided by C_n permutations of length n, where C_n is the Catalan number $\frac{1}{n+1}\binom{2n}{n}$.

By extension, we say that two patterns p and q are *strongly Wilf equivalent* if they have the same *distribution* on the set of permutations of length n for each n, that is, if for each nonnegative integer k the number of permutations of length n with exactly k occurrences of p is the same as that for q. It is easy to see that the symmetry equivalences mentioned above imply strong Wilf equivalence. For example, $p = 1$-3-2 is strongly Wilf equivalent to $q = 2$-3-1, since the bijection defined by reversing a permutation turns an occurrence of p into an occurrence of q and conversely. On the other hand, 1-3-2 and 1-2-3 are not strongly Wilf equivalent, although they are Wilf equivalent. For example, the permutation 1234 has four occurrences of 1-2-3, but there is no permutation of length 4 with four occurrences of 1-3-2.

Just as finding the distribution of occurrences of a pattern is in general
harder than finding the number of avoiders, so there are still few results
about strong Wilf equivalence. In fact, the only nontrivial such results I
am aware of are recent results of Kasraoui [30], who gives an infinite family
of strong Wilf equivalences for non-classical vincular patterns, including,
for example, the equivalence of 3-421 and 421-3.

The mesh patterns, which may seem overly general at first sight, turn
out to be the right level of generalisation for expressing a seemingly deep
algebraic relationship (see Section 6) that encompasses all the kinds of
patterns defined above. They also allow for simple expressions of some
more cumbersome definitions, such as some of the so-called barred patterns, and they provide elegant expressions for some well known statistics
on permutations, such as the number of left-to-right maxima and the number of components in a permutation π, that is, the maximum number of
terms in a direct sum decomposition of π, to be defined in Section 4.

3   Enumeration and asymptotics of avoiders: The case of 1324

For information about the state of the art in enumeration of permutations avoiding given patterns, refer to [31]. Here I will essentially only
treat one unresolved case, that of the only classical pattern of length 4, up
to Wilf equivalence, for which neither exact enumeration nor asymptotics
have been determined. Of course, there is an endless list of problems left to
solve when it comes to avoidance, until we find general theorems. Whether
that will ever happen is likely to remain unknown for quite a while, given
the slow progress so far. The situation is similar for vincular patterns,
a short survey on which appears in [45], and even less has, understand-
ably, been done when it comes to bivincular and mesh patterns. Although
we are up against a major obstacle in furthering the knowledge about
classical pattern avoidance, there are probably some reasonably easy, and
interesting, results left to be found for the vincular, bivincular and mesh
patterns.

Observe that in the remainder of the paper, unless otherwise noted, I
am talking about classical patterns, which will be written in the classical
way, that is, without any dashes to separate their letters (as should be
done were they being considered as vincular patterns).

In 2004, Marcus and Tardos [36] proved the Stanley-Wilf conjecture,
stating that, for any classical pattern p, we have $\mathrm{Av}_p(n) < C^n$ for some
constant C depending only on p, where $\mathrm{Av}_p(n)$ is the number of permutations of length n avoiding p. It had been shown earlier by Arratia [5] that

this was equivalent to the existence of the limit

$$\mathrm{SW}(p) = \lim_{n \to \infty} \sqrt[n]{\mathrm{Av}_p(n)},$$

which is called the *Stanley-Wilf limit for p*. It should be noted that this exponential growth does not apply to vincular patterns in general. For example, the avoiders of length n of the pattern 1-23 are enumerated by the Bell numbers B_n [22], which count set partitions and grow faster than any exponential function. Namely, when n goes to infinity, the quotient B_n/B_{n-1} grows like $n/\log n$ [32, Prop. 2.6]. In fact, it seems likely (see [45, Section 8]) that there are no vincular patterns of length greater than 3 with exponential growth.

The Stanley-Wilf limit is 4 for all patterns of length 3, which follows from the fact that the number of avoiders of any one of these is the n-th Catalan number C_n, as mentioned above. This limit is known to be 8 for the pattern 1342 (see Bóna's paper [12]). For the pattern 1234 the limit is 9. This is a special case of a result of Regev [39] (see also the more recent [38]), who provided a formula for the asymptotic growth of the number of standard Young tableaux with at most k rows, pairs of which are in bijection, via the Robinson-Schensted correspondence, with permutations avoiding an increasing pattern of length $k + 1$. This limit can also be derived from Gessel's general result [26] for the number of avoiders of an increasing pattern of any length.

The only Wilf class of patterns of length 4 for which the Stanley-Wilf limit is unknown is represented by 1324. A lower bound of 9.47 was established by Albert et al. [1], who used an interesting technique, the *insertion encoding* of a permutation, introduced in [3]. This encoding was used to analyse, in an efficient fashion, how a 1324-avoider can be built up by inserting the letters $1, 2, \ldots, n$ in increasing order. They then used this to show that a certain subset of 1324-avoiders has insertion encodings that are accepted by a particular finite automaton, from which they could deduce the lower bound.

Successively improved upper bounds have been established in several steps. The first reasonably small one, was given in [23], where 1324-avoiders were shown to inject into the set of pairs of permutations where one avoids 132 and the other avoids 213; this led to an upper bound of $4 \cdot 4 = 16$. Refining the method used in [23], Bóna [10] was able to reduce this bound to $7 + 4 \cdot \sqrt{3} \approx 13.9282$.

In [23] it is conjectured that $\mathrm{SW}(1324) < e^{\pi\sqrt{2/3}} \approx 13.001954$. This would follow from another conjecture in [23], which says that the number of 1324-avoiders of length n with a fixed number k of inversions[3] is increasing

---

[3]An inversion in a permutation $a_1 a_2 \ldots a_n$ is a pair (i, j) such that $i < j$ and $a_i > a_j$.

as a function of n. It is further conjectured that this holds for avoiders of any classical pattern other than the increasing ones, and there is some evidence that this in fact applies to all non-increasing vincular patterns.

Using Markov chain Monte Carlo methods to generate 1324-avoiders at random, Madras and Liu [34] estimated that, with high likelihood, the Stanley-Wilf limit SW(1324) lies in the interval $[10.71, 11.83]$. Recent computer simulations[4] I have done (unpublished), building on the random generation method of Madras and Liu, but estimating SW(1324) in a different way, point to the actual limit being close to 11. Given how hard it seems to determine SW(1324) makes it understandable that Doron Zeilberger is claimed to have said [25] that "Not even God knows the number of 1324-avoiders of length 1000." I'm not sure how good Zeilberger's God is at math, but I believe that some humans will find this number in the not so distant future.

4 Are layered patterns the most easily avoided?

A *layered permutation* is a permutation that is a concatenation of decreasing sequences, each containing smaller letters than in any of the following sequences. An example of a layered permutation is 32145768, whose layers are displayed by $321 - 4 - 5 - 76 - 8$. A layered pattern is thus a *direct sum* of decreasing permutations. The direct sum $\sigma \oplus \tau$ of two permutations σ and τ is obtained by appending τ to σ after adding the length of σ to each letter of τ. The *skew sum* $\sigma \ominus \tau$ of σ and τ is obtained by prepending σ to τ after adding the length of τ to each letter of σ. Thus, for example, we have $3142 \oplus 231 = 3142675$ and $3142 \ominus 231 = 6475231$.

Of course, reversing a layered pattern p, or taking its complement, gives a pattern that is Wilf equivalent to p, and such a pattern/permutation might be called *up-layered*, since each layer is increasing. Clearly, an up-layered pattern is the skew sum of increasing permutations. To simplify the discussion in this section, without changing the traditional definition

---

[4]The simulations were done in the following way: Using the method of [34, Sections 2 and 4], I first generated, in each of roughly 155 independent processes, 10^9 1324-avoiders of length 1000 (the initialisation phase in the terminology of [34]). Using each of these 155 generated seed permutations, each in an independent process, I then generated a total of approx. $2.57 \cdot 10^{12}$ further avoiders, (the data collection phase). Using one in every ten thousand of those avoiders, I found the right end of the leftmost occurrence of the pattern 132. The number of that place equals the number of different places where $n + 1 = 1001$ can be inserted to obtain an avoider of length 1001 from one of length 1000. In the limit where $n \to \infty$ the average of the number of this place over all avoiders of a given length n would give the S-W limit. The average I obtained was approximately 11.01146. However, the convergence may be very slow, so, as pointed out to me by Josef Cibulka [21], this may be close to the correct value for $n = 1000$ although the actual limit might be significantly greater.

of layered permutations, I will abuse notation by letting "layered" refer both to layered and up-layered patterns.

Evidence going back at least to Julian West's thesis [50, Section 3.3] supports the conjecture that among all patterns of length k the pattern avoided by most permutations of a sufficiently large length n is a layered pattern. This conjecture has been around for a long time, and several variations on it, ranging from asymptotic dominance to the conjecture that $\mathrm{Av}_\sigma(n) \leq \mathrm{Av}_\tau(n)$ for all n if σ and τ have the same length and τ is layered but σ is not. The latter conjecture here is false, as pointed out by Vít Jelínek [28], who noted that, by [13, Theorem 4.2], the non-layered pattern obtained as the direct sum $p \oplus q$ of the layered pattern p with layer sizes $(1, 2, 1, 2, \ldots, 1, 2, 1)$ and $q = 231$ has a larger Stanley-Wilf limit than the layered pattern $12 \ldots k$, if k is sufficiently large. One of the weakest of these conjectures, widely believed to be true, is made explicit in [23, Conjecture 1] (although it may have been made before) and says that among all patterns of a given length k, the largest Stanley-Wilf limit is attained by some layered pattern.

The first published instances of such conjectures I am aware of appear in Burstein's thesis [18, Conjectures 9.5], where several conjectures are listed, both by Burstein and others, some of which have been refuted, while others have yet to be proved right or wrong. I think these questions are quite interesting, apart from their intrinsic interest, because any results about them are likely to be accompanied by quite general results about the enumeration and asymptotics of pattern avoidance.

Although the evidence is strong in support of the conjecture that the most easily avoided pattern of any given length is a layered pattern, there is currently no general conjecture that fits all the known data about the particular layered patterns with the most avoiders. However, there are some ideas about what form the most avoided layered patterns ought to have, and specific conjectures that have not been shown to be false.

As is mentioned in Burstein's thesis [18, see Conjectures 9.10, 9.11 and 9.12] Kézdy and Snevily had made some conjectures about this, and Burstein did too. In all these cases, which fall into three classes, depending on the congruence class modulo 4 of k, the patterns conjectured to have the maximal number of avoiders have small layers, and are highly symmetric. For example, a conjecture of Kézdy and Snevily [18, Conj. 9.10] that has not been refuted is that for patterns of even length k the most avoided pattern is the one with layers $(1, 2, 2, \ldots, 2, 2, 1)$. Also, Burstein's Conjecture 9.11 in [18] that for $k \equiv 1 \pmod 4$ the most avoided pattern of length k has symmetric layers $(1, 2, \ldots, 2, 3, 2, \ldots, 2, 1)$ still stands. Data computed by Vít Jelínek [28] support these conjectures. Jelínek provided all such data mentioned in this section. These computations range over all patterns of lengths up to 10 and permutations of lengths up to 14, although they are not exhaustive within these ranges.

For patterns of length $k \equiv 3 \pmod 4$, however, the situation is not quite that simple. Burstein's Conjecture 9.12 in [18], has the symmetric pattern $(1,2,\ldots,2,1,2,\ldots,2,1)$ as most avoided, but among permutations of lengths 11 and 12 the most avoided pattern of length 7 is 1432657, with layers $(1,3,2,1)$. The next pattern in this respect is 2143657, with layers $(2,2,2,1)$, and only in third place comes 1324657, with the symmetric layers $(1,2,1,2,1)$. The respective numbers of avoiders of these patterns, for permutations of length 12, are 457 657 176, 457 656 206 and 457 655 768, differing in their last four digits.

But, strange things do occasionally happen in this field, so Burstein's pretty conjecture for $k \equiv 3 \pmod 4$ cannot yet be written off entirely as far as asymptotic growth is concerned. Namely, Stankova and West [42, Figure 9] unearthed a somewhat irregular behaviour in pattern avoidance that gives pause here. They computed data that show, among other examples, that although $\mathrm{Av}_p(n)(53241) < \mathrm{Av}_p(n)(43251)$ when $7 \leq n \leq 12$, the inequality is switched for $n = 13$. It is still unknown whether such a switch occurs twice or more for any pair of patterns, but Stankova and West conjecture [42, Conjecture 2] that this does not happen. They also make the weaker conjecture that for any pair of patterns σ and τ there is an N such that either $\mathrm{Av}_\sigma(n) < \mathrm{Av}_\tau(n)$ for all $n > N$ or else the opposite inequality holds for all $n > N$. In other words, that two patterns do not switch places infinitely often.

It was shown in [23, Corollary 7] that the Stanley-Wilf limit of a layered pattern of length k is at most $4k^2$. In a recent preprint [11] Bóna, building on results from [23] and [11], proves that this bound can be reduced to $2.25k^2$. If the conjecture is true that the most avoided pattern of length k, for any given k, is a layered pattern then this would be an upper bound for the Stanley-Wilf limit of any pattern of length k, but it is not known whether this bound is the best possible.

5   The pattern poset, its Möbius function and topology

The set of all permutations forms a poset $\mathcal{P}$ with respect to classical pattern containment. That is, a permutation σ is smaller than π, denoted $\sigma \leq \pi$, if σ occurs as a pattern in π. This poset is the underlying object of all studies of pattern avoidance and containment. In the following two subsections I treat the Möbius function of $\mathcal{P}$, perhaps the most studied invariant of a poset, and then the topological aspects of the simplicial complexes naturally associated to intervals in $\mathcal{P}$.

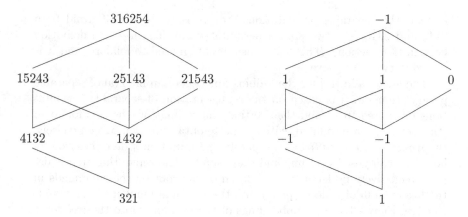

Figure 3: The interval $\mathcal{I} = [321, 316254]$ in $\mathcal{P}$ and the values of $\mu(321, \tau)$ on its elements τ.

5.1 The Möbius function of the permutation poset

An *interval* $[x, y]$ in a poset P is the set of all elements $z \in P$ such that $x \leq z \leq y$. An interval is thus a subposet with unique minimum and maximum elements, namely x and y, respectively, except when $x \not\leq y$, in which case $[x, y]$ is empty. The *Möbius function* of an interval is defined recursively as follows: for all x, we set $\mu(x, x) = 1$ and

$$\mu(x, y) = - \sum_{x \leq z < y} \mu(x, z).$$

Thus, the Möbius function is uniquely defined by setting its sum over any interval $[x, y]$ to 1 if $x = y$ and to 0 otherwise. Figure 3 gives an example of an interval from $\mathcal{P}$, and the values of the Möbius function on this interval, computed from bottom to top using the above recursive definition, showing that $\mu(321, 316254) = -1$.

The Möbius function of intervals in the permutation poset exhibits a great variety in values, even for permutations of small length, and is seemingly very hard to determine in the general case. The first such results were found by Sagan and Vatter [40], who found a formula for intervals of layered permutations.

In [46] the problem was solved for some special cases, and in [19] an effective (polynomial time) formula was given for all *separable* permutations. These are the permutations that can be built from singletons by combinations of direct sums and skew sums. The permutation 31254 is separable, since $31254 = (1 \ominus (1 \oplus 1)) \oplus (1 \ominus 1)$, whereas 3142 is not.

In fact, the separable permutations are precisely those that avoid both 3142 and 2413 (the two *simple* permutations of length 4; see definition later in this section). This fact is usually attributed to folklore, and it is straightforward to prove.

The formula in [19] for the Möbius function of an interval of separable permutations $[\sigma, \tau]$ is based on the representation of separable permutations by rooted trees, and then certain embeddings of (the tree for) σ in (the tree for) τ are counted, with a sign, to obtain $\mu(\sigma, \tau)$. More precisely, the *reduced separating tree* of a separable permutation π describes exactly how π is composed by sums and skew sums. The separating tree is defined recursively by letting the children of the root be the summands in the (skew) sum of π decomposed into the maximal number of summands (see [19, Figure 1]). The embeddings of the tree for τ into the tree for σ that are counted to compute the Möbius function are the so called *normal embeddings*, defined by a rather technical condition that we won't go into here. Each embedding corresponds to a unique occurrence of σ in τ. The formula for the Möbius function $\mu(\sigma, \tau)$ is then given by

$$\mu(\sigma, \tau) = \sum_{f \in N(\sigma, \tau)} \text{sgn}(f), \tag{5.1}$$

where $N(\sigma, \tau)$ is the set of normal embeddings of σ in τ and $\text{sgn}(f)$ is a sign associated to the embedding f, depending on which leaves of the tree for τ belong to the embedding of σ in τ. Again, this is determined by a rather technical condition that we omit, but we refer the reader to [19, Section 4].

Formula (5.1) implies, among many other things, that the absolute value of $\mu(\sigma, \tau)$ cannot exceed the number of occurrences of σ in τ (which is far from true in the general case), because each of the embeddings in question corresponds to a unique occurrence of σ in τ.

This formula also allows for many easy computations of values of the Möbius function, such as this example: if $\pi_i = 1, 3, 5, \ldots, 2i-1, 2i, \ldots, 4, 2$, then

$$\mu(\pi_k, \pi_n) = \binom{n + k - 1}{n - k}.$$

For example, $\mu(\pi_2, \pi_4) = \mu(1342, 13578642) = \binom{4+2-1}{4-2} = 10$.

It is conjectured in [19, Conjecture 30] that the maximum value of $\mu(\sigma, \pi)$ for any separable π of length $n \geq 3$ is obtained by a permutation of this form, for k that is roughly $n/2$ (but whose exact formula depends on the parity of the length of π).

In [19] a recursive formula for the Möbius function was also given in the case of *decomposable* permutations, those that can be written non-trivially as sums or skew sums $(241365 = 2413 \oplus (1 \ominus 1)$ is decomposable, whereas 2413 is not). This reduces the problem to the indecomposable permutations, for which it seems unlikely there will be a general formula anytime soon. However, solutions for large classes of these are reasonable to hope for. Moreover, indications are that the results for separable permutations, which are characterised by avoiding 2413 and 3142, can be extended to other more complicated permutation classes. That would allow us to find the maximum (absolute) value of the Möbius function on infinite permutation classes by computing it on a small finite set of permutations in the class.

An occurrence of a *consecutive* (vincular) pattern p in a permutation π is an occurrence of p in π whose letters are consecutive in π, such as the occurrence 364 of the pattern 132 in 5136472. An effective formula for computing the Möbius function of the poset of permutations ordered by containment as consecutive patterns was given in [8]. Sagan and Willenbring [41] later provided another proof, and used discrete Morse theory to determine the homotopy type of intervals of this poset. This poset is rather simple; its Möbius function is restricted to 0, 1 and -1, and it was shown in [41] that its intervals are either contractible or else homotopy equivalent to a single sphere (see the next subsection, on topology of the permutation poset).

A recent paper by McNamara and Sagan [37] established similar and further results in the more wide ranging case of the poset of generalised subword order. In particular, they determined the Möbius function for all intervals of this poset, and exhibited a family of intervals that are homotopic to wedges of spheres. Given the similarity in definition of this poset to the poset $\mathcal{P}$ of permutations with the classical pattern containment order, it seems reasonable to hope that the methods developed in [37] can be adapted to obtain results for large classes of intervals in $\mathcal{P}$.

Let **1** be the permutation of length 1. It is shown in [19, Corollary 24] that for any separable permutation π the only possible values of $\mu(\mathbf{1}, \pi)$ are 0, 1 and -1. Although nobody has bothered proving this yet, it appears certain that for arbitrary permutations π the absolute value $\mu(\mathbf{1}, \pi)$ is unbounded. For example, it seems a safe guess (based on computed data) that

$$\mu(\mathbf{1}, 2468\ldots(2n)135\ldots(2n-1)) = -\binom{n+1}{2}.$$

As an example, $\mu(\mathbf{1}, 24681357) = -\binom{4+1}{2} = -10$.

The maximum absolute value of $\mu(\mathbf{1}, \pi)$ is known for all π of length at most 11, as mentioned in [19, Section 5]. Recent computations I have done, although not exhaustive, suggest that for $n = 12$ the maximum is attained only by the permutation $\pi = 4\ 7\ 2\ 10\ 5\ 1\ 12\ 8\ 3\ 11\ 6\ 9$, and its symmetric equivalents, with $\mu(\mathbf{1}, \pi) = -261$. That would match the results for $n < 12$, mentioned in [19], namely that the maximum is in each case attained by only one permutation (up to trivial symmetries), and that that permutation is simple.[5] A permutation $\pi = a_1 a_2 \ldots a_n$ is *simple* if it has no segment $a_i a_{i+1} \ldots a_{i+k}$, where $0 < k < n - 1$, that consists of a segment of values, that is, such that $\{a_i, a_{i+1}, \ldots, a_{i+k}\} = \{\ell, \ell+1, \ldots, \ell+k\}$ for some ℓ. For example, 315264 is simple, but 461325 is not, since 132 is a segment of consecutive values that is nontrivial, that is, neither a singleton nor the entire permutation. A nice survey on simple permutations, which are important in the study of permutation classes, is found in [16].

It is worth noting here that separable permutations and simple permutations are in some imprecise sense each others' opposites; the separable ones decompose very nicely, while being simple is an obstruction to such decomposition. Thus, it would not be surprising if the maximum of $|\mu(\mathbf{1}, \pi)|$ over all π of length n turns out to be attained by a simple permutation. It seems less certain that there is, for all n, a unique permutation, up to trivial symmetries, that attains the maximum value for each n. The sequence of values of $\mu(\mathbf{1}, \pi)$ for which $|\mu(\mathbf{1}, \pi)|$ is maximised (for each length n, starting at $n = 1$) begins with

$$1, -1, 1, -3, 6, -11, 15, -27, -50, -58, 143, -261, \ldots$$

(the last entry still conjectural, as mentioned above) but no nontrivial upper bound on its n-th term is known.

Although many families of intervals $[\sigma, \tau]$ with $\mu(\sigma, \tau) = 0$ are described in [19, 46], it is an open problem to characterise such intervals completely. It might also be interesting to characterise those intervals for which $|\mu(\sigma, \pi)|$ equals the number of occurrences of σ in π.

Another question raised in [19] is whether it is possible to find a bound on $|\mu(\sigma, \tau)|$ that depends only on the number of occurrences of σ in τ. As mentioned above, it has been shown that for separable σ and τ, the value $|\mu(\sigma, \tau)|$ cannot exceed the number of occurrences of σ in τ, but nothing similar is known for the general case.

In conclusion, although a general formula for the Möbius function of an interval will likely remain untractable for a while, it seems reasonable to expect much further progress. In particular, the fact that there are families of intervals whose Möbius function has a "nice" formula, such as binomial coefficients, gives hope that these intervals have a structure that

---

[5]Except when $n = 3$, for which there are no simple permutations.

can be understood and exploited, and used to elicit the topology of these intervals, which is treated in the next section.

5.2   Topology of the permutation poset

Another important aspect of the poset of permutations, as for any combinatorially defined poset, is the topology of (the *order complexes* of) its intervals. (For terminology not defined in this section, see [43].) The order complex of a poset P, denoted $\Delta(P)$, is the abstract simplicial complex consisting of the *chains* of P. A chain in a poset P is a set of elements in P that are pairwise comparable, and thus totally ordered. Any subset of a chain forms a chain, so the set of chains is closed in this respect, which is a defining property of simplicial complexes. To study the topology of an interval $\mathcal{I}$ we remove the maximum and minimum element of $\mathcal{I}$, and take the order complex of the remaining "interior" of $\mathcal{I}$, which we denote by $\bar{\mathcal{I}}$.

There are some indications that large classes of intervals in the permutation poset $\mathcal{P}$ may have a "nice" topology, meaning that the topology can be simply described, and thus that the homology, and the homotopy type, of these intervals can be well understood. Gaining such understanding may well lead to significant progress in answering other questions, for example about the Möbius function. This is because, in some sense, the topological properties of a complicated interval give a much clearer picture of the important overall structure, sweeping aside irrelevant details that obscure the view. One well known connection to the Möbius function is that the Möbius function of a poset is equal to the *reduced Euler characteristic* of its order complex, which is a topological invariant (see [44, Proposition 3.8.6]).

Figure 4 shows the interval $\mathcal{I} = [321, 316254]$, and the order complex of $\bar{\mathcal{I}}$. This complex is homotopy equivalent to a sphere, since contracting the edge between 1432 and 21543 leaves a 1-dimensional sphere (topologically speaking) consisting of the edges of the rectangle in the figure. Note that $\mu(321, 21543) = 0$ and so removing 21543 from $\mathcal{I}$ does not affect the Möbius function of $\mathcal{I}$. The reduced Euler characteristic of a 1-dimensional sphere is -1, which, of course, equals $\mu(321, 316254)$. An obvious question (see below) is whether it is common for order complexes arising in this way from intervals of $\mathcal{P}$ to have similarly nice properties, such as being homotopy equivalent to wedges of spheres of the same dimension.

The elements (which are sets) in a simplicial complex are called *faces*. A *facet* of a simplicial complex is a face that is maximal with respect to containment. A simplicial complex is *pure* if all its facets have the same dimension. A property that implies a pure simplicial complex is homotopy equivalent to a wedge of spheres is being *shellable*. Informally, this means

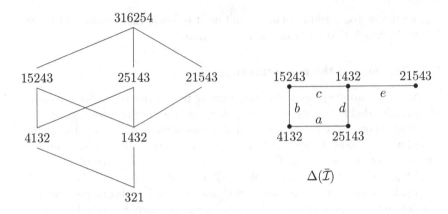

Figure 4: The interval $\mathcal{I} = [321, 316254]$ in $\mathcal{P}$ and the order complex of $\bar{\mathcal{I}}$. Note that this is a deceptively simple example, since intervals of a higher rank have order complexes of higher dimension, which are not so easy to depict.

that the complex can be built up, one facet at a time, such that each facet that is added, apart from the first one, intersects the union of the previous ones in a pure subcomplex of the maximum possible dimension (which is one less than the dimension of the facets). The complex in Figure 4 is shellable. One shelling order of its facets (edges) is a, b, c, d, e, whereas beginning with $a, c, \ldots$ can not give a shelling since the edge c does not intersect a. As mentioned before, a shellable complex is necessarily homotopy equivalent to a wedge of spheres, and showing shellability is probably the most common tool used to determine the topology of combinatorially defined complexes.

As mentioned in the previous section, Sagan and Willenbring [41] used discrete Morse theory to determine the homotopy type of intervals of the poset of consecutive patterns, whose intervals are either contractible or else homotopy equivalent to a single sphere, and it is reasonable to expect that results using similar techniques will show large classes of intervals in $\mathcal{P}$ to be homotopy equivalent to wedges of spheres. The Möbius function of such intervals is, up to a sign, just the number of spheres in the wedge.

In [19, Question 31] the following questions were raised, where, given an interval $\mathcal{I} = [\sigma, \pi] \in \mathcal{P}$, we let $\Delta(\sigma, \pi)$ be the order complex of $\bar{\mathcal{I}}$:

1. For which σ and π does $\Delta(\sigma, \pi)$ have the homotopy type of a wedge of spheres?

2. Let Γ be the subcomplex of $\Delta(\sigma, \pi)$ induced by those elements τ of $[\sigma, \pi]$ for which $\mu(\sigma, \tau) \neq 0$. Is Γ a pure complex, that is, do

all its maximal simplices (with respect to inclusion) have the same dimension?

3. If σ occurs precisely once in π, and $\mu(\sigma, \pi) = \pm 1$, is $\Delta(\sigma, \pi)$ homotopy equivalent to a sphere?

4. For which σ and π is $\Delta(\sigma, \pi)$ shellable?

An example where $\Delta(\sigma, \pi)$ is not shellable is given by $\sigma = 231$ and $\pi = 231564$, since $\Delta(\sigma, \pi)$ in this case consists of two disjoint components, each of which is contractible. If, however, we remove from $[231, 231564]$ all those elements τ for which $\mu(231, \tau) = 0$, we get a shellable complex, namely a boolean algebra of rank 2. For questions 2 and 3 above we don't know any counterexamples. In fact, we know no counterexamples to question 3 even without the condition of just one occurrence. However, since we have so far only examined intervals of small rank, our evidence is weak.

6 Other algebraic aspects

In addition to the Möbius function (and the underlying incidence algebra of $\mathcal{P}$) there are some algebraic aspects of permutation patterns that have been little studied, but which might harbour some interesting things. I mention two here: the ring of functions of vincular patterns, and Brändén and Claesson's reciprocity theorem for mesh patterns.

Vincular patterns can be regarded as functions from the set of all permutations to the ring of integers, counting occurrences of themselves in a permutation. For example, 2-31(416253) = 2, corresponding to the 462 and 453, which are all the occurrences of 2-31 in 416253. Linear combinations of vincular patterns were used in [6] to classify the Mahonian permutation statistics, which are those that are equidistributed with the number of inversions. Other such combinations played a crucial role in [47], where they were used to record the distribution of various statistics on the filled Young tableaux treated there.

No further work seems to have been done along these lines, although it is almost certain that there are many equivalences to be found of the kind discussed in [6]. A promising indication in this context is that computer experiments suggest that the distribution of two linear combinations of patterns is the same for all n provided that it is the same for all n smaller than some (small) constant depending only on the length of the patterns involved. As an example, exhaustive computer search shows that if two linear combinations of three vincular patterns of length 3 diverge for $n < 10$, then they diverge already for $n = 6$. For vincular patterns of length 4 all such combinations with different distributions for $n = 9$ differ already for $n = 8$. Of course, it is possible that divergence will occur again for greater n, but this seems unlikely. A general theorem to this effect, guaranteeing that checking such equidistributions for small n is sufficient to establish equidistribution for all n, would be a major breakthrough, provided the values that need to be checked are small enough, since that would give automatic proofs of various theorems. More importantly, proving such a theorem would undoubtedly require a general understanding we lack today, and thus would likely lead to significant other progress.

Seen as functions, as described above, the set of vincular patterns constitutes a ring of functions. (It is a tedious but straightforward exercise to verify that the product of two vincular patterns can be expressed as a sum of vincular patterns.) One relation is known in this ring, namely the *upgrading* mentioned in [6, Equation (2)], an example of which is

$$(21\text{-}3) = (21\text{-}43) + (21\text{-}34) + (31\text{-}24) + (32\text{-}14) + (213).$$

Since this ring contains all linear combinations of vincular patterns, it might be worthwhile to study its algebraic structure further. In particular, it would be interesting to know if there are other relations in this ring.

The set of mesh patterns also forms a ring of functions that should be further investigated for its properties. It is in this ring that the striking Reciprocity Theorem of Brändén and Claesson lives [15]. The Reciprocity Theorem expresses any mesh pattern (including the classical patterns) as a (possibly infinite) linear combination of classical patterns whose coefficients are obtained from values of the *dual* pattern on permutations. To express that theorem, let $p = (\pi, R)$ be a mesh pattern, and let R^c be the complement of R, that is, $R^c = [0,n]^2 \setminus R$, where n is the length of π. We then define the dual pattern of p as $p^* = (\pi, R^c)$, and define $\lambda(\sigma)$ by $\lambda(\sigma) = (-1)^{n-k} p^*(\sigma)$, where k is the length of σ. The Reciprocity Theorem is then the following identity, where the sum is over all classical patterns σ:

$$p = \sum_{\sigma \in \mathcal{S}} \lambda(\sigma)\sigma.$$

It seems likely that much can be gained from this theorem, due to its universal nature.

7   Growth rates of permutation classes

A *permutation class* is a set of permutations that is closed with respect to containment. That is, if $\pi \in \mathcal{C}$ for a class $\mathcal{C}$, and σ occurs as a pattern in π, then $\sigma \in \mathcal{C}$. The set of permutations avoiding any classical pattern, or set of such patterns, is easily seen to be a class (which is *not* true for vincular, bivincular or mesh patterns) and every permutation class is characterised by a unique antichain of permutations that are avoided by all elements of $\mathcal{C}$. That antichain is called the *basis* of $\mathcal{C}$. Note that there are infinite antichains of permutations (see Brignall [17] for the most general construction to-date), so bases can be infinite.

Studies of the poset of permutations have in recent years yielded many results about the diverse collection of permutation classes, parallelling work being done on other types of object (surveyed in Bollobás [9]). One of the most active and successful avenues of investigation has been into the *growth rates* of permutation classes. Given a class $\mathcal{C}$, where $\mathcal{C}_n$ is a set of permutations in $\mathcal{C}$ of length n, the growth rate of $\mathcal{C}$ is defined as

$$\mathrm{gr}(\mathcal{C}) = \limsup_{n \to \infty} \sqrt[n]{|\mathcal{C}_n|}.$$

To connect this terminology with that of Section 3, note that the Stanley-Wilf limit of the (classical) pattern p is the growth rate of the class of p-avoiding permutations. Thus the Stanley-Wilf Conjecture in this context states that all proper permutation classes have finite growth rates. One of the most natural open questions is whether the limit superior above in the definition can be replaced by a limit; this is known to be possible in the case of singleton-based classes by Arratia [5].

The line of research on growth rates attempts to determine both which growth rates are possible and where notable phase transitions take place in this spectrum. The first answers were provided by Kaiser and Klazar [29], who characterised the growth rates up to 2. At the smallest end of the scale, it is clear that 0 and 1 are growth rates of permutation classes, and that no classes have growth rates between these two numbers. Kaiser and Klazar showed that the next growth rate is the golden ratio and established the stronger result that if $|\mathcal{C}_n| < F_n$ (the nth Fibonacci number) for *any* n, then $|\mathcal{C}_n|$ is eventually polynomial (the structural properties of such classes were later explored by Huczynska and Vatter [27]). Between the golden ratio and 2, Kaiser and Klazar showed that all growth rates are roots of $x^k - x^{k-1} - \cdots - x - 1$ for some k; note that this makes 2 the least accumulation point of growth rates.

Vatter [49] extended the characterisation of growth rates up to $\kappa \approx 2.21$, the unique positive root of $x^3 - 2x^2 - 1$. Moreover, κ represents a sharp phase transition: There are only countably many permutation classes of growth rate less than κ, but because infinite antichains of permutations begin to appear at this growth rate, there are uncountably many permutation classes of growth rate κ. Viewed on the number-line of growth rates, κ also lies in an interesting place, as it is the least accumulation point of accumulation points of growth rates. Recent work by Albert, Ruškuc, and Vatter [4] has established another threshold at κ: every permutation class of growth rate less than κ has a rational generating function, while there are (by an elementary counting argument using the existence of infinite antichains) permutation classes of growth rate κ whose generating functions are not even holonomic. (A function on the natural numbers is holonomic if it satisfies a linear homogeneous recurrence relation with polynomial coefficients.)

Working in the closely related context of ordered graphs, Balogh, Bollobás, and Morris [7] characterised the growth rates up to 2 and made two conjectures which would have implied that growth rates of permutation classes are always algebraic integers and that the set of growth rates contains no accumulation points from above. These conjectures were both disproved by Albert and Linton [2], who constructed an uncountable set of growth rates. It remains open if, as suggested by Klazar [33], the conjectures of Balogh, Bollobás, and Morris hold when restricted to finitely based permutation classes.

Building on the work of Albert and Linton, Vatter [48] showed that every real number greater than or equal to $\lambda \approx 2.48$, the unique positive root of $x^5 - 2x^4 - 2x^2 - 2x - 1$, is the growth rate of a permutation class, and conjectured that λ is best possible.

Thus a striking problem remains: To characterise the growth rates between κ and λ. While it may well be impossible to describe the set of growth rates once it becomes uncountable (and before it consists of all real numbers), one could perhaps hope to describe it up to this point. The work of Vatter [48] implies that this happens at or before $\xi \approx 2.32$, the unique positive root of $x^5 - 2x^4 - x^2 - x - 1$. Thus, just as κ represents the transition from countably many to uncountably many permutation classes, ξ may represent the transition from countably many to uncountably many growth rates.

8 Acknowledgements

I am deeply grateful to Vince Vatter, who provided invaluable help, in particular with the section on growth rates. I thank Alex Burstein, Miklós Bóna, Vít Jelínek and Josef Cibulka for helpful information about the subject of Section 4. Jelínek also generously provided all the computations mentioned in that section. I am also indebted to a meticulous and insightful referee who made many good suggestions, pointed out several errors and spurred me to make the paper somewhat more extensive, all of which has improved it significantly from its original version.

References

[1] Michael H. Albert, M. Elder, A. Rechnitzer, P. Westcott, and M. Zabrocki, *On the Wilf-Stanley limit of 4231-avoiding permutations and a conjecture of Arratia*, Adv. in Appl. Math. **36** (2006), no. 2, 95–105.

[2] Michael H. Albert and Steve Linton, *Growing at a perfect speed*, Combin. Probab. Comput. **18** (2009), 301–308.

[3] Michael H. Albert, Steve Linton, and Nik Ruškuc, *The insertion encoding of permutations*, Electron. J. Combin. **12** (2005), Research Paper 47, 31 pp. (electronic).

[4] Michael H. Albert, Nikola Ruškuc, and Vincent Vatter, *Inflations of geometric grid classes of permutations*, arXiv:1202.1833v1 [math.CO].

[5] Richard Arratia, *On the Stanley-Wilf conjecture for the number of permutations avoiding a given pattern*, Electron. J. Combin. **6** (1999), Note 1, 4 pp.

[6] Eric Babson and Einar Steingrímsson, *Generalized permutation patterns and a classification of the Mahonian statistics*, Sém. Lothar. Combin. **44** (2000), Article B44b, 18 pp.

[7] József Balogh, Béla Bollobás, and Robert Morris, *Hereditary properties of ordered graphs*, Topics in discrete mathematics (M. Klazar, J. Kratochvíl, M. Loebl, J. Matoušek, R. Thomas, and P. Valtr, eds.), Algorithms Combin., vol. 26, Springer, Berlin, 2006, 179–213.

[8] Antonio Bernini, Luca Ferrari, and Einar Steingrímsson, *The Möbius function of the consecutive pattern poset*, Electron. J. Combin. **18** (2011), no. 1, Paper 146, 12.

[9] Béla Bollobás, *Hereditary and monotone properties of combinatorial structures*, Surveys in Combinatorics 2007 (Anthony Hilton and John Talbot, eds.), London Mathematical Society Lecture Note Series, no. 346, Cambridge University Press, 2007, 1–39.

[10] Miklós Bóna, *A new upper bound for 1324-avoiding permutations*, arXiv:1207.2379v1 [math.CO].

[11] Miklós Bóna, *On the best upper bound for permutations avoiding a pattern of a given length*, arXiv:1209.2404v1 [math.CO].

[12] Miklós Bóna, *Exact enumeration of* 1342-*avoiding permutations: a close link with labeled trees and planar maps*, J. Combin. Theory Ser. A **80** (1997), no. 2, 257–272.

[13] Miklós Bóna, *New records in Stanley-Wilf limits*, European J. Combin. **28** (2007), no. 1, 75–85.

[14] Mireille Bousquet-Mélou, Anders Claesson, Mark Dukes, and Sergey Kitaev, $(2+2)$-*free posets, ascent sequences and pattern avoiding permutations*, J. Combin. Theory Ser. A **117** (2010), no. 7, 884–909.

[15] Petter Brändén and Anders Claesson, *Mesh patterns and the expansion of permutation statistics as sums of permutation patterns*, Electron. J. Combin. **18** (2011), no. 2, Paper 5, 14.

[16] Robert Brignall, *A survey of simple permutations*, Permutation Patterns (Steve Linton, Nik Ruškuc, and Vincent Vatter, eds.), London Math. Soc. Lecture Note Ser., vol. 376, Cambridge Univ. Press, Cambridge, 2010, 41–65.

[17] Robert Brignall, *Grid classes and partial well order*, J. Combin. Theory Ser. A **119** (2012), 99–116.

[18] Alexander Burstein, *Enumeration of words with forbidden patterns*, ProQuest LLC, Ann Arbor, MI, 1998, Ph.D thesis, University of Pennsylvania.

[19] Alexander Burstein, Vít Jelínek, Eva Jelínková, and Einar Steingrímsson, *The Möbius function of separable and decomposable permutations*, J. Combin. Theory Ser. A **118** (2011), no. 8, 2346–2364.

[20] William Y. C. Chen, Alvin Y. L. Dai, Theodore Dokos, Tim Dwyer, and Bruce E. Sagan, *On* 021-*avoiding ascent sequences*, arXiv:1206.2849v2 [math.CO].

[21] Josef Cibulka, Personal communication.

[22] Anders Claesson, *Generalized pattern avoidance*, European J. Combin. **22** (2001), no. 7, 961–971.

[23] Anders Claesson, Vít Jelínek, and Einar Steingrímsson, *Upper bounds for the Stanley-Wilf limit of 1324 and other layered patterns*, J. Combin. Theory Ser. A **119** (2012), no. 8, 1680–1691.

[24] Paul Duncan and Einar Steingrímsson, *Pattern avoidance in ascent sequences*, Electron. J. Combin. **18** (2011), no. 1, Paper 226, 17.

[25] Murray Elder and Vincent Vatter, *Problems and conjectures presented at the Third International Conference on Permutation Patterns, University of Florida, March 7–11, 2005*, arXiv:math.CO/0505504 [math.CO].

[26] Ira M. Gessel, *Symmetric functions and P-recursiveness*, J. Combin. Theory Ser. A **53** (1990), no. 2, 257–285.

[27] Sophie Huczynska and Vincent Vatter, *Grid classes and the Fibonacci dichotomy for restricted permutations*, Electron. J. Combin. **13** (2006), Research paper 54, 14 pp.

[28] Vít Jelínek, Personal communication.

[29] Tomáš Kaiser and Martin Klazar, *On growth rates of closed permutation classes*, Electron. J. Combin. **9** (2003), no. 2, Research paper 10, 20 pp.

[30] Anisse Kasraoui, *New Wilf-equivalence results for vincular patterns*, European J. Combin. **34** (2012), no. 2, 322–337.

[31] Sergey Kitaev, *Patterns in permutations and words*, Monographs in Theoretical Computer Science, Springer-Verlag, 2011.

[32] Martin Klazar, *Counting set systems by weight*, Electron. J. Combin. **12** (2005), Research Paper 11, 8 pp. (electronic).

[33] Martin Klazar, *Overview of some general results in combinatorial enumeration*, Permutation Patterns (Steve Linton, Nik Ruškuc, and Vincent Vatter, eds.), London Mathematical Society Lecture Note Series, vol. 376, Cambridge University Press, 2010, 3–40.

[34] Neal Madras and Hailong Liu, *Random pattern-avoiding permutations*, Algorithmic probability and combinatorics, Contemp. Math., vol. 520, Amer. Math. Soc., Providence, RI, 2010, 173–194.

[35] Toufik Mansour and Mark Shattuck, *Some enumerative results related to ascent sequences*, arXiv:1207.3755v1 [math.CO].

[36] Adam Marcus and Gábor Tardos, *Excluded permutation matrices and the Stanley-Wilf conjecture*, J. Combin. Theory Ser. A **107** (2004), no. 1, 153–160.

[37] Peter McNamara and Bruce E. Sagan, *The Möbius function of generalized subword order*, Adv. Math. **229** (2012), 2741–2766.

[38] Amitai Regev, *Asymptotics of Young tableaux in the strip, the d-sums*, arXiv:1004.4476 [math.CO].

[39] Amitai Regev, *Asymptotic values for degrees associated with strips of Young diagrams*, Adv. in Math. **41** (1981), no. 2, 115–136.

[40] Bruce E. Sagan and Vincent Vatter, *The Möbius function of a composition poset*, J. Algebraic Combin. **24** (2006), no. 2, 117–136.

[41] Bruce E. Sagan and R. Willenbring, *Discrete Morse theory and the consecutive pattern poset*, J. Algebraic Combin. (to appear).

[42] Zvezdelina Stankova and Julian West, *A new class of Wilf-equivalent permutations*, J. Algebraic Combin. **15** (2002), no. 3, 271–290.

[43] Richard P. Stanley, *Combinatorics and commutative algebra*, second ed., Progress in Mathematics, vol. 41, Birkhäuser Boston Inc., Boston, MA, 1996.

[44] Richard P. Stanley, *Enumerative combinatorics. Vol. 1*, Cambridge Studies in Advanced Mathematics, vol. 49, Cambridge University Press, Cambridge, 1997.

[45] Einar Steingrímsson, *Generalized permutation patterns – a short survey*, Permutation patterns, London Math. Soc. Lecture Note Ser., vol. 376, Cambridge Univ. Press, Cambridge, 2010, 137–152.

[46] Einar Steingrímsson and Bridget Eileen Tenner, *The Möbius function of the permutation pattern poset*, J. Comb. **1** (2010), no. 1, 39–52.

[47] Einar Steingrímsson and Lauren K. Williams, *Permutation tableaux and permutation patterns*, J. Combin. Theory Ser. A **114** (2007), no. 2, 211–234.

[48] Vincent Vatter, *Permutation classes of every growth rate above 2.48188*, Mathematika **56** (2010), 182–192.

[49] Vincent Vatter, *Small permutation classes*, Proc. Lond. Math. Soc. (3) **103** (2011), 879–921.

[50] Julian West, *Permutations with forbidden subsequences and stack-sortable permutations*, ProQuest LLC, Ann Arbor, MI, 1990, Ph.D thesis, Massachusetts Institute of Technology.

[51] Sherry H. F. Yan, *Ascent sequences and 3-nonnesting set partitions,* arXiv:1208.1915 [math.CO].

Department of Computer and Information Sciences
University of Strathclyde
Glasgow G1 1XH, UK
einar@alum.mit.edu

The world of hereditary graph classes viewed through Truemper configurations

Kristina Vušković

Abstract

In 1982 Truemper gave a theorem that characterizes graphs whose edges can be labeled so that all chordless cycles have prescribed parities. The characterization states that this can be done for a graph G if and only if it can be done for all induced subgraphs of G that are of a few specific types, that we will call Truemper configurations. Truemper was originally motivated by the problem of obtaining a co-NP characterization of bipartite graphs that are signable to be balanced (i.e. bipartite graphs whose node-node incidence matrices are balanceable matrices).

The configurations that Truemper identified in his theorem ended up playing a key role in understanding the structure of several seemingly diverse classes of objects, such as regular matroids, balanceable matrices and perfect graphs. In this survey we view all these classes, and more, through the excluded Truemper configurations, focusing on the algorithmic consequences, trying to understand what structurally enables efficient recognition and optimization algorithms.

1 Introduction

Optimization problems such as coloring a graph, or finding the size of a largest clique or stable set are NP-hard in general, but become polynomially solvable when some configurations are excluded. On the other hand they remain difficult even when seemingly quite a lot of structure is imposed on an input graph. For example, determining whether a graph is 3-colorable remains NP-complete for triangle-free graphs with maximum degree 4 [92]. The approximation approach to these problems does not help either, since for example unless P=NP, there does not exist a polynomial time algorithm that can find a $2\chi(G)$-coloring of a graph G [73]. So if $\mathcal{C}$ is a class of graphs for which there exist polynomial time algorithms that find the chromatic number, $\mathcal{C}$ must have some "strong structure". Understanding structural reasons that enable efficient algorithms is our primary interest in this survey.

In 1982 Truemper [121] gave a theorem (Theorem 2.1) that characterizes graphs whose edges can be labeled so that all chordless cycles have prescribed parities. The characterization states that this can be done for a graph G if an only if it can be done for all induced subgraphs of G that are of a few specific types (depicted in Figure 1), that we will call *Truemper configurations*, and will describe precisely in Section 2. Truemper was originally motivated by the problem of obtaining a co-NP characterization of bipartite graphs that are signable to be balanced (i.e. bipartite graphs whose node-node incidence matrices are balanceable matrices, a class of matrices that have important polyhedral properties).

The configurations that Truemper identified in his theorem ended up playing a key role in understanding the structure of several seemingly diverse classes of objects, such as regular matroids, balanceable matrices and perfect graphs. A powerful technique called the decomposition method, which we describe in Section 3, was used in structural analysis of all these classes. In these decomposition theorems Truemper configurations appear both as excluded structures that are convenient to work with, and as structures around which the actual decomposition takes place.

In this survey, in trying to understand what structurally enables efficient recognition and optimization algorithms, we will view different classes of objects and their associated decomposition theorems, through excluded Truemper configurations. We survey the above mentioned classes, as well as other classes closed under taking graph minors (such as cycle-free graphs, outerplanar graphs, series-parallel graphs, etc.) and those closed under taking induced subgraphs (such as hole-free graphs, claw-free graphs, bull-free graphs, even-hole-free graphs, odd-hole-free graphs, graphs that do not contain cycles with a unique chord, ISK4-free graphs, etc.).

Most generally all of the above mentioned classes of objects can be viewed as hereditary graph classes, i.e. classes of graphs closed under taking induced subgraphs. We say that a graph G *contains* a graph F, if F is isomorphic to an induced subgraph of G, and it is *F-free* if it does not contain F. For a family of graphs $\mathcal{F}$ we say that G is *$\mathcal{F}$-free* if G is F-free for every $F \in \mathcal{F}$. So for every hereditary graph class $\mathcal{C}$ there is a family $\mathcal{F}$ of graphs such that $\mathcal{C}$ is precisely the set of graphs that are $\mathcal{F}$-free.

Throughout the paper all graphs are finite and simple. A *hole* in a graph is an induced cycle of length at least 4, and it is *even* or *odd* depending on the parity of its length. A *clique* is a graph in which every pair of nodes are adjacent. A *stable set* (or *independent set*) S in a graph G is a subset of the vertex set of G such that no pair of vertices of S are adjacent. For $S \subseteq V(G)$, $G[S]$ denotes the subgraph of G induced by S, $N(S)$ denotes the set of nodes in $V(G) \setminus S$ with at least one neighbor in S, and $N[S] = N(S) \cup S$.

The remainder of this survey is organised in the following sections.

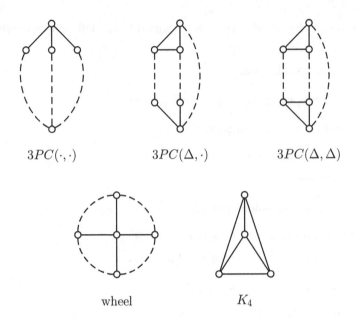

$$3PC(\cdot, \cdot) \qquad\qquad 3PC(\Delta, \cdot) \qquad\qquad 3PC(\Delta, \Delta)$$

wheel K_4

Figure 1: Truemper configurations and K_4.

2   Truemper's Theorem

Theorem 2.1 (Truemper [121]) *Let β be a $\{0,1\}$ vector whose entries are in one-to-one correspondence with the chordless cycles of a graph G. Then there exists a subset F of the edge set of G such that $|F \cap C| \equiv \beta_C$ (mod 2) for all chordless cycles C of G, if and only if for every induced subgraph G' of G that is a Truemper configuration or K_4 (see Figure 1), there exists a subset F' of the edge set of G' such that $|F' \cap C| \equiv \beta_C$ (mod 2), for all chordless cycles C of G'.*

Truemper configurations are depicted in Figure 1, where a solid line denotes an edge and a dashed line denotes a chordless path containing one or more edges. We now define these configurations.

The first three configurations in Figure 1 are referred to as *3-path configurations* ($3PC$'s). They are structures induced by three paths P_1, P_2 and P_3, in such a way that the nodes of $P_i \cup P_j$, $i \neq j$, induce a hole. More specifically, a $3PC(x, y)$ is a structure induced by three paths that connect two nonadjacent nodes x and y; a $3PC(x_1x_2x_3, y)$, where $x_1x_2x_3$ is a triangle, is a structure induced by three paths having endnodes x_1, x_2 and x_3 respectively and a common endnode y; a $3PC(x_1x_2x_3, y_1y_2y_3)$, where $x_1 x_2 x_3$ and $y_1 y_2 y_3$ are two node-disjoint triangles, is a structure induced by three paths P_1, P_2 and P_3 such that, for $i = 1, 2, 3$, path P_i has endnodes x_i and y_i. We say that a graph G contains a $3PC(\cdot, \cdot)$ if it contains a $3PC(x, y)$ for some $x, y \in V(G)$, a $3PC(\Delta, \cdot)$ if it contains a $3PC(x_1 x_2 x_3, y)$ for some $x_1, x_2, x_3, y \in V(G)$, and it contains a $3PC(\Delta, \Delta)$ if it contains a $3PC(x_1 x_2 x_3, y_1 y_2 y_3)$ for some $x_1, x_2, x_3, y_1, y_2,$ $y_3 \in V(G)$. Note that the condition that nodes of $P_i \cup P_j$, $i \neq j$, must induce a hole, implies that all paths of a $3PC(\cdot, \cdot)$ have length greater than one, and at most one path of a $3PC(\Delta, \cdot)$ has length one. In the literature $3PC(\cdot, \cdot)$ is also referred to as *theta* [23], $3PC(\Delta, \cdot)$ as *pyramid* [22], and $3PC(\Delta, \Delta)$ as *prism* [23].

A *wheel* (H, x) consists of a hole H, called the *rim*, and a node x, called the *center*, that has at least three neighbors in H. Finally, a K_4 is a clique on four vertices. We note that in [121] K_4's are also referred to as wheels, but in this paper we choose to separate these two structures. In this survey we will refer to 3-path-configurations and wheels as *Truemper configurations*.

Truemper's interest in this theorem at the time was to obtain a co-NP characterization of balanceable matrices, that are a generalization of regular matrices. An alternative simple proof of Theorem 2.1 is given by Conforti, Gerards and Kapoor in [52], where they also give some of its consequences, such as an easy way to obtain Tutte's characterization of regular matrices.

2.1 Recognizing Truemper configurations

A natural question to ask is whether Truemper configurations can be recognized in polynomial time. These questions in fact arose when people were studying how to construct polynomial time recognition algorithms for even-hole-free graphs and perfect graphs. Observe that if a graph contains a $3PC(\Delta, \Delta)$ or a $3PC(\cdot, \cdot)$ then it must contain an even hole, and if it contains a $3PC(\Delta \cdot)$ then it must contain an odd hole. Even-hole-free graphs and perfect graphs are further discussed in Section 9. We now briefly describe different general techniques that were developed when trying to recognize whether a graph contains a particular Truemper configuration.

In [22] it is shown that detecting whether a graph contains a $3PC(\Delta, \cdot)$ can be done in $\mathcal{O}(n^9)$ time. This algorithm is based on the *shortest-paths detector* technique developed by Chudnovsky and Seymour. The idea of their algorithm is as follows. If G has a $3PC(\Delta, \cdot)$, then it has a $\Sigma = 3PC(\Delta, \cdot)$ with fewest number of nodes. The algorithm "guesses" some vertices of Σ, and then finds shortest paths in G between the guessed vertices that are joined by a path in Σ. If the graph induced by the union of these paths is a $3PC(\Delta, \cdot)$, then clearly G contains a $3PC(\Delta, \cdot)$. If it is not, then it turns out that G is $3PC(\Delta, \cdot)$-free.

Chudnovsky and Seymour [29] show that detecting whether a graph contains a $3PC(\cdot, \cdot)$ can be done in $\mathcal{O}(n^{11})$ time. For this detection problem, the shortest-paths detector technique does not work. The detection of $3PC(\cdot, \cdot)$'s relies on being able to solve a more general problem called the *three-in-a-tree problem* defined as follows: given a graph G and three specified vertices a, b and c, the question is whether G contains a tree that passes through a, b and c. It is shown in [29] that this problem can be solved in $\mathcal{O}(n^4)$ time. What is interesting is that the algorithm for the three-in-a-tree problem is based on an explicit construction of the cases when the desired tree does not exist, and that this construction can be directly converted into the algorithm. As we shall see in this survey, this direct connection between structure and algorithm does not occur so frequently for graph classes closed under taking induced subgraphs. The three-in-a-tree algorithm is quite general, and can be used to solve different detection problems, including the detection of a $3PC(\cdot, \cdot)$, and a $3PC(\Delta, \cdot)$ (this time in $\mathcal{O}(n^{10})$ time).

It turns out that detecting whether a graph contains a $3PC(\Delta, \Delta)$ is NP-complete, as shown by Maffray and Trotignon [94]. Detecting whether a graph contains a wheel remains an open problem.

A number of related detection problems will be looked at throughout this survey. The reader is also referred to [86] for more on detection of induced subgraphs problems.

3   The decomposition method

In the past few decades a number of important results were obtained through the use of decomposition theory, such as a polynomial time recognition algorithm for regular matroids [115] and the proof of the Strong Perfect Graph Conjecture (SPGC) [26] (discussed further in Sections 4 and 9). The power of decomposition is that it allows us to understand complex structures by breaking them down into simpler ones. Once these simpler structures are understood, this knowledge is propagated back to the original structure by understanding how their composition behaves. Decomposition is a general concept that applies to different classes of objects. Here we start by introducing the method in the context of graphs.

In a connected graph G, a subset S of nodes and/or edges is a *cutset* if its removal disconnects G. If S consists only of nodes then it is referred to as a *node cutset*, and if it consists only of edges then it is referred to as an *edge cutset*. A decomposition theorem for a class of graphs C is of the following form.

Decomposition Theorem: *If G belongs to C then G is either "basic" or G has a cutset S for $S \in \mathcal{S}$.*

Depending on what one wants to prove about the class of graphs C using the Decomposition Theorem, "basic" graphs and cutsets in $\mathcal{S}$ have to have adequate properties. For example, the SPGC was proved using the decomposition theorem for Berge graphs [26], by ensuring that "basic" graphs were simple in the sense that the SPGC could be easily proved for them directly, and the cutsets in $\mathcal{S}$ had the property that no minimal imperfect graph could contain them (or if it did it would have to be an odd hole or an odd antihole).

To use a decomposition theorem to recognize a class of graphs C, "basic" graphs need to be simple in the sense that they can be easily recognized, and the cutsets $S \in \mathcal{S}$ need to have the following property. The removal of a cutset S from a graph G disconnects G into two or more connected components. From these components *blocks of decomposition* are constructed by adding some more nodes and edges. A decomposition is C-*preserving* if it satisfies the following: G belongs to C if and only if all the blocks of decomposition belong to C. A recognition algorithm takes a graph G as input and decomposes it using C-preserving decompositions into undecomposable blocks, which are then checked whether they belong to C (which according to the decomposition theorem reduces to checking whether they are basic). The decomposition can be represented with a *decomposition tree* T, whose root is the input graph, and for every non-leaf node H of T, its children in T are the blocks of decomposition of H. In order for such an algorithm to have polynomial complexity we need to ensure that T can be constructed in polynomial time (which in particular means that we can find the cutsets in polynomial time and that we can ensure that the decomposition tree is polynomial in size) and that checking whether a graph is basic can be done in polynomial time.

This is an ideal scenario, which works, for example, for obtaining a recognition algorithm for regular matroids [115]. On the other hand, it does not work, for example, for obtaining a recognition algorithm for perfect graphs. The problem is that for the cutsets from the decomposition theorem in [26], one does not know how to construct the blocks of decomposition that would, at the same time, be class-preserving as well as guarantee polynomiality of the decomposition tree. This problem was first encountered when trying to construct a polynomial time recognition algorithm for balanced matrices. At that time a technique called "cleaning" (i.e. preprocessing the input graph, so that later when the decomposition is applied it would be class-preserving) was developed by Conforti and Rao [54] that enabled them to recognize, in polynomial time, linear balanced matrices. This technique was further developed and used in obtaining decomposition based polynomial time recognition algorithms for balanced matrices [47], balanced $0, \pm 1$ matrices [44], even-hole-free graphs [46, 62], and it was the key to obtaining a recognition algorithm for perfect (in fact Berge) graphs [22].

Decomposition can also be used to construct optimization algorithms. The general paradigm would be as follows: given a decomposition tree T for a graph G obtained by using S-decompositions, for $S \in \mathcal{S}$ (referring to the general decomposition theorem stated above), with the property that for every leaf L of T one can solve an optimization problem (such as coloring or finding the size of the largest clique or a stable set), can we construct an algorithm to solve the problem on G? This general paradigm sometimes works nicely, but most of the time it is difficult to apply to classes whose decomposition theorems use "powerful cutsets".

We next illustrate the ideal scenarios, discussed above, for using a decomposition theorem for constructing a recognition algorithm as well as for obtaining combinatorial optimization algorithms, on the class of triangulated graphs. We close this section by introducing some cutsets that commonly appear in the decomposition theorems we will discuss in this survey.

3.1   Triangulated graphs

A graph is *triangulated* (or *chordal* or *hole-free*) if it does not contain a hole. On this class we will illustrate different techniques for obtaining recognition and combinatorial optimization algorithms. Although, as we shall see, there are more efficient methods for obtaining algorithms for triangulated graphs, we will start by describing the use of the general decomposition method, because it is ideally illustrated on this class and it generalizes to other more complex classes of objects. First, it is simple to obtain the following decomposition theorem for triangulated graphs. A node set $S \subseteq V(G)$ is a *clique cutset* of G, if S is a node cutset of G and it induces a clique in G.

Theorem 3.1 (Dirac [65]) *If G is a connected triangulated graph, then G is either a clique or G has a clique cutset.*

Proof Suppose that G is not a clique. Then G clearly has a node cutset. Let S be a minimal node cutset of G, and let C_1 and C_2 be two connected components of $G \setminus S$. Suppose S is not a clique and let u and v be two non-adjacent vertices of S. Since S is minimal, both u and v have a neighbor in both C_1 and C_2. Hence, for $i = 1, 2$, there exists a chordless path P_i from u to v whose interior vertices belong to C_i. But then $P_1 \cup P_2$ induces a hole, a contradiction. □

Let S be a clique cutset of a graph G, and let $C_1, \ldots, C_k$ be the connected components of $G \setminus S$. We define the *blocks of decomposition by a clique cutset* S to be graphs $G_i = G[C_i \cup S]$, for $i = 1, \ldots, k$. It is now easy to see that this definition of blocks is class-preserving for the class of triangulated graphs.

Theorem 3.2 G *is triangulated if and only if all the blocks of decomposition by a clique cutset are triangulated.*

Proof Since the blocks of decomposition are all induced subgraphs of G, if G is triangulated then so are all the blocks. Now suppose that all the blocks $G_1, \ldots, G_k$ are triangulated, but that G contains a hole H. Since H cannot be contained in any of the blocks, it must contain nodes of at least two connected components $C_1, \ldots, C_k$. Consequently H contains at least two nodes of S that are not consecutive on H, which contradicts the assumption that S is a clique. □

Theorems 3.1 and 3.2 actually give us a complete structure theorem for the class of triangulated graphs, i.e. they show how (connected) triangulated graphs can be built starting from cliques, gluing them together through cliques (clique composition), and all graphs built this way are triangulated. Such structure theorems are stronger than the usual decomposition theorems, and are quite rare for classes of graphs closed under taking induced subgraphs.

We now turn to using a decomposition theorem to construct algorithms. We construct a *decomposition tree* T using clique cutsets as follows: the root of T is our input graph G; for every internal node G' of T, the children of G' are the blocks of decomposition of G' by some clique cutset; and the leaves of T are graphs that have no clique cutset. An $\mathcal{O}(nm)$ algorithm is given in [126] for finding a clique cutset in a graph, and a simple counting argument shows that the number of nodes in T is bounded by $\mathcal{O}(n^2)$, giving an $\mathcal{O}(n^3 m)$ algorithm for constructing T. As we shall see one can actually do better than that.

First observe that, by Theorem 3.1 and Theorem 3.2, the input graph G is triangulated if and only if all the leaves of T are cliques. One can now recognize triangulated graphs (in the same time it takes to construct

T) as follows: construct a decomposition tree T using clique cutsets, check whether all the leaves of T are cliques, if yes then G is triangulated, and otherwise it is not.

Clique cutsets have another interesting property, that is quite useful for constructing algorithms. We say that S is an *extreme clique cutset* if for some i, the block of decomposition $G_i = G[C_i \cup S]$ has no clique cutset. We say that G_i is an *extreme block*. It turns out that every graph that has a clique cutset has an extreme clique cutset. This is a very useful property, that not many types of cutsets have.

Lemma 3.3 *If a graph G has a clique cutset, then it has an extreme clique cutset.*

Proof Let S be a clique cutset of G such that out of all clique cutsets of G a connected component C of $G \setminus S$ is smallest possible. Suppose that $G' = G[C \cup S]$ has a clique cutset S'. Since S is a clique, there is a connected component C' of $G' \setminus S'$ such that $C' \cap S = \emptyset$. Clearly, $S' \cap C \neq \emptyset$, and hence C' is a proper subset of C. In particular $|C'| < |C|$. Also C' is a connected component of $G \setminus S'$, contradicting our choice of S and C. So G' has no clique cutset, and hence S is an extreme clique cutset of G. □

We will now use extreme clique cutsets to decompose. Suppose that S is an extreme clique cutset with G_i being an extreme block. This time we will construct only two blocks of decomposition: $G_B = G_i = G[C_i \cup S]$ and $G_A = G \setminus C_i$. We now construct an *extreme decomposition tree* T using clique cutsets as follows: the root of T is our input graph G; for every internal node G' of T, the children of G' are the blocks of decomposition G'_A and G'_B of G' by some extreme clique cutset; and the leaves of T are graphs that have no clique cutset. Note that every G'_B is a leaf, so T is a binary tree in which every internal node has a child that is a leaf.

It turns out that such an extreme decomposition tree using clique cutsets can be built in $\mathcal{O}(nm)$ time [117]. This relies on being able to find a particular ordering of vertices, called a minimal elimination ordering, in $\mathcal{O}(nm)$ time, which is done in [112] using lexicographic breadth-first search (Lex-BFS).

For a graph G, let T be an associated extreme decomposition tree using clique cutsets, and let $L_1, \ldots, L_t$ be the leaves of T. We now consider how T can be used to construct combinatorial optimization algorithms for maximum weight clique, vertex coloring and maximum weight independent set problems, assuming that these problems can be efficiently solved on the leaves of T (see [117]). For any graph G, let $\omega(G)$ denote the weight of a maximum weighted clique of G, $\chi(G)$ the chromatic number of G, and $\alpha(G)$ the weight of a maximum weighted stable set of G.

Since any clique of G is contained in one of the blocks of decomposition by a clique cutset, it follows that $\omega(G) = \max\{\omega(L_1), \ldots, \omega(L_t)\}$. And hence the problem of finding a maximum weight clique reduces to doing it on the leaves. Similarly, the coloring problem reduces to coloring the leaves, since any k-colorings of the blocks of decomposition by a clique cutset S can be combined into a k-coloring of the graph by renaming the colors in the blocks so that they agree on S. In particular $\chi(G) = \max\{\chi(L_1), \ldots, \chi(L_t)\}$. For both of these problems it is not essential that T is an extreme decomposition tree, but it gives a better time bounds if it is.

For solving the maximum weight independent set problem in polynomial time (assuming this is possible to do on the leaves of T) it actually does matter that T is an extreme decomposition tree. Let H be an interior node of T and let H_A and H_B be its children (where H_B is a leaf of T), obtained by decomposing H with clique cutset S. Let w be the weight function defined on the nodes of H. For every $u \in S$ redefine the weight of u in H_A to be $w(u) + \alpha(H[V(H_B) \setminus N_{H_B}(u)]) - \alpha(H_B \setminus S)$. Let H_A' be the resulting weighted graph. Then it is easy to see that $\alpha(H) = \alpha(H_A') + \alpha(H_B \setminus S)$. So the independent set problem for H reduces to recursively solving the independent set problem on block H_A' (with newly defined weights). Note that computing the weights for H_A' and computing $\alpha(H_B \setminus S)$ amounts to solving $|S| + 1$ independent set problems on H_B. Since H_B is a leaf this is not a problem since it requires no further recursion, but if we were not using an extreme decomposition tree this method could lead to an exponential explosion.

Note that if the input graph G is triangulated, then all the leaves of T are cliques, and hence all of the above mentioned problems can be solved on G in the same time it takes to construct T.

Triangulated graphs are in fact characterized by having very special types of minimal elimination orderings that can be found more efficiently. A *perfect elimination ordering* is an ordering of vertices $v_1, \ldots, v_n$ such that v_i is simplicial (where a vertex is *simplicial* if its neighborhood induces a clique) in $G[v_i, \ldots, v_n]$.

Theorem 3.4 (Dirac [65]) *G is triangulated if and only if G has a perfect elimination ordering.*

Proof Suppose G has a perfect elimination ordering $v_1, \ldots, v_n$, but is not triangulated. Let H be a hole of G, and let v_i be a smallest indexed vertex of H. Then clearly v_i has two nonadjacent neighbors in $G[v_i, \ldots, v_n]$, a contradiction. To prove the converse, assume G is triangulated. If G is a clique then any ordering of vertices is a perfect elimination ordering. Otherwise by Theorem 3.1, G has a clique cutset, and by Lemma 3.3 it has an extreme clique cutset S. So for some connected component C of $G \setminus S$, $G' = G[C \cup S]$ has no clique cutset. By Theorem 3.1, G' is a clique, and hence any vertex $u \in C$ is simplicial in G' and hence in G as well. Let $u = v_1$ and inductively construct the remainder of a perfect elimination ordering.   □

Note that if $v_1, \ldots, v_n$ is a perfect elimination ordering of G and G is not a clique, then $N(v_1)$ is an extreme clique cutset of G separating a single vertex v_1 from the rest of the graph.

In [112] it is shown that a perfect elimination ordering of a triangulated graph can be found in linear time using Lex-BFS, and more generally, by using Lex-BFS to construct this particular ordering and checking that in fact the ordering constructed is a perfect elimination ordering one gets a linear time recognition algorithm for triangulated graphs. It follows that all of the above mentioned optimization problems can be solved in linear time for triangulated graphs. Note that triangulated graphs can also be optimally colored, in linear time, by applying a greedy coloring algorithm to the vertices in the reverse of a perfect elimination ordering.

3.2 Common cutsets

Here we introduce some cutsets that commonly appear in decompositions of graph classes closed under taking induced subgraphs.

We start by introducing several edge cutsets. First observe that a disconnected graph can be defined as a graph that has a partition (X_1, X_2) of its vertex set satisfying: there are no edges between X_1 and X_2; and for $i = 1, 2$, $|X_i| \geq 1$. This concept can be generalized by controlling the kinds of edges that go across the partition.

A partition (X_1, X_2) of the vertex set of a graph G is a *general join* if, for $i = 1, 2$, there exist disjoint $A_i, B_i, C_i \subseteq X_i$ satisfying the following: every vertex of A_1 is adjacent to every vertex of A_2, every vertex of B_1 is adjacent to every vertex of B_2, every vertex of C_1 is adjacent to every vertex of $A_2 \cup B_2$, every vertex of C_2 is adjacent to every vertex of $A_1 \cup B_1$, and there are no other edges between X_1 and X_2. Sets X_1 and X_2 are the two *sides* of the general join. We say that $(X_1, X_2, A_1, B_1, C_1, A_2, B_2, C_2)$ is a *split* of a general join (X_1, X_2). For $i = 1, 2$ sets A_i, B_i, C_i are called the *special sets* of general join (X_1, X_2).

A *general k-join*, for $k = 0, 1$, is a general join with split $(X_1, X_2, A_1, B_1, C_1, A_2, B_2, C_2)$ such that for $i = 1, 2$ exactly k of the sets A_i, B_i, C_i are nonempty, there are at least k edges going from X_1 to X_2, and $|X_i| \geq k+1$. A *general 2-join* is a general join with split $(X_1, X_2, A_1, B_1, C_1, A_2, B_2, C_2)$ such that for $i = 1, 2$ at least 2 of the sets A_i, B_i, C_i are nonempty, and $|X_i|$ is greater than the number of nonempty sets among A_i, B_i, C_i. A general 2-join was first introduced in [41]. General joins generalize some of the previously introduced edge cutsets.

A general 0-join corresponds to a disconnected graph. A general 1-join is exactly the *1-join* (or *join* or *split decomposition*) as introduced by Cunningham and Edmonds [59]. A related notion is that of a *homogeneous set* (or *module*) of a graph G, that is a proper subset S of $V(G)$ of at least two vertices such that every vertex not in S is adjacent to either all or none of the vertices in S. Note that if $V(G) \setminus S \geq 2$, then homogeneous set corresponds to a 0-join, or 1-join with split $(X_1, X_2, A_1, \emptyset, \emptyset, A_2, \emptyset, \emptyset)$ such that $X_1 \setminus A_1 = \emptyset$.

A general 2-join with $C_1 = C_2 = \emptyset$ and all the other special sets nonempty is called a *2-join* and it was first introduced by Cornuéjols and Cunningham [56]. A general 2-join with $B_1 = C_2 = \emptyset$ (or equivalently $A_1 = A_2 = \emptyset$) and all the other special sets nonempty is called a *N-join*. A general 2-join with $C_2 = \emptyset$ (or equivalently $B_2 = \emptyset$) and all the other special sets nonempty is called a *M-join*. A general 2-join with all special sets nonempty is called a *6-join* and it was first introduced by Conforti, Cornuéjols, Kapoor and Vušković [44].

General joins also generalize the notion of a *homogeneous pair* introduced by Chvátal and Sbihi [39]. A homogeneous pair in a graph G is a pair $\{Q_1, Q_2\}$ of disjoint sets of vertices of G such that: every vertex of $V(G) \setminus (Q_1 \cup Q_2)$ is adjacent to either all vertices of Q_1 or to no vertex of Q_1; every vertex of $V(G) \setminus (Q_1 \cup Q_2)$ is adjacent to either all vertices of Q_2 or to no vertex of Q_2; $|Q_1| \geq 2$ or $|Q_2| \geq 2$; and $|V(G) \setminus (Q_1 \cup Q_2)| \geq 2$. Note that a homogeneous pair in a graph with no homogeneous set is a special case of a general 2-join, where $A_1, B_1 \neq \emptyset$ and $X_1 \setminus (A_1 \cup B_1) = \emptyset$.

Furthermore, there is a correspondence between k-separations, $k = 1, 2, 3$, in binary matroids and general joins. A 1-separations corresponds to a general 0-join, a 2-separation corresponds to a general 1-join and a 3-separation corresponds to a general 2-join. This correspondence is discussed in Section 4.

We now consider some commonly appearing node cutsets. Let S be a node cutset of a graph G. S is a *k-node cutset* if $|S| = k$. We say that S is a *small node cutset* if $|S|$ is bounded by some fixed integer k. Recall that S is a *clique cutset* if S induces a clique in G.

A node set $S \subseteq V(G)$ is a *k-star* if S is comprised of a clique C (the *clique center* of S) of size k and nodes with at least one neighbor in C, so $S \subseteq N[C]$. A *k-star cutset* is a k-star S that is a node cutset. A 1-star cutset is also referred to as a *star cutset*, a 2-star cutset as a *double star cutset*, and a 3-star cutset as a *triple star cutset*.

Here is another generalization of a star cutset, that is a special case of a double star cutset. A node cutset S is a *skew cutset* if there exists a partition (S_1, S_2) of S such that every node of S_1 is adjacent to every node of S_2. Star cutsets and skew cutsets were first introduced by Chvátal [38].

In trying to understand why these cutsets appear "naturally" in decomposition theorems, we first observe that with clique cutsets one can only separate vertices that are not contained in a hole. When we need to break a hole, we can either use a node that has neighbors on this hole as a center of a star cutset (or more generally a k-star cutset), or when no such node exists we can hope for example that two edges of this hole will extend to a 2-join that separates the hole.

4   Regular and balanced matrices

A matrix is *totally unimodular* if every square submatrix has determinant equal to $0, \pm 1$. In particular, all entries of a totally unimodular matrix are $0, \pm 1$. A $0, 1$ matrix is *balanced* if it does not contain a square submatrix of odd order with two 1's per row and per column. This notion was introduced by Berge [6], and it was extended to $0, \pm 1$ matrices by Truemper [121]. A $0, \pm 1$ matrix is *balanced* if, in every square submatrix with exactly two nonzero entries per row and column, the sum of the entries is a multiple of 4. Note that the class of $0, \pm 1$ balanced matrices properly includes totally unimodular matrices. All these matrices have important polyhedral properties, see for example [55]. In this section we describe decomposition theorems that were the key to obtaining polynomial time recognition algorithms for all these classes of matrices.

4.1   Decomposition of regular matroids

What enabled the structural understanding of totally unimodular matrices, which led to their polynomial time recognition, was the translation of the property into the realm of matroids, and the use of existent powerful tools from matroid theory.

A $0, 1$ matrix is *regular* if its nonzero entries can be signed $+1$ or -1 so that the resulting matrix is totally unimodular. Camion [10] observed that this signing is unique up to multiplying rows and columns by -1, and gave a simple signing algorithm, from which it follows that the recognition of totally unimodular matrices reduces to the recognition of regular matrices. This shift to regular matrices allows for the focus on the structure of the pattern of zero/nonzero entries.

$$\begin{pmatrix} 1 & 1 & 0 & 1 \\ 1 & 0 & 1 & 1 \\ 0 & 1 & 1 & 1 \end{pmatrix}$$

Figure 2: Partial representation matrix of the Fano matroid.

Let M be a binary matroid and $X \subseteq V(M)$ a base of M. The *partial representation* of M with respect to X is the 0,1 matrix $A(M)$ with rows indexed by the elements of X, columns indexed by the elements of $Y = V(M) \setminus X$, and $a_{xy} = 1$ if and only if x belongs to the unique circuit contained in $X \cup \{y\}$. Note that if $A(M)$ is a partial representation of a binary matroid M, then $(I, A(M))$ is a binary representation of M.

A binary matroid is *regular* if all of its partial representation matrices are regular. Let A be a partial representation matrix of a binary matroid, i.e. rows of A are indexed by a base of the matroid. One can always go from one partial representation of a binary matroid to another by using GF(2)-pivoting and row and column permutations. Pivoting over GF(2) consist in replacing $A = \begin{pmatrix} 1 & y \\ x & D \end{pmatrix}$ by $\tilde{A} = \begin{pmatrix} 1 & y \\ x & D + xy \end{pmatrix}$. It can be shown that if A is a regular matrix, then so is $\tilde{A}$, and hence it is the case that for a binary matroid either all or none of its partial representation matrices are regular.

The matrix in Figure 2 is not regular, and is a partial representation matrix of the *Fano matroid* F_7. The transpose of this matrix is a partial representation of the dual of the Fano matroid F_7^*. Let M be a binary matroid and A a partial representation of M. Any submatrix of A is a partial representation of a binary matroid M'. A matroid M' obtained from M in this way is a *minor* of M. A convenient way to work with regular matroids is provided by the following excluded minors characterization.

Theorem 4.1 (Tutte [124]) *A binary matroid is regular if and only if it has no minor isomorphic to F_7 or F_7^*.*

What enabled the polynomial time recognition of regular matroids, and hence totally unimodular matrices, is the following decomposition theorem. Let M be a matroid defined by a finite ground set $V(M)$ and a family $E(M)$ of subsets of $V(M)$ that are the independent sets of M. The *rank* $r(U)$ of a set $U \subseteq V(M)$ is the maximum cardinality of an independent set contained in U. A *k-separation* of M is a partition (U_1, U_2) of $V(M)$ such that $|U_1| \geq k$, $|U_2| \geq k$ and $r(U_1) + r(U_2) \leq r(V(M)) + k - 1$.

Theorem 4.2 (Seymour [115]) *A regular matroid is either graphic, co-graphic or R_{10} (a certain 10-element matroid), or it has a k-separation, for $k = 1, 2, 3$.*

This theorem leads to a decomposition based polynomial time recognition algorithm for regular matroids in the following way. First of all, 1-, 2-and 3-separations can be found in polynomial time (see [123] Section 8.4). For 1-, 2- and 3-separation, blocks of decomposition can be constructed that are regularity-preserving and lead to a linear size of the decomposition tree (see for example [55]). Finally, by Theorem 4.2, it just remains to check whether the leaves of this decomposition tree are R_{10}, graphic or co-graphic matroids, which can be done in polynomial time (see for example [123] Section 10.6). By Camion's signing algorithm it follows that totally unimodular matrices can be recognized in polynomial time. We note that before Seymour's decomposition approach no polynomial time recognition algorithm for totally unimodular matrices was known. The fastest known algorithm for testing total unimodularity is the $\mathcal{O}(n + m)^3$ algorithm of Truemper [122], where the input is an $n \times m$ real matrix. The algorithm uses Seymour's decomposition theorem, but does not blindly search for 3-separations as described above. Instead it searches for 3-separations by starting with particular minors that have 3-separation that should extend to the entire matroid.

This decomposition approach was later extended to recognition algorithms for other classes of matrices and graphs (such as balanced matrices and perfect graphs), but as we shall see, with many more complications. To relate these results, we close this section by translating the work described above into graphs.

Let A be a $0, 1$ matrix. A can be thought of as a node-node incidence matrix of a bipartite graph, which we denote with $G(A)$ and call the *bipartite graph representation of A*. We say that a bipartite graph $G(A)$ is *regular* if A is regular. Pivoting on an entry a_{ij} of A corresponds to the following operation on $G(A)$: let ij be the edge of $G(A)$ that corresponds to the pivot element, then $G(\tilde{A})$ is obtained from $G(A)$ by complementing the edges between $N(i) \setminus \{j\}$ and $N(j) \setminus \{i\}$. We refer to this operation as *pivoting on the edge ij*. Note that the bipartite representation of the matrix in Figure 2 is a wheel whose rim is of length 6 (in particular, the center of the wheel has three neighbors on the rim and they are all on the same side of the bipartition). Let us call this wheel *Fano wheel*. Theorem 4.1 now translates into the following:

A bipartite graph is regular if and only if it cannot be transformed into a Fano wheel by a sequence of edge pivots and/or node deletions.

Let G be a bipartite graph. Since G is bipartite, the only Truemper configurations that G can possibly have are $3PC(\cdot,\cdot)$'s and wheels. Let (H,x) be a wheel. Suppose that x has more than 3 neighbors on H. It can easily be seen that if we pivot on an edge xx_i, where x_i is a neighbor of x on H, we get a wheel (H',x) such that x has fewer neighbors on H' than on H. Now suppose that a sector S of (H,x) is of length greater than 2, and let uv be an interior edge of that sector. If we pivot on uv we get a wheel (H',x) that has all the sectors of (H,x) except for S, and the sector S' of (H',x) that corresponds to S in (H,x) is shorter than S. So clearly, a wheel (H,x) such that x has an odd number of neighbors on H, can be transformed into a Fano wheel by a sequence of edge pivots and node deletions. Let us call such a wheel in a bipartite graph an *odd bipartite wheel*. Similarly it can be seen that a $3PC(u,v)$ where u and v are on opposite sides of the bipartition, can be transformed into a Fano wheel. Let us call such a $3PC(u,v)$ a *3-odd-path configuration*. Therefore regular bipartite graphs cannot contain odd bipartite wheels nor 3-odd-path configurations (as well as all the other configurations that can be transformed into a Fano wheel). In other words, out of all the Truemper configurations, regular bipartite graphs may contain only wheels (H,x) such that x has an even number of neighbors on H, and $3PC(u,v)$'s such that u and v are on the same side of the bipartition.

Let M be a binary matroid, and consider a k-separation (U_1,U_2) where $r(U_1)+r(U_2) = r(|V(M)|)+k-1$. Let X_2 be a maximal independent subset of U_2, and enlarge X_2 by subset X_1 of U_1 to a base of M. The partial representation matrix A of M w.r.t. base $X_1 \cup X_2$ is $A = \begin{array}{c} X_1 \\ X_2 \end{array} \left(\begin{array}{cc} A_1 & 0 \\ D & A_2 \end{array} \right)$ where the sum of rows and columns of A_i is at least k, for $i = 1, 2$, and the rank of D over $GF(2)$ is $k-1$. Observe that when $k = 1$, then $D = 0$, and hence $G(A)$ corresponds to a disconnected graph, or a general 0-join. By similarly analyzing the possibilities for matrix D, it turns out that a 2-separation corresponds to a general 1-join in $G(A)$, and a 3-separation corresponds to a general 2-join in $G(A)$. (As mentioned in Section 3.2 different forms of general joins were introduced by different authors, interestingly without being aware of this correlation. They were notions that emerged naturally when dealing with different graph classes.) Therefore, Theorem 4.2 translated states that *regular bipartite graphs can be decomposed by general k-joins, for $k = 0, 1, 2$.*

4.2   Decomposition of balanced matrices

We immediately switch from $0,1$ matrices to their bipartite graph representations. So a bipartite graph is *balanced* if it does not contain a hole of length 2 (mod 4). A *signed bipartite graph* is a bipartite graph with edge weights $+1$ and -1. A signed bipartite graph is *balanced*, if it does not contain a hole of weight 2 (mod 4). A bipartite graph is *balanceable* if there exists a signing of its edges so that the resulting signed bipartite graph is balanced.

If a graph is a balanceable bipartite graph, there exists an easy signing algorithm that signs it into a balanced signed bipartite graph (since if such a signing exists, it is essentially unique and easy to find by Camion's signing algorithm [10], see also [50]). So the recognition of signed balanced bipartite graphs reduces to the recognition of balanceable bipartite graphs.

Clearly the class of balanceable bipartite graphs is closed under taking induced subgraphs, but it is not closed under edge pivoting. Consider for example a graph G that consists of a $3PC(x, y)$ where all the paths have length 3 together with an edge xy. This graph is balanceable, but if we pivot on the middle edge of any of the paths, edge xy disappears and we get the original 3-odd-path configuration $3PC(x, y)$, which is not balanceable (since no matter how we sign its edges two of the paths will have weights that are congruent (mod 4) and would hence induce a hole of weight 2 (mod 4)). Observe that it also follows that G is not regular. So balanceable bipartite graphs properly contain regular bipartite graphs.

The following theorem characterizes balanceable bipartite graphs in terms of excluded induced subgraphs, and provides a convenient way to work with this class.

Theorem 4.3 (Truemper [121]) *A bipartite graph is balanceable if and only if it does not contain an odd bipartite wheel nor a 3-odd-path configuration.*

The first known polynomial time recognition algorithm for balanced matrices (or equivalently, balanced bipartite graphs) is given by Conforti, Cornuéjols and Rao [47], and it is based on the following decomposition theorem.

Theorem 4.4 (Conforti, Cornuéjols, Rao [47]) *If a bipartite graph is balanced but not totally unimodular, then it has a double star cutset.*

These results were later extended to balanceable bipartite graphs. The first known polynomial time recognition algorithm for balanceable bipartite graphs is given by Conforti, Cornuéjols, Kapoor and Vušković [44], and it is based on the following decomposition theorem.

Theorem 4.5 (Conforti, Cornuéjols, Kapoor, Vušković [44]) *A connected balanceable bipartite graph is either strongly balanceable or R_{10} (a certain 10-element graph), or has a 2-join, a 6-join or a double star cutset.*

We observe that the 2-joins in the above theorem are in fact of a special type that we call connected non-path 2-joins and describe in Section 9. The major difficulty in using Theorems 4.4 and 4.5 to construct decomposition based recognition algorithms is the double star cutsets. For the 2-join and 6-join it is possible to construct blocks of decomposition that are balancedness-preserving and keep the decomposition tree polynomial in size (see [44]), but it is not clear how to do that for the double star cutset. The double star cutsets in Theorems 4.4 and 4.5 are actually more structured, but that does not help, the problem appears even when trying to use just the star cutsets. Consider for example an odd wheel (H, x) whose every sector is of length 2. This wheel can be decomposed with a star cutset $S = N[x]$. If we construct blocks of decomposition as we did for the clique cutset decomposition in Section 3.1, we get that all the blocks of decomposition are balanced (or balanceable), but (H, x) is not. (We observe that it was precisely for the decomposition of wheels that double star cutsets are needed in the proofs of Theorems 4.4 and 4.5.) One might add some more information to the blocks to make the decomposition balancedness-preserving, but then the decomposition tree blows up in size. To deal with this problem, a technique called *cleaning* was developed by Conforti and Rao [54], which enabled them to recognize linear balanced matrices in polynomial time. This technique was further developed and used in obtaining decomposition based polynomial time recognition algorithms for balanced matrices [47], balanced $0, \pm 1$ matrices [44], and a new level of cleaning had to be developed for recognition of even-hole-free graphs [46, 62], that was also used in the cleaning for recognition of perfect graphs [22].

We now describe the cleaning procedure in the context of its use for recognizing balanced bipartite graphs. A hole of length 2 (mod 4) is called an *unbalanced hole*. Given an input graph G, the cleaning procedure produces, in polynomial time, a clean graph G', such that G is balanced if and only if G' is balanced, and if G contains an unbalanced hole then G' contains a clean unbalanced hole (i.e. an unbalanced hole for which there are no nodes outside the hole that have problematic neighbors on the hole, which can be used as centers of star cutsets to break the hole). This is done by studying the structure of a smallest unbalanced hole in a graph, showing that such a hole contains a fixed number of nodes that see all the problematic neighbors of the hole, and using that information to remove them. Once we have a clean graph G', decomposition by (double) star cutsets can be applied safely, since it will now be balancedness-preserving, as well as lead to a polynomial decomposition tree.

Using Theorem 4.5 Chudnovsky and Seymour prove the following decomposition theorem for balanceable bipartite graphs, resolving a conjecture from [44]. We observe that this decomposition theorem does not help with the recognition algorithm, since the double star cutsets are still used.

Theorem 4.6 (Chudnovsky and Seymour [28]) *A balanceable bipartite graph that is not regular has a double star cutset.*

The following conjecture is the last unresolved conjecture about balanced (balanceable) bipartite graphs in Cornuéjols' book [55].

Conjecture 4.7 (Conforti and Rao [53]) *Every balanced bipartite graph contains an edge that is not the unique chord of a cycle.*

This conjecture was proved recently for linear balanced bipartite graphs and balanced bipartite graphs whose maximum degree is at most 3 in [4] using the idea of extreme decomposition (in fact in this paper the analogous form of this conjecture for balanceable bipartite graphs is proved for 4-hole-free balanceble bipartite graphs and subcubic balanceable bipartite graphs).

5   Classes closed under minor taking

In this section we briefly consider graph classes that are not only closed under deletion of vertices, but also under deletion and contraction of edges, i.e. classes of graphs that are closed under minor taking. Some important examples of such classes are cycle-free graphs (or forests), series-parallel graphs, planar graphs or more generally classes of graphs embeddable in any fixed surface.

A graph H is a *minor* of a graph G, if it is isomorphic to a graph that can be produced from G by a sequence of contracting edges, and deleting vertices and edges. A class of graphs $\mathcal{G}$ is *minor-closed*, if for every $G \in \mathcal{G}$, every minor of G also belongs to $\mathcal{G}$. Trivially, every minor-closed class of graphs can be characterized by a list of excluded minors, by just listing all the graphs that are not in the class. Wagner conjectured that this can always be done by a *finite* list of excluded minors. This famous conjecture was proved by Robertson and Seymour in their monumental work on revealing the structure of minor-closed families of graphs, with far reaching algorithmic consequences.

Theorem 5.1 (Robertson and Seymour [111]) *Every minor-closed class of graphs can be characterized by a finite family of excluded minors.*

The proof of this theorem is based on the following structural characterization: if a minor-closed class of graphs does not contain all graphs, then every graph in it is "glued" together in a tree-like fashion from graphs that can almost be embedded in a fixed surface. To be more specific we need to introduce the concept of tree-decomposition [108]. A *tree-decomposition* of a graph G is a pair (T, W), where T is a tree and $W = (W_t : t \in V(T))$ is such that:

(i) $\cup_{t \in V(T)} W_t = V(G)$, and every edge of G has both endnodes in some W_t, and

(ii) if $t, t', t'' \in V(T)$ and t' lies on the path from t to t'' in T, then $W_t \cap W_{t''} \subseteq W_{t'}$.

The *width* of (T, W) is the $\max\{|W_t| - 1 : t \in V(T)\}$, and the *tree-width* of G is the least integer k such that G has a tree-decomposition of width k.

Theorem 5.2 (Robertson and Seymour [109]) *For every planar graph H there is an integer $k > 0$ such that if a graph is H-minor-free, then its tree-width is at most k.*

In other words, if a graph does not contain some planar graph as a minor, then it has bounded tree-width, and hence it can be constructed from bounded sized graphs by "gluing" them together in a tree-like structure. In [110] an analogous construction is given for H-minor-free graphs in general, starting with graphs embedded in a connected closed surface with genus at most k, adding more nodes in a specified way, and "gluing" such pieces together in a tree-like fashion. This time the pieces that are "glued on" are not necessarily of bounded size, but the parts that are being glued over are.

This structural characterization leads to an $\mathcal{O}(n^3)$ algorithm to test whether a graph G is H-minor-free (although there is a constant factor that depends superpolynomially on the size of G). Together with Theorem 5.1 we get the following algorithmic consequence.

Theorem 5.3 *Every minor-closed class of graphs can be recognized in polynomial time.*

This theoretically beautiful result, has its practical shortcomings. Unless a minor-closed class of graphs is given by its finite list of excluded minors, from Theorem 5.1 we only get an existence of a polynomial time algorithm.

There are further algorithmic consequences for graph classes that have tree-decompositions of bounded tree-width, as is the case for example with any minor-closed family that does not include all planar graphs (by Theorem 5.2). Many problems that are NP-hard in general, such as the independent set problem or the coloring problem, can be solved by dynamic programming in linear time when the input graph has bounded tree-width. In fact, each problem that can be formulated in Monadic Second Order Logic can be solved in linear time on graphs of bounded tree-width [57].

In the terminology used in this survey, a tree-decomposition of width k corresponds to decomposing a graph into blocks of size at most $k+1$ by a sequence of "non-crossing" node cutset decompositions, where the cutsets are all of size at most k. Indeed, let (T, W) be a tree-decomposition of a graph G, let $t_1 t_2$ be an edge of T, and for $i = 1, 2$, let T_i be the subgraph of $G \setminus t_1 t_2$ that contains t_i. Then $W_{t_1} \cap W_{t_2}$ is a cutset of G that separates $\cup_{t \in V(T_1)} \setminus (W_{t_1} \cap W_{t_2})$ from $\cup_{t \in V(T_2)} \setminus (W_{t_1} \cap W_{t_2})$. Clearly the size of all such cutsets is at most the width of (T, W). Let us now say that for $A, B \subseteq V(G)$, (A, B) is a *separation* of G if $A \cup B = V(G)$ and there are no edges between $A \setminus B$ and $B \setminus A$. Two separations (A, B) and (C, D) *do not cross* if one of the following holds: $A \subseteq C$ and $B \supseteq D$, or $A \subseteq D$ and $B \supseteq C$, or $A \supseteq C$ and $B \subseteq D$, or $A \supseteq D$ and $B \subseteq C$. So a tree-decomposition corresponds to a family of cross-free separations of a graph G.

As we shall see in this survey, we cannot hope for such strong structure results, with sweeping algorithmic consequences, for graph classes that are closed just under vertex deletion. Their structure is a lot more general, so that much stronger cutsets are needed for their decomposition which makes it a lot more difficult to make use of in algorithms. On the other hand Geelen, Gerards and Whittle have worked on generalizing results and techniques of Robertson and Seymour's Graph Minor Theory to matroids representable over finite fields, see [74]. They have shown that binary matroids are well-quasi-ordered by minors, and that any minor-closed property can be tested in polynomial time for binary matroids.

6 $(3PC(\cdot, \cdot), 3PC(\Delta, \cdot), 3PC(\Delta, \Delta), \text{wheel})$-**free graphs**

Cycle-free graphs are an example of a graph class that does not contain any of the Truemper configurations. This class of graphs is closed under minor taking and is in fact the class of K_3-minor-free graphs. The graphs in this class have tree-width at most 1. Outerplanar graphs (or $(K_4, K_{2,3})$-minor-free graphs), generalize cycle-free graphs and also do not contain any of the Truemper configurations. Their tree-width is at most 2, meaning that they can be decomposed by a sequence of non-crossing node cutsets of size at most 2 into cliques of size at most 3. Triangulated, or hole-free graphs, are another generalization of cycle-free graphs, that are not closed under minor taking, but still have the property of not containing any of the Truemper configurations. They can be decomposed by clique cutsets into cliques (as we have discussed in Section 3.1). The following class introduced in [42] generalizes all these classes of graphs.

Let γ be a $\{0,1\}$ vector whose entries are in one-to-one correspondence with the holes of a graph G. G is *universally signable* if for all choices of vector γ, there exists a subset F of the edge set of G such that $|F \cap H| \equiv \gamma_H$ (mod 2), for all holes H of G. From Theorem 2.1 it is easy to obtain the following characterization of universally signable graphs in terms of forbidden induced subgraphs.

Theorem 6.1 ([42]) *A graph is universally signable if and only if it is* $(3PC(\cdot,\cdot), 3PC(\Delta,\cdot), 3PC(\Delta,\Delta), wheel)$-*free.*

This characterization of universally signable graphs is then used to obtain the following decomposition theorem.

Theorem 6.2 (Conforti, Cornuéjols, Kapoor, Vušković [42]) *A connected* $(3PC(\cdot,\cdot), 3PC(\Delta,\cdot), 3PC(\Delta,\Delta), wheel)$-*free graph is either a clique or a hole or has a clique cutset.*

By the discussion in Section 3.1 it is clear how Theorem 6.2 can be used to construct efficient decomposition based algorithms for the recognition of the class, for finding the size of a largest clique, or independent set, and coloring the class. From Theorem 6.2 it is easy to deduce that every universally signable graph has a vertex that is simplicial or of degree 2. Recently it was shown in [2] that using LexBFS one can find in linear time an ordering of vertices $v_1, \ldots, v_n$ of a universally signable graph G such that for $i = 1, \ldots, n$, v_i is simplicial or of degree 2 in $G[\{v_1, \ldots, v_i\}]$. This implies a linear time robust algorithm for the maximum weight clique and coloring problems.

As we shall see in the following sections, once Truemper configurations are allowed to appear in a graph class, one will need more complex cutsets to decompose the class.

7 $(3PC(\Delta,\cdot), 3PC(\Delta,\Delta), \text{wheel})$-**free graphs**

A multigraph is called *series-parallel* if it arises from a forest by applying the following operations: adding a parallel edge or subdividing an edge. A *series-parallel graph* is a series-parallel multigraph with no parallel edges. Series-parallel graphs are an example of a graph class that is $(3PC(\Delta,\cdot), 3PC(\Delta,\Delta), \text{wheel})$-free. This class of graphs is closed under minor taking and is in fact the class of K_4-minor-free graphs. Their tree-width is at most 2. In this section we describe three more classes that are $(3PC(\Delta, \cdot), 3PC(\Delta, \Delta),$ wheel)-free, but are also closed under taking induced subgraphs.

7.1   (ISK4, wheel)-free graphs

A *subdivision* of a graph G is obtained by subdividing edges of G into paths of arbitrary length (at least 1). An *ISK4* is a subdivision of a K_4. Note that graphs that have no ISK4 as a subgraph are precisely series-parallel graphs. ISK4-free graphs are studied by Lévêque, Maffray and Trotignon in [87]. They prove a decomposition theorem for this class that uses double star cutsets. Unfortunately this does not lead to a recognition algorithm for ISK4-free graphs, which remains an open problem. In [87] a complete structural characterization of (ISK4, wheel)-free graphs is given, which we now describe.

A node cutset $S = \{a, b\}$ of a connected graph G is a *proper 2-cutset* if a and b are nonadjacent and both of degree at least 3, and such that $V(G) \setminus S$ can be partitioned into X and Y so that: $|X| \geq 2, |Y| \geq 2$; there are no edges between X and Y; and both $G[X \cup S]$ and $G[Y \cup S]$ contain an ab-path, but neither is a chordless ab-path.

Given a graph G, an induced subgraph K of G, and a set C of vertices of $G \setminus K$, the *attachment* of C over K is $N(C) \cap V(K)$. When a set $S = \{u_1, u_2, u_3, u_4\}$ induces a square (i.e. a 4-hole) in a graph G, with u_1, u_2, u_3, u_4 in this order along the square, a *link* of S is an induced pp'-path P of G such that either $p = p'$ and $N_S(p) = S$, or $N_S(p) = \{u_1, u_2\}$ and $N_S(p') = \{u_3, u_4\}$, or $N_S(p) = \{u_1, u_4\}$ and $N_S(p') = \{u_2, u_3\}$, and no interior vertex of P has a neighbor in S. A link with ends p, p' is said to be *short* if $p = p'$, and *long* otherwise. A *rich square* (resp. *long rich square*) is a graph K that contains a square S such that $K \setminus S$ has at least two components and every component of $K \setminus S$ is a link (resp. long link) of S.

A graph is *chordless* if all its cycles are chordless. It is easy to check that a line graph $G = L(R)$ is wheel-free if and only if R is chordless.

Theorem 7.1 (Lévêque, Maffray, Trotignon [87]) *An (ISK4, wheel)-free graph either has a clique cutset or a proper 2-cutset, or is one of the following types:*

- *a series-parallel graph,*

- *a complete bipartite graph,*

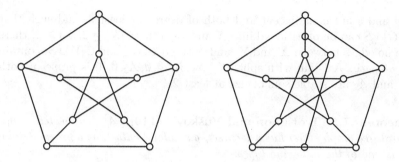

Figure 3: Petersen and Heawood graph.

- *line graph of a chordless graph with maximum degree at most 3, or*

- *a long rich square.*

The structure of chordless graphs is given by the following theorem (that was implicitly proved in [119] and explicitly stated and proved in [87]). A graph G is *sparse* if every edge of G has an endnode that is of degree at most 2. Note that chordless graphs were first studied in the 1960s by Dirac [66] and Plummer [104]. A description of their work can also be found in [3].

Theorem 7.2 ([119, 87]) *A connected chordless graph is either sparse or has a 1-cutset or proper 2-cutset.*

Theorems 7.1 and 7.2 are used in [87] to recognize (ISK4, wheel)-free graphs in $\mathcal{O}(n^2m)$ time, as well as to show that (ISK4, wheel)-free graphs are 3-colorable and to give an $\mathcal{O}(n^2m)$ time coloring algorithm.

7.2 Unichord-free graphs

The class of graphs that do not contain a cycle with a unique chord, also known as *unichord-free graphs*, is studied by Trotignon and Vušković in [119], where they obtain the following structure theorem for this class.

The Petersen and Heawood graphs are the two famous graphs, depicted in Figure 3, that were discovered at the end of the XIXth century in the research on the four color conjecture ([103], [79]), and they also appear as basic classes of unichord-free graphs. A graph is *strongly 2-bipartite* if it is 4-hole-free and bipartite with bipartition (X, Y) where X is the set of all degree 2 vertices of G and Y is the set of all nodes of G of degree at least 3. A node cutset $S = \{a, b\}$ of a connected graph G is a *special 2-cutset* if a and b are nonadjacent and both of degree at least 3, and such that $V(G) \setminus S$ can be partitioned into X and Y so that: $|X| \geq 2, |Y| \geq 2$; there are no edges between X and Y; and both $G[X \cup S]$ and $G[Y \cup S]$ contain an ab-path. A 1-join with split $(X_1, X_2, A_1, \emptyset, \emptyset, A_2, \emptyset, \emptyset)$ is *proper* if both A_1 and A_2 are stable sets of size at least 2.

Theorem 7.3 (Trotignon and Vušković [119]) *A connected unichord-free graph either has a 1-cutset, a special 2-cutset, or a proper 1-join, or is one of the following types:*

- *a clique,*

- *a hole of length at least 7,*

- *a strongly 2-bipartite graph, or*

- *an induced subgraph of the Petersen or the Heawood graph.*

We note that the decomposition theorem above in fact implies a complete structure theorem for unichord-free graphs. The decompositions in Theorem 7.3 can be reversed into compositions in such a way that every unichord-free graph can be built starting from basic graphs, that can be explicitly constructed, and gluing them together with prescribed composition operations, and all graphs built this way are unichord-free. This also implies a straightforward decomposition based recognition algorithm for this class that runs in $\mathcal{O}(nm)$ time.

Since unichord-free graphs are diamond-free, any edge of a unichord-free graph is contained in a unique maximal clique. Hence to find a maximum clique it is enough to look for common neighbors of every edge. This leads to an $\mathcal{O}(nm)$ time algorithm. In [119] an $\mathcal{O}(n+m)$ algorithm is given for the maximum clique problem for unichord-free graphs. It is based on the fact that a connected unichord-free graph that contains a triangle is either a clique or has a 1-cutset. Also in [119] it is shown how Theorem 7.3 can be used to obtain an $\mathcal{O}(nm)$ coloring algorithm for unichord-free graphs. It turns out that every unichord-free graph G is either 3-colorable or has an $\omega(G)$-coloring, and in particular $\chi(G) \leq \omega(G) + 1$. The problem of finding a maximum stable set of a unichord-free graph is NP-hard (follows from 2-subdivisions [105]).

Another characterization of unichord-free graphs is given by McKee in [97]: unichord-free graphs are precisely the graphs whose all minimal separators are stable sets (where a *separator* in a graph G is a set $S \subseteq V(G)$ such that $G \setminus S$ has more connected components than G).

7.3 Propeller-free graphs

Motivated by trying to understand the structure of wheel-free graphs, whose recognition remains an open problem, Aboulker, Radovanović, Trotignon and Vušković studied in [3] a subclass of wheel-free graphs known as *propeller-free graphs*. A *propeller* is a a graph that consists of a cycle C and a node x that has at least two neighbors on C. Let $\mathcal{C}_0$ be the class of graphs that have no node that has at least two neighbors of degree at least 3, $\mathcal{C}_1$ the class of graphs that have no propeller as a subgraph, and $\mathcal{C}_2$ the class of propeller-free graphs. Clearly $\mathcal{C}_0 \subsetneq \mathcal{C}_1 \subsetneq \mathcal{C}_2$.

First let us point out that by considering a longest path it is easy to show that graphs in $\mathcal{C}_2$ must always have a node of degree at most 2, and hence they are 3-colorable, see [3]. Observe that since a clique on 4 nodes is a propeller finding the size of a largest clique in a propeller-free graph can easily be done in polynomial time. On the other hand, finding a maximum independent set of a propeller-free graph is NP-hard (follows easily from [105], see also [119]).

The following decomposition theorems are given in [3], and used to obtain an $\mathcal{O}(nm)$ recognition algorithm for class $\mathcal{C}_1$, and an $\mathcal{O}(n^2 m^2)$ recognition algorithm for class $\mathcal{C}_2$.

A 2-cutset $\{a, b\}$ of a graph G is an S_2-*cutset* (resp. K_2-*cutset*) if ab is not an edge (resp. is an edge). An S_2-cutset is *proper* if nodes of $G \setminus \{a, b\}$ can be partitioned into sets X and Y so that no node of X is adjacent to a node of Y, and neither $G[X \cup \{a, b\}]$ nor $G[Y \cup \{a, b\}]$ is a chordless ab-path. A K_2-cutset is *proper* if $G \setminus \{a, b\}$ contains no node adjacent to both a and b. A 3-cutset $\{u, v, w\}$ of a graph G is an I-*cutset* if $G[\{u, v, w\}]$ contains exactly one edge.

Theorem 7.4 (Aboulker, Radovanović, Trotignon, Vušković [3])
A connected graph in $\mathcal{C}_1$ is either in $\mathcal{C}_0$ or it has a 1-cutset, a proper K_2-cutset or a proper S_2-cutset.

Theorem 7.5 (Aboulker, Radovanović, Trotignon, Vušković [3])
A graph in $\mathcal{C}_2$ is either in $\mathcal{C}_1$ or it has an I-cutset.

Furthermore, it is shown in [3] that propeller-free graphs admit an extreme decomposition, which is used to prove that 2-connected propeller-free graphs must always have an edge whose endnodes are of degree 2. This implies that propeller-free graphs can also be edge-colored in polynomial time.

8 $(3PC(\Delta,\cdot),\text{proper wheel})$-**free**

Let (H,x) be a wheel. A *sector* of (H,x) is a minimal subpath of H, of length at least one, whose endnodes are neighbors of x on H. A sector is *short* if it is of length one, and *long* otherwise. (H,x) is a *triangle-free wheel* if it has no short sectors. (H,x) is a *universal wheel* if x is adjacent to all nodes of H, i.e. it has no long sectors. (H,x) is a *line wheel* if it has four sectors, exactly two of which are short, and the short sectors have no common node. (H,x) is a *fanned wheel* if it has exactly one long sector. A *proper wheel* is a wheel that is not a triangle-free wheel, a universal wheel, a line wheel, or a fanned wheel with 2 or 3 short sectors. We now consider three different subclasses of the class of $(3PC(\Delta,\cdot),\text{proper wheel})$-free graphs.

8.1 Cap-free graphs

A *cap* is a hole together with a node that is adjacent to exactly two adjacent nodes on the hole. Cap-free graphs were studied in [43], where the focus was on obtaining polynomial time algorithms for recognizing whether a cap-free graph contains an odd (respectively even) hole. Note that the only Truemper configurations that cap-free graphs can contain are $3PC(\cdot,\cdot)$'s and wheels that are either triangle-free, universal or fanned with exactly two short sectors. The following decomposition theorem is obtained in [43] for this class, generalizing the decomposition theorem for Meyniel graphs obtained by Burlet and Fonlupt in [7].

A graph G contains a *1-amalgam* (or *amalgam*) (X_1, X_2, K, A_1, A_2) if $V(G) = X_1 \cup X_2 \cup K$, where X_1, X_2 and K are disjoin sets, $|X_1| \geq 2$, $|X_1| \geq 2$ and the nodes of K induce a clique in G (possibly K is empty). Furthermore, for $i = 1, 2$, $\emptyset \neq A_i \subseteq X_i$; every node of A_1 is adjacent to every node of A_2, and these are the only edges between X_1 and X_2; and every node of K is adjacent to every node of $A_1 \cup A_2$. Amalgams were first introduced in [7], and they generalize 1-joins, as a 1-amalgam with $K = \emptyset$ corresponds to a 1-join.

A *basic cap-free graph* G is either a triangulated graph or a biconnected triangle-free graph with at most one additional node, that is adjacent to all other nodes of G

Theorem 8.1 (Conforti, Cornuéjols, Kapoor, Vušković [43])
A cap-free graph is either basic or it has a 1-amalgam.

Cap-free graphs can easily be recognized in polynomial time directly, but in [43] Theorem 8.1 is used to obtain decomposition based recognition algorithms for cap-free even-signable and cap-free odd-signable graphs (even-signable graphs are a generalization of odd-hole-free graphs, and odd-signable graphs are a generalization of even-hole-free graphs; they are formally defined in Section 9).

Since triangle-free graphs are cap-free (basic), it follows that the problems of coloring and finding the size of a largest independent set are both NP-hard for cap-free graphs. On the other hand, it is easy to see how to use Theorem 8.1 to obtain a polynomial time algorithm to solve the maximum weight clique problem for cap-free graphs, see [51].

Theorem 8.1 is a generalization of analogous result obtained by Burlet and Fonlupt [7] for *Meyniel* graphs, which are exactly (cap, odd-hole)-free graphs. The decomposition of Meyniel graphs by 1-amalgams is used in [7] to obtained the first known polynomial time recognition algorithm for this class. Subsequently, Roussel and Rusu [113] obtained a faster algorithm for recognizing Meyniel graphs (of complexity $\mathcal{O}(m^2)$), that is not decomposition based.

Hertz [80] gives an $\mathcal{O}(nm)$ algorithm for coloring and obtaining a largest clique of a Meyniel graphs. This algorithm is based on contractions of even pairs. Roussel and Rusu [114] give an $\mathcal{O}(n^2)$ algorithm that colors a Meyniel graph without using even pairs. This algorithm "simulates" even pair contractions and it is based on lexicographic breadth-first search and greedy sequential coloring. In [9] Cameron, Lévêque and Maffray give another $\mathcal{O}(n^2)$ algorithm for coloring Meyniel graphs, which takes as input any graph and finds either a clique and a coloring of the same size or a Meyniel obstruction (i.e. an odd cycle of length at least 5 with at most one chord).

Conforti and Gerards [51] show how to obtain a polynomial time algorithm for solving maximum weight independent set problem, on any class of graphs that is decomposable by amalgams into basic graphs for which one can solve the maximum weight independent set problem in polynomial time. In particular, using Theorem 8.1, they obtain a polynomial time algorithm for solving the maximum weight independent set problem for (cap, odd-hole)-free graphs (i.e. Meyniel graphs) and (cap, even-hole)-free graphs (and more generally, cap-free odd-signable graphs). Whether (cap, even-hole)-free graphs can be colored in polynomial time remains an open problem.

8.2 Claw-free graphs

A *claw* is a complete bipartite graph $K_{1,3}$. Observe that claw-free graphs are $3PC(\cdot,\cdot)$-free and the only wheels they may contain are line wheels, universal wheels whose rim is of length 4 or 5, and fanned wheels with 2 or 3 short sectors.

Structural study of claw-free graphs started in the context of perfect graphs. Claw-free Berge graphs have been shown to be perfect by Parthasarathy and Ravindra [101] by exploiting properties of minimally imperfect graphs. Another proof, based on properties of minimally imperfect graphs, is given by Giles, Trotter and Tucker in [75]. The first insight into the structure of claw-free Berge graphs is given by Chvátal and Sbihi [15]. They use the following characterization of claw-free Berge graphs to obtain a polynomial time recognition algorithm for this class (based on decomposition by clique cutsets). A graph is *elementary* if its edges can be colored by two colors in such a way that edges xy and yz have distinct colors whenever x and z are nonadjacent. It is easy to see that every elementary graph is claw-free Berge.

Theorem 8.2 (Chvátal and Sbihi [15]) *A claw-free graph G with no clique cutset is Berge if and only if it has at least one of the following properties:*

(i) G is elementary,

(ii) $\alpha(G) \geq 3$ and G contains no hole of length at least 5.

The following strengthening of Theorem 8.2 describes graphs that satisfy (ii) more precisely. A *cobipartite* graph G is the complement of bipartite graph, and *cobipartition* of G is its vertex partition (X, Y) such that X and Y are cliques. A graph is *peculiar* if it can be obtained as follows: take three, pairwise vertex-disjoint, cobipartite graphs (A_1, B_1), (A_2, B_2), (A_3, B_3) such that each of them has at least one pair of nonadjacent vertices; add all edges between every two of them; then take three cliques K_1, K_2, K_3 that are pairwise disjoint and disjoint from the A_i's and B_i's; add all the edges between K_i and $A_j \cup B_j$ for $j \neq i$; there is no other edge in the graph.

Theorem 8.3 (Chvátal and Sbihi [15]) *A claw-free Berge graph either has a clique cutset or it is elementary or peculiar.*

Maffray and Reed [93] further strengthen Theorem 8.3 by giving complete description of the structure of elementary graphs.

An edge is *flat* if it does not lie in a triangle. Let xy be a flat edge of a graph G and let B be a cobipartite graph, disjoin from G, with cobipartition (X, Y) such that there is at least one edge between X and Y in B. Let G' be the graph obtained from $G \setminus \{x, y\}$ and B by adding all possible edges between X and $N(x) \setminus \{y\}$ and between Y and $N(y) \setminus \{x\}$. We say that G' is obtained from G by *augmenting along xy with augment B*.

Now let $x_1y_1, \ldots, x_hy_h$ be pairwise non-incident flat edges of G, and let $B_1, \ldots, B_h$ be pairwise disjoint cobipartite graphs, that are also disjoint from G, with cobipartitions $(X_1, Y_1), \ldots, (X_h, Y_h)$. Let G' be the graph obtained from G by augmenting respectively each edge x_iy_i with augment B_i. Graph G' is called an *augmentation* of G.

A *line graph* of a graph G, denoted by $L(G)$, is a graph whose vertices are edges of G, and two vertices of $L(G)$ are adjacent if and only if the corresponding edges of G have a common vertex.

Theorem 8.4 (Maffray and Reed [93]) *G is elementary if and only if G is an augmentation of a line graph of bipartite multigraph.*

Theorem 8.3 and Theorem 8.4 yield a new proof of the perfection of claw-free Berge graphs: by directly showing that peculiar graphs are perfect, using the above structural characterization of elementary graphs to show directly that they are perfect, and the fact that composing along clique cutsets preserves perfection. We now show that Theorems 8.3 and 8.4 in fact imply the following decomposition theorem for claw-free Berge graphs.

Corollary 8.5 *If G is a connected claw-free Berge graph, then either G has a clique cutset, a 1-join (whose one side is a homogeneous set that is a cobipartite graph) or a 2-join (whose one side is a homogeneous pair of cliques), or G is cobipartite or a line graph of a bipartite multigraph.*

It is easy to see that peculiar graphs have a 1-join whose one side is in fact a homogeneous set that induces a cobipartite graph. Let G be a connected claw-free graph with flat edge xy, and let G' be an augmentation of G along xy with augment B with cobipartition (X, Y). Let $N_x = N(x) \setminus \{y\}$ and $N_y = N(y) \setminus \{x\}$. We first observe that since G is claw-free, it follows that both N_x and N_y are cliques. If $V(G) = N[x]$ then G' is a cobipartite graph. In particular, a cobipartite graph itself can be viewed as an augmentation of the line graph of a bipartite graph consisting of just two adjacent vertices. If $N_y = \emptyset$ and $V(G) \setminus N[x] \neq \emptyset$ then N_x is a clique cutset. Let us now assume that G' is not cobipartite, $G' \neq G$, and that G' does not have a clique cutset. Then it follows that $(X \cup Y, V(G) \setminus \{x, y\}, X, Y, \emptyset, N_x, N_y, \emptyset)$ is a split of a 2-join. Given a 2-join of a graph H with split $(X_1, X_2, A_1, B_1, \emptyset, A_2, B_2, \emptyset)$, let us construct the blocks of decomposition by this 2-join as follows: block H_1 (resp H_2) is the graph obtained from $H[X_1]$ (resp. $H[X_2]$) by adding an edge a_2b_2 (resp. a_1b_1), all edges between a_2 and A_1 (resp. a_1 and A_2), and all edges between b_2 and B_1 (resp. b_1 and B_2). We observe that augmenting along a flat edge is the reverse of decomposing along a 2-join with such construction of the blocks of decomposition.

Given a graph G we say that the two 2-joins (X_1, X_2) and (Y_1, Y_2) of G are *non-crossing* if $X_1 \subseteq Y_1$ or $Y_1 \subseteq X_1$. Theorem 8.4 in fact shows that connected elementary graphs can be decomposed by a sequence of non-crossing 2-joins into cobipartite graphs and a line graph of a bipartite multigraph. Furthermore, if the graph G contains a line graph of a bipartite multigraph H (and H is a maximal such graph in G), then at each step of the decomposition of G a cobipartite graph is split off from the skeleton of H. Moreover, it is shown in [93] how to efficiently find such a sequence of 2-joins.

Minty [98] (corrected by Nakamura and Tamura [99]) showed that there is a polynomial algorithm to find a stable set of maximum weight in a claw-free graph by generalizing the algorithm of Edmonds [67, 68] for finding a maximum weighted matching in a graph. The problems of finding a largest clique and a minimum coloring are both NP-hard for claw-free graphs [83] (finding α in a triangle-free graph is NP-hard, and hence so is ω in a claw-free graph; edge-coloring on general graphs can be reduced to vertex coloring on claw-free graphs and hence the HP-hardness of that problem).

Hsu and Nemhauser [84] gave a combinatorial polynomial time algorithm which finds a maximum weighted clique and a minimum coloring of claw-free perfect graphs. Note that finding a maximum weighted clique in a claw-free perfect graph follows easily from the fact that the neighborhood of any vertex in such a graph is cobipartite. Li and Zang [88] gave a combinatorial polynomial time algorithm which finds a minimum weighted coloring of claw-free perfect graphs and is based on Theorems 8.3 and 8.4.

Chudnovsky and Seymour, in a series of papers [30, 31, 32, 33, 34, 35, 36], extend the above structural characterization of claw-free Berge graphs, to claw-free graphs in general. They show how all claw-free graphs can be obtained through explicit constructions starting from a few basic classes that can all be described explicitly. The full structural characterization they obtain is too complicated to explain. Here we state the decomposition theorem they obtain in [33] and then use in [34] for describing the construction. *Basic claw-free graphs* consist of seven subclasses, some of which are line graphs (of multigraphs), induced subgraphs of icosahedron, circular interval graphs and antiprismatic graphs (claw-free graphs in which every four vertices induce a subgraph with at least two edges). Antiprismatic graphs are further studied in [30] and [31]. We state the decomposition theorem in a weakened form and then explain the strong form actually obtained in [33].

Theorem 8.6 (Chudnovsky and Seymour [33]) *A connected claw-free graph is either basic claw-free or it has a 1-join, 2-join, M-join or 6-join.*

The decomposition theorem in [33] is stronger than the statement we gave above in two ways of key importance for being able to reverse the decomposition into a construction. First, the general joins used in the decomposition theorem in [33] have a particular structure, some of which is directly implied by the fact that the graph is assumed to be connected claw-free, but some is not. For example the 1-joins used in [33] have the following property: if neither side of the split of a 1-join is a homogeneous set then both special sets are cliques (this follows from being connected claw-free), and otherwise at least one of the sides is a homogeneous set and in this case at least one of the two possible homogeneous sets is a clique (note that this is not directly implied by being claw-free when both sides are homogeneous sets). For the 2-join the requirement is that at least one side has special sets that are cliques, and if neither side is a homogeneous pair then all special sets must be cliques. For the M-join, either one side is a homogeneous pair of cliques (that are neither complete nor anticomplete to each other) or all special sets are cliques. For the 6-join, all special sets are cliques and the two sides consist only of special sets.

The second important strengthening of Theorem 8.6 is that in [33] the decomposition theorem is actually proved for claw-free trigraphs. A *trigraph* is an object that generalizes a graph: in a graph every pair of vertices is either adjacent or nonadjacent, and in a trigraph every pair of vertices is either adjacent, or nonadajacent or semi-adjacent. A general join with split (X_1, X_2) in a trigraph has exactly the same requirements for adjacent and nonadjacent pairs of vertices from different sides of the split as in the graph version. In other words, if a pair of vertices are semi-adjacent they must both belong to the same side of the split. The decomposition theorem for claw-free trigraphs is used to strengthen the structure of needed general 2-joins to the point that decompositions can be reversed into compositions.

It is interesting to observe that out of all types of general 2-joins, the only one that does not appear in this decomposition theorem is the N-join. Here is why. Suppose that G is a claw-free graph that has an N-join but does not have any of the cutsets described in Theorem 8.6. Then it is easy to see that G must have a clique cutset S such that $V(G)$ can be partitioned into sets S, V_1, V_2 with $|V_i| \geq 2$, for $i = 1, 2$. If G has such a clique cutset, then (as shown in [33]) it follows that G must be a linear interval graph, i.e. one of the basic graphs (since every linear interval graph is also a circular interval graph).

Let us point out that the NP-hardness of the coloring problem and the maximum clique problem on claw-free graphs stems from the NP-hardness of these problems on the basic subclasses. For example, coloring line graphs is NP-hard, and finding a maximum clique in the class of graphs with no stable set of size 3 (a subclass of antiprismatic graphs) is already NP-hard. On the other hand, the chromatic number of a claw-free graph is bounded by the function of the size of its largest clique: it is easy to see that for a claw-free graph G, $\chi(G) \leq \omega(G)^2$, and that this is not far from being best possible since every graph with no stable set of size 3 is claw-free. One consequence of the structure theory for claw-free graphs is the following boundedness of the chromatic number for claw-free graphs that do contain a stable set of size 3.

Theorem 8.7 (Chudnovsky and Seymour [36]) *If G is a connected claw-free graph with $\alpha(G) \geq 3$, then $\chi(G) \leq 2\omega(G)$ (and this is asymptotically best possible).*

8.3   Bull-free graphs

A *bull* is a graph with five vertices a, b, c, d, e and five edges ab, bc, cd, be, ce. Bull-free graphs cannot contain $3PC(\Delta, \cdot)$'s, the only $3PC(\Delta, \Delta)$'s they can have are $\bar{C}_6$'s (i.e. the complements of holes of length 6), and the only wheels they can have are triangle-free wheels, universal wheels, fanned wheels with 2 short sectors, and wheels whose rim is a 5-hole and whose centre has 4 neighbors on the rim.

The study of bull-free graphs also started in the context of perfect graphs. First Chvátal and Sbihi [39] proved that bull-free Berge graphs are perfect by obtaining the following decomposition theorem.

Theorem 8.8 (Chvátal and Sbihi [39]) *A connected bull-free Berge graph is either bipartite or cobipartite, or it has a homogeneous pair or a star cutset in the graph or its complement.*

Since bipartite and cobipartite graphs are perfect and minimal imperfect graphs cannot have homogeneous pairs [39], nor star cutsets [38], nor star cutsets in the complement (which follows from the Perfect Graph Theorem: a graph is perfect if and only if its complement is perfect [89]), it follows that bull-free Berge graphs must be perfect.

Reed and Sbihi [107] showed how bull-free perfect graphs can be recognized in polynomial time by decomposing them with homogeneous sets and finding vertices whose removal does not change whether the graph is Berge or not, and hence avoiding decomposition by star cutsets.

De Figueiredo and Maffray [63] give a combinatorial strongly polynomial time algorithm for solving the maximum weighted clique problem on bull-free perfect graphs. Since this class is self-complimentary, this algorithm implies combinatorial polynomial time algorithms for maximum weighted stable set problem, minimum weighted coloring problem and minimum weighted clique covering problem. Their algorithm is based on the following decomposition theorem. A graph is *weakly triangulated* if it is (hole, antihole)-free. A graph is *transitively orientable* if it admits a *transitive orientation*, i.e. an orientation of its edges with no circuit and with no P_3 abc with the orientation $\vec{ab}$ and $\vec{bc}$. Such graphs are also called *comparability graphs*.

Theorem 8.9 (De Figueiredo and Maffray [63]) *A connected bull-free Berge graph is either weakly triangulated, transitively orientable, complement of a transitively orientable graph, or it has a homogeneous set or a homogeneous pair.*

A maximum weighted clique of a weakly triangulated graph can be found in strongly polynomial time by the algorithm in [81], of a transitively orientable graph by the algorithm in [82], and of the complement of a transitively orientable graph by the algorithm in [8].

The complete structural characterization of bull-free graphs in general is done by Chudnovsky in a series of papers [18, 19, 20, 21]. This characterization is too difficult to explain precisely here, but we give some flavor of it. First let us consider some examples of bull-free graphs. Triangle-free graphs are clearly bull-free, and since a bull is a self-complementary structure, so are their complements. Note that from these two classes of graphs it follows that the maximum clique problem, maximum stable set problem and the vertex coloring problem are all NP-hard for bull-free graphs. Another example of a bull-free graph is an *ordered split graph*: a graph G whose vertex set is a union of a clique $\{k_1, \ldots, k_n\}$ and a stable set $\{s_1, \ldots, s_n\}$, and s_i is adjacent to k_j if and only if $i + j \leq n + 1$. A larger bull-free graph can be created from smaller ones using the operation of *substitution*: input are two bull-free graphs G_1 and G_2 with disjoint vertex sets, and vertex $v \in V(G_1)$; output is a new graph G whose vertex set is $V(G_1) \cup V(G_2) \setminus \{v\}$ and whose edge set is $E(G_1 \setminus \{v\}) \cup E(G_2) \cup \{xy : x \in V(G_1) \setminus \{v\}, y \in V(G_2), \text{ and } xv \in E(G_1)\}$. We observe that this composition operation is the reverse of the homogeneous set decomposition.

Chudnovsky's construction of all bull-free graphs starts from three explicitly constructed classes of *basic bull-free graphs*: $\mathcal{T}_0, \mathcal{T}_1$ and $\mathcal{T}_2$. $\mathcal{T}_0$ is a class of graphs with few nodes, the graphs in $\mathcal{T}_1$ are built from a triangle-free graph F and a collection of disjoint cliques with prescribed attachments in F (so triangle-free graphs are in this class, and also ordered split graphs), and $\mathcal{T}_2$ generalizes graphs G that have a pair uv of vertices, so that uv is dominating both in G and $\bar{G}$. Furthermore, each graph G in $\mathcal{T}_1 \cup \mathcal{T}_2$ comes with a list $\mathcal{L}_G$ of "expandable edges". Chudnovsky shows that every bull-free graph that is not obtained by substitution from smaller ones, can be constructed from a basic bull-free graph by expanding the edges in $\mathcal{L}_G$ (where edge expansion is an operation corresponding to "reversing the homogeneous pair decomposition"). To prove this result, again it was convenient to work on trigraphs, and the first step is to obtain the following decomposition theorem for bull-free trigraphs.

Theorem 8.10 (Chudnovsky [18, 19, 20, 21]) *If G is a bull-free trigraph, then either G or $\bar{G}$ is basic bull-free, or G has a homogeneous set or a homogeneous pair.*

The fact that the theorem is proved on trigraphs makes it possible to put enough structure on the homogeneous pairs actually needed in the decomposition to allow for the reversal of the decomposition into a composition.

Recall that the way homogeneous sets and homogeneous pairs are defined for trigraphs, is the same as for graphs when it comes to adjacent and nonadjacent pairs that go across the split, the semi-adjacent pairs are only allowed to be fully contained in a side of a split. So in some sense the above decomposition theorem is saying that there is a sequence of non-crossing decompositions by homogeneous sets and homogeneous pairs that can break the graph down to a basic graph. We can see this by thinking of semi-adjacent edges in the trigraph as marker edges used in the construction of blocks of decomposition by homogeneous pairs.

One consequence of Chudnovsky's characterization of bull-free graphs is that the Erdős-Hajnal conjecture holds for them.

Conjecture 8.11 (Erdős and Hajnal [69]) *For every graph H, there exists $f(H) > 0$, such that if G is H-free, then G contains either a clique or a stable set of size at least $|V(G)|^{f(H)}$*

Theorem 8.12 (Chudnovsky and Safra [27]) *If G is a bull-free graph then G contains a stable set or a clique of size at least $|V(G)|^{\frac{1}{4}}$.*

The proof of Theorem 8.12 is actually based on the following decomposition theorem.

Theorem 8.13 (Chudnovsky [18, 27]) *If G is a bull-free graph that contains a hole H of length at least 5, and vertices $c, a \in V(G) \setminus V(H)$ such that c is complete to $V(H)$ and a is anticomplete to $V(H)$, then G has a homogeneous set.*

The structure theorem for bull-free graphs is also used to derive a structure theorem for bull-free perfect graphs [25], which is then used in [102] to derive combinatorial polynomial time algorithm for maximum weighted clique problem on bull-free perfect graphs that is a bit faster than the algorithm in [63].

9   Excluding some wheels and some 3-path-configurations

The class of regular bipartite graphs and the class of balanceable bipartite graphs, that generalizes it, were discussed in Section 4. As we have seen, the only Truemper configurations that balanceable bipartite graphs can have are bipartite wheels whose center has an even number of neighbors on the rim, and $3PC(u, v)$'s where u and v are on the same side of the bipartition. In this section we discuss three more well studied classes of graphs where some 3-path-configurations and some wheels are excluded, but enough is left in to make them structurally quite complex, namely the classes of even-hole-free graphs, odd-hole-free graphs and perfect graphs.

9.1   Even-hole-free graphs

The class of even-hole-free graphs is structurally quite similar to the class of perfect graphs, which was the key initial motivation for their study. The first major structural study of even-hole-free graphs was done by Conforti, Cornuéjols, Kapoor and Vušković in [45] and [46]. They were focused on showing that even-hole-free graphs can be recognized in polynomial time (a problem that at that time was not even known to be in NP), and their primary motivation was to develop techniques which can then be used in the study of perfect graphs. In [45] a decomposition theorem is obtained for even-hole-free graphs, based on which the first known polynomial time recognition algorithm for even-hole-free graphs is constructed in [46]. This research kick-started a number of other studies of even-hole-free graphs which we survey in this section. A more detailed survey of even-hole-free graphs is given in [125].

The class of even-hole-free graphs is also of independent interest due to its relationship to β-perfect graphs introduced by Markossian, Gasparian and Reed [96]. For a graph G, let $\delta(G)$ be the minimum degree of a vertex in G. Consider the following total order on $V(G)$: order the vertices by repeatedly removing a vertex of minimum degree in the subgraph of vertices not yet chosen and placing it after all the remaining vertices but before all the vertices already removed. Coloring greedily on this order gives the upper bound $\chi(G) \leq \beta(G)$, where $\beta(G) = \max\{\delta(G') + 1 : G'$ is an induced subgraph of $G\}$. A graph is β-perfect if for each induced subgraph H of G, $\chi(H) = \beta(H)$. It is easy to see that β-perfect graphs belong to the class of even-hole-free graphs, and that this containment is proper.

The essence of even-hole-free graphs is actually captured by their generalization to signed graphs. A graph is *odd-signable* if there exists an assignment of $0, 1$ weights to its edges that makes every chordless cycle of odd weight. We say that a wheel (H, x) is *even* if x has an even number of neighbors on H. The following characterization of odd-signable graphs can be easily derived from Theorem 2.1.

Theorem 9.1 ([43]) *A graph is odd-signable if and only if it does not contain an even wheel, a $3PC(\cdot, \cdot)$ nor a $3PC(\Delta, \Delta)$.*

All decomposition theorems for even-hole-free graphs which we now describe are in fact proved for 4-hole-free odd-signable graphs, and the above characterization of odd-signable graphs is repeatedly used in the proofs.

A 2-join with split $(X_1, X_2, A_1, B_1, \emptyset, A_2, B_2, \emptyset)$ is *connected* if for $i = 1, 2$, $G[X_i]$ contains a path whose one endnode is in A_i and the other in B_i. It is a *path 2-join* if for some $i \in \{1, 2\}$, $G[X_i]$ is a chordless path whose one endnode is in A_i and the other in B_i. A *non-path 2-join* is a 2-join that is not a path 2-join. A graph is a *clique tree* if each of its maximal 2-connected components is a clique. A graph is an *extended clique tree* if it can be obtained from a clique tree by adding at most two vertices.

Theorem 9.2 (Conforti, Cornuéjols, Kapoor, Vušković [45])
A connected even-hole-free graph is either an extended clique tree, or it has a k-star cutset for $k \leq 3$ or a connected non-path 2-join.

This theorem was strong enough to be used in the construction of a polynomial time recognition algorithm for even-hole-free graphs in [46], but even at that time it was suspected that a stronger decomposition theorem was possible. The strengthening of Theorem 9.2 was eventually given in [62].

Theorem 9.3 (da Silva and Vušković [62]) *A connected even-hole-free graph is either an extended clique tree, or it has a star cutset or a connected non-path 2-join.*

We observe that in the decomposition theorems in [45] and [62], the basic graphs are defined in a more specific way, but for the purposes of the algorithms the statements of Theorems 9.2 and 9.3 suffice. As in the case of the decomposition based recognition algorithm for balanced bipartite graphs, described in Section 4.2, the problem in using the above theorems for constructing a recognition algorithm for even-hole-free graphs are the star cutsets. For 2-joins it is possible to construct the blocks of decomposition that are class-preserving for the class of even-hole-free graphs (by replacing a side of a 2-join by a path of appropriate length, which clarifies the usefulness of connected non-path 2-joins in the above decomposition theorems), and lead to a polynomial decomposition tree. To use the decomposition by star cutsets, one first needs to *clean* the graph (as described in Section 3). The decomposition based recognition algorithm for even-hole-free graphs in [46] is of complexity of about $\mathcal{O}(n^{40})$. In [24] an $\mathcal{O}(n^{31})$ recognition algorithm for even-hole-free graphs is given, that first cleans the graph and then directly looks for an even hole (using the shortest-paths detector technique described in Section 2.1). In [62] an $\mathcal{O}(n^{19})$ decomposition based algorithm is obtained. Finally, by using Theorem 9.3 Chang and Lu [11] obtain an $\mathcal{O}(n^{11})$ recognition algorithm for even-hole-free graphs. They improve the complexity by introducing a new idea of a "tracker" that allows for fewer graphs that need to be recursively decomposed by star cutsets, and they improve the complexity of the cleaning procedure by first looking for certain structures, using the three-in-a-tree algorithms from [29], before applying the cleaning. We observe that detecting whether a graph contains a $3PC(\cdot, \cdot)$ or a $3PC(\Delta, \Delta)$ can be done in $\mathcal{O}(n^{35})$ time [23]. The high complexity of all these algorithms is due to the cleaning procedure.

The following intermediate result is used as one of the steps in the proof of Theorem 9.3, and as we shall see later, it is of an independent interest. A *diamond* is the graph obtained from a clique on 4 nodes by removing an edge. A *bisimplicial cutset* is a node cutset that either induces a clique or two cliques with exactly one common node. Note that a bisimplicial cutset is a very special type of a star cutset.

Theorem 9.4 (Kloks, Müller, Vušković [85]) *A connected (even-hole, diamond)-free graph is either an extended clique tree, or it has a bisimplicial cutset or a connected non-path 2-join.*

We now survey known results related to combinatorial optimization on even-hole-free graphs. In Section 10 we shall see that 2-joins can be used in a decomposition based optimization algorithm, but it is not clear how to use star cutsets. The decomposition by star cutsets can, on the other hand, be used to obtain local structural properties that can then be used in algorithms.

The complexities of finding a maximum independent set and an optimal coloring are not known for even-hole-free graphs. One can find a maximum clique of an even-hole-free graph in polynomial time, since as observed by Farber [70], 4-hole-free graphs have $\mathcal{O}(n^2)$ maximal cliques and hence one can list them all in polynomial time. The following structural characterization of even-hole-free graphs leads to a faster algorithm for computing a maximum clique in an even-hole-free graph.

Theorem 9.5 (da Silva and Vušković [61]) *Every even-hole-free graph has a node whose neighborhood is triangulated.*

This result follows from the fact that universal wheels in even-hole-free graphs can be decomposed in a particular way by star cutsets. For any node x in a graph G, a maximal clique belongs to $G[N[x]]$ or $G \setminus \{x\}$. Therefore Theorem 9.5 reduces the problem of finding a maximum clique in an even-hole-free graph to the problem of finding a maximum clique in a triangulated graph, which as we have seen in Section 3.1 can be done efficiently. Observe that in order to find a maximum clique, it is not necessary that we know that the input graph is even-hole-free. The algorithm proceeds by attempting to construct an ordering of vertices $x_1, \ldots, x_n$ of the input graph G such that, for every $i = 1, \ldots, n$, the neighborhood of x_i in $G_i = G[\{x_i, \ldots x_n\}]$ is triangulated. If it cannot complete the sequence, then it follows from Theorem 9.5 that the input graph is not even-hole-free. Otherwise, we get a sequence of triangulated graphs $G_1, \ldots, G_n$ such that every maximal clique of G belongs to exactly one of them. It follows that there are at most $n + 2m$ maximal cliques in an even-hole-free graph and all of them can be generated in $\mathcal{O}(n^2 m)$ time (and hence in the same time a maximum weighted clique in a weighted even-hole-free graph can be found). In [2] it is shown how LexBFS can be used to find the above ordering of vertices, reducing the complexity of finding a maximum weighted clique to $\mathcal{O}(nm)$. Again, the algorithms discussed are robust in the sense that they either correctly compute the desired clique or they correctly identify the input graph as not being even-hole-free.

Here is another property of even-hole-free graphs that shows that this class is χ-bounded (i.e. the chromatic number is bounded by a function of the size of a largest clique). A *bisimplicial vertex* is a vertex whose set of neighbors induces a graph that is a union of two cliques.

Theorem 9.6 (Addario-Berry, Chudnovsky, Havet, Reed, Seymour [1]) *Every even-hole-free graph has a bisimplicial vertex.*

It is interesting to observe that Theorem 9.6 is also obtained using decomposition, although in [1] not all even-hole-free graphs are decomposed, but enough structures are decomposed using special double star cutsets (star cutsets and cutsets that become double star cutsets after some edges are added) to obtain the desired result. It clearly implies the following corollary.

Corollary 9.7 ([1]) *If G is even-hole-free then $\chi(G) \leq 2\omega(G) - 1$.*

Recall that β-perfect graphs are a subclass of even-hole-free graphs that can be efficiently colored, by coloring greedily on a particular easily constructable ordering of vertices. Unfortunately it is not known whether β-perfect graphs can be recognized in polynomial time. In [96] it is shown that (even-hole, diamond, cap)-free graphs are β-perfect, and in [64] it is shown that (even-hole, diamond, cap-on-6-vertices)-free graphs are β-perfect. These results are further generalized in [85] where it is shown that (even-hole, diamond)-free graphs are β-perfect, and hence can be both recognized and colored in polynomial time. This result follows from the following property of (even-hole, diamond)-free graphs, that is obtained by using Theorem 9.4. A vertex is *simplicial* if its neighborhood set induces a clique, and it is a *simplicial extreme* if it is either simplicial or of degree 2.

Theorem 9.8 (Kloks, Müller, Vušković [85]) *Every (even-hole, diamond)-free graph has a simplicial extreme.*

Theorem 9.8 and the following property of minimal β-imperfect graphs, imply that (even-hole, diamond)-free graphs are β-perfect.

Lemma 9.9 (Markossian, Gasparian, Reed [96]) *A minimal β-imperfect graph that is not an even hole, contains no simplicial extreme.*

Corollary 9.10 ([85]) *Every (even-hole, diamond)-free graph is β-perfect.*

Note that the fact that (even-hole, diamond)-free graphs have simplicial extremes implies that for such graphs G, $\chi(G) \leq \omega(G) + 1$.

9.2 Perfect graphs and odd-hole-free graphs

A graph G is *perfect* if for every induced subgraph H of G, $\chi(H) = \omega(H)$. In 1961 Berge [5] made a conjecture that characterizes perfect graphs in terms of excluded induced subgraphs in the following way: *a graph is perfect if and only if it does not contain an odd hole nor an odd antihole* (where an *antihole* is a complement of a hole). The graphs that

do not contain an odd hole nor an odd antihole are known as *Berge* graphs.
It is easy to see that perfect graphs must be Berge, so the essence of the
conjecture is to show that Berge graphs must be perfect. This famous conjecture, known as the Strong Perfect Graph Conjecture (SPGC), sparked
an enormous amount of diverse research until it was finally proved in 2002
by Chudnovsky, Robertson, Seymour and Thomas [26], and is now known
as the Strong Perfect Graph Theorem.

The approach that eventually worked for proving the SPGC is the
decomposition method. This approach entails proving a decomposition
theorem for Berge graphs, in such a way that the undecomposable (basic)
graphs are simple enough so that the SPGC can be proved directly for
them, and the cutsets used have the property that no minimum counter-
example to the conjecture can have them. This approach was used to prove
the SPGC for a number of subclasses of graphs (i-triangulated graphs
using clique cutsets [72], weakly triangulated graphs using star cutsets
[78], bull-free graphs using homogeneous pairs and star cutsets [14]), but
the one subclass that came closest to revealing the structure of Berge
graphs in general is the class of 4-hole-free Berge graphs. In [48] Conforti,
Cornuéjols and Vušković prove the SPGC for 4-hole-free graphs by the
following decomposition theorem, and the fact that it was already known
that no minimal imperfect graph can have a star cutset [38], and that if a
a minimal imperfect graph has a 2-join then it must be an odd hole [56].

Theorem 9.11 (Conforti, Cornuéjols, Vušković [48]) *A 4-hole-free
Berge graph is either bipartite or line graph of a bipartite graph, or it
has a star cutset or a connected non-path 2-join.*

Before we describe the decomposition theorem for Berge graphs in general, we state the decomposition theorem for odd-hole-free graphs (a su-
perclass of Berge graphs), that also preceded the work in [26].

Theorem 9.12 (Conforti, Cornuéjols, Vušković [49]) *An odd-hole-
free graph is either bipartite, line graph of a bipartite graph or complement of a line graph of a bipartite graph, or it has a double star cutset or
a connected non-path 2-join.*

We observe that, as in the study of even-hole-free graphs, a convenient setting for the study of odd-hole-free graphs is their generalization
to signed graphs. A graph is *even-signable* if there exists an assignment of
$0, 1$ weights to its edges that makes every triangle of odd weight and every
hole of even weight. An *odd wheel* is a wheel that induces an odd number
of triangles. The following characterization of even-signable graphs can
easily be derived from Theorem 2.1.

Theorem 9.13 ([43]) *A graph is even-signable if and only if it does not contain an odd wheel nor a $3PC(\Delta, \cdot)$.*

We now describe the decomposition theorems for Berge graphs, by first introducing the specific cutsets used and the basic graphs.

A 2-join with split $(X_1, X_2, A_1, B_1, \emptyset, A_2, B_2, \emptyset)$ is a P_3-*path 2-join* if for some $i \in \{1, 2\}$, $G[X_i]$ induces a path on three nodes whose one endnode is in A_i and the other in B_i. A *non-P_3-path 2-join* is a 2-join that is not a P_3-path 2-join.

The homogeneous pair as defined in Section 3.2 was first introduced by Chvátal and Sbihi in [39], where it was also shown that no minimal imperfect graph has a homogeneous pair. The definition that we give here is a slight variation that is used in [26]. A *homogeneous pair* is a partition of $V(G)$ into six non-empty sets (A, B, C, D, E, F) such that:

- every vertex in A has a neighbor in B and a non-neighbor in B, and vice versa;

- the pairs (C, A), (A, F), (F, B), (B, D) are complete;

- the pairs (D, A), (A, E), (E, B), (B, C) are anticomplete.

If S is a skew cutset in a graph G, then $(S, V(G) \setminus S)$ is also called a *skew partition* of G. A *balanced skew partition* is a skew partition (S, T) with the additional property that every induced path of length at least 2 in G with ends in S and interior in T has even length, and every induced path of length at least 2 in $\overline{G}$ with ends in T and interior in S has even length. If (S, T) is a balanced skew partition we say that the skew cutset S is *balanced*. Balanced skew partitions were first defined in [26] where it was also shown that no minimum counter-example to the strong perfect graph conjecture admits a balanced skew partition.

A *double split graph* is a graph G constructed as follows. Let $m, n \geq 2$ be integers. Let $A = \{a_1, \ldots, a_m\}$, $B = \{b_1, \ldots, b_m\}$, $C = \{c_1, \ldots, c_n\}$, $D = \{d_1, \ldots, d_n\}$ be four disjoint sets. Let G have vertex set $A \cup B \cup C \cup D$ and edges in such a way that:

- a_i is adjacent to b_i for $1 \leq i \leq m$. There are no edges between $\{a_i, b_i\}$ and $\{a_{i'}, b_{i'}\}$ for $1 \leq i < i' \leq m$;

- c_j is non-adjacent to d_j for $1 \leq j \leq n$. There are all four edges between $\{c_j, d_j\}$ and $\{c_{j'}, b_{j'}\}$ for $1 \leq j < j' \leq n$;

- there are exactly two edges between $\{a_i, b_i\}$ and $\{c_j, d_j\}$ for $1 \leq i \leq m$, $1 \leq j \leq n$ and these two edges are disjoint.

Note that $C \cup D$ is a non-balanced skew cutset of G and that $\overline{G}$ is a double split graph. Note that in a double split graph, vertices in $A \cup B$ all have degree $n + 1$ and vertices in $C \cup D$ all have degree $2n + m - 2$. Since $n \geq 2, m \geq 2$ implies $2n - 2 + m > 1 + n$, it is clear that given a double split graph the partition $(A \cup B, C \cup D)$ is unique. Hence, we call *matching edges* the edges that have an end in A and an end in B.

A graph is said to be *basic* if one of $G, \overline{G}$ is either a bipartite graph, the line-graph of a bipartite graph or a double split graph.

The following theorem was first conjectured in a slightly different form in [48]. A corollary of it is the Strong Perfect Graph Theorem.

Theorem 9.14 (Chudnovsky, Robertson, Seymour, Thomas [26])
Let G be a Berge graph. Then either G is basic or G has a homogeneous pair or a balanced skew partition, or one of $G, \overline{G}$ has a connected non-P_3-path 2-join.

The theorem that we state now is due to Chudnovsky who proved it from scratch, that is without assuming Theorem 9.14. Her proof uses the notion of a *trigraph*. The theorem shows that homogeneous pairs are not necessary to decompose Berge graphs. Thus it is a result stronger than Theorem 9.14.

Theorem 9.15 (Chudnovsky, [17, 16]) *Let G be a Berge graph. Then either G is basic, or one of $G, \overline{G}$ has a connected non-P_3-path 2-join or a balanced skew partition.*

In [22] Chudnovsky, Cornuéjols, Liu, Seymour and Vušković show that Berge graphs (and hence perfect graphs) can be recognized in polynomial time. As expected, cleaning (that, recall from Section 3, was developed in order to be able to use cutsets such as star cutsets in decomposition based recognition algorithms) was the key to the work in [22]. What was surprising, as Chudnovsky and Seymour observed, was that once the cleaning is performed, one does not need the decomposition based algorithm, one can simply look for the odd hole directly (using the shortest-paths detector technique described in Section 2.1). In [22] two recognition algorithms for Berge graphs are given: an $\mathcal{O}(n^9)$ Chudnovsky/Seymour style algorithm that uses the direct method, and an $\mathcal{O}(n^{18})$ decomposition based algorithm (that uses Theorem 9.12). Whether odd-hole-free graphs can be recognized in polynomial time remains an open problem.

Finding a maximum clique, a maximum independent set and an optimal coloring can all be done in polynomial time for perfect graphs. This result of Grötschel, Lovász and Schrijver uses the ellipsoid method and a polynomial time separation algorithm for a certain class of positive semidefinite matrices related to Lovász's upper bound on the Shannon capacity of a graph [90]. The question remains whether these optimization problems can be solved for perfect graphs by purely combinatorial polynomial time algorithms, avoiding the numerical instability of the ellipsoid method. Some partial results in this direction are described in Section 10. We observe that for a number of classes we have seen so far, the decomposition theorem is used to prove the existence of a vertex with a particular neighborhood, which is then used to obtain some optimization algorithms. Such a result is not yet known for perfect graphs.

The complexities of finding a maximum independent set and an optimal coloring for odd-hole-free graphs are not known. Finding a maximum clique is NP-hard for odd-hole-free graphs (follows from 2-subdivisions [105]: if G' is the graph obtained from G by subdividing every edge twice then $\alpha(G') = \alpha(G) + |E(G)|$; also all holes of G' are of length at least 9, and hence $\overline{G'}$ does not contain a hole of length at least 5).

10   Combinatorial optimization with 1-joins and 2-joins

In this section we consider how decompositions with 1-joins and 2-joins can be used for construction of different optimization algorithms.

10.1   1-Joins

1-Join decompositions (also known as split decompositions) were used for circle graph recognition [71, 116] and parity graph recognition [40, 60]. Cunningham [58] showed how 1-join decompositions can be used for the independent set problem, and Cicerone and Di Stefano [40] showed how this algorithm can be applied to parity graphs. Rao [106] shows how to use 1-join decompositions for the clique and coloring problems. To describe these results we first need to describe the blocks of decomposition by a 1-join.

Let $(X_1, X_2, A_1, \emptyset, \emptyset, A_2, \emptyset, \emptyset)$ be a 1-join of a graph G. The *block of decomposition by a 1-join* are graphs $G_1 = G[X_1 \cup \{m_2\}]$ (where m_2 is any vertex of A_2) and $G_2 = G[X_2 \cup \{m_1\}]$ (where m_1 is any vertex of A_1). It turns out that if a graph has a 1-join then it has an *extreme 1-join*, i.e. a 1-join where one of the blocks of decomposition does not have a 1-join. So one can construct an extreme decomposition tree by 1-joins, similarly to the extreme decomposition tree by clique cutsets described in Section 3.1. Dahlhaus shows that this decomposition tree can in fact be constructed in linear time ([60], see also [13]). (The first algorithm for decomposing a graph by 1-joins, of complexity $\mathcal{O}(n^3)$, was given in

[58], this was later improved to an $\mathcal{O}(nm)$ algorithm in [71], and to an $\mathcal{O}(n^2)$ algorithm in [91]). The fact that one can compute an extreme decomposition tree by 1-joins is quite useful when constructing optimization algorithms.

To solve the independent set, clique and coloring problems using 1-join decomposition we need to move to the weighted versions of these problems. Let $w : V(G) \longrightarrow \mathcal{N}$ be a weight function for a graph G. When H is an induced subgraph of G, $w(H)$ denotes the sum of the weights of vertices in H. By $\alpha_w(G)$ we denote the weight of a maximum weighted independent set of G, and by $\omega_w(G)$ we denote the weight of a maximum weighted clique of G. From the discussion above and the following lemmas it is easy to see how to obtain polynomial time algorithms to solve the weighted independent set and the weighted clique problems for graphs that are decomposable by 1-joins into basic graphs for which these problems can be solved in polynomial time. Similarly, one can also solve the weighted chromatic number problem [106]. For $a \in \mathcal{N}$, denote by $w|_{v \to a}$ the function on domain $V(G) \cup \{v\}$ such that $w|_{v \to a}(v) = a$ and $w|_{v \to a}(u) = w(u)$ for all $u \in V(G) \setminus \{v\}$.

Lemma 10.1 ([58]) *Let* $a = \alpha_w(G_2 \setminus N_{G_2}[m_1])$ *and* $a' = \alpha_w(G_2 \setminus m_1)$. *Then* $\alpha_w(G) = \alpha_{w|_{m_2 \to a' - a}}(G_1) + a$.

Lemma 10.2 ([106]) *Let* $a = \omega_w(G_2[N_{G_2}[m_1]])$. *Then* $\omega_w(G) = \max\{\omega_w(G_2 \setminus m_1), \omega_{w|_{m_2 \to a}}(G_1)\}$.

10.2   2-Joins

To use 2-joins in a decomposition based optimization algorithm is a lot more difficult. In [120] Trotignon and Vušković focused on developing techniques for combinatorial optimization with 2-joins, by considering two classes of graphs decomposable by 2-joins into basic graphs for which we know how to solve the respective optimization problems in polynomial time. They give combinatorial polynomial time algorithms for finding the size of a largest independent set in even-hole-free graphs with no star cutset; as well as finding the size of a largest independent set, the size of a largest clique and an optimal coloring for Berge graphs with no skew cutset, no 2-join in the complement and no homogeneous pair. The coloring algorithm can be implemented to run in $\mathcal{O}(n^7)$ time, and all the other ones in $\mathcal{O}(n^6)$ time. Coloring of Berge graphs actually follows from being able to compute the size of a largest independent set and

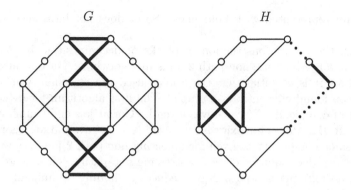

Figure 4: A graph G with no extreme 2-join.

largest clique ([76, 77]), so these two problems are the focus of the work in [120].

Using 2-joins in combinatorial optimization algorithms requires building blocks of decomposition and asking at least two questions for at least one block, while for recognition algorithms one question suffices. Applying this process recursively can lead to an exponential blow-up even when the decomposition tree is linear in size of the input graph. In [120] this problem is bypassed by using *extreme 2-joins*, i.e. 2-joins whose one block of decomposition is basic. Graphs in general do not have extreme 2-joins, this is a special property of 2-joins in graphs with no star cutset.

Consider the following way of constructing blocks of decomposition by 2-joins. Let $(X_1, X_2, A_1, B_1, \emptyset, A_2, B_2, \emptyset)$ be a 2-join of the graph G. The *blocks of decomposition by 2-join* are graphs $G_1^{k_1}$ and $G_2^{k_2}$ defined as follows. $G_1^{k_1}$ is obtained by replacing X_2 by a *marker path* P_2, of length k_1, from a vertex a_2 complete to A_1, to a vertex b_2 complete to B_1 (the interior of P_2 has no neighbor in X_1). The block $G_2^{k_2}$ is obtained similarly by replacing X_1 by a marker path P_1 of length k_2. It is easy to see that in an even-hole-free graph or an odd-hole-free graph, all paths from a node in A_i to a node in B_i with interior in $X_i \setminus (A_i \cup B_i)$ have the same parity. So if we are careful about the parity of the marker paths, the blocks of decomposition will be class-preserving for the classes of even-hole-free graphs and odd-hole-free graphs.

The graph G in Figure 4 has exactly two 2-joins, one is represented with bold lines, and the other is equivalent to it. Both of the blocks of decomposition are isomorphic to graph H (where dotted lines represent paths of arbitrary length, possibly of length 0), and H has a 2-join whose edges are represented with bold lines. So G does not have an extreme 2-join.

With the above construction of blocks of decomposition by 2-joins, clearly one needs to use non-path 2-joins in algorithms. For optimization algorithms it is essential that the 2-joins used are extreme. Can these 2-joins be found efficiently in a graph? The first algorithm for detecting 2-joins, of complexity $\mathcal{O}(n^3m)$, was given by Cornuéjols and Cunningham in [56]. In [12] this complexity is improved to $\mathcal{O}(n^2m)$, and an algorithm of the same complexity is given for detecting non-path 2-joins, as well as an $\mathcal{O}(n^3m)$ algorithm for detecting extreme non-path 2-joins. Note that finding an extreme non-path 2-join reduces to finding a minimally-sided non-path 2-join.

We now give a method from [120] that can be used to solve the maximum weighted clique problem for any class of graphs that can be decomposed with extreme (non-path) 2-joins into basic graphs for which the problem can be solved efficiently. To be able to apply arguments inductively one needs to switch to the weighted version of the problem. Let G be a weighted graph with a weight function $w : \mathcal{N}^+ \longrightarrow V(G)$. Let $\omega(G)$ denote the weight of a maximum weighted clique of G.

Let G_1 and G_2 be the blocks of decomposition by a 2-join of G. Let us also assume that the lengths of the marker paths are at least 3 (this is important in [120] because there it is not just important that the parity of holes is preserved in the blocks, but also the property of not having a star cutset). Let $P_1 = a_1, x_1, \ldots, x_k, b_1$ be the marker path of G_2, where a_1 is adjacent to all of A_2 and b_1 is adjacent to all of B_2. The weights of vertices of G_2 are modified as follows:

- for every $u \in X_2$, $w_{G_2}(u) = w_G(u)$;

- $w_{G_2}(a_1) = \omega(G[A_1])$;

- $w_{G_2}(b_1) = \omega(G[B_1])$;

- $w_{G_2}(x_1) = \omega(G[X_1]) - \omega(G[A_1])$;

- $w_{G_2}(x_i) = 0$, for $i = 2, \ldots, k$.

With such modification of weights it can be shown that $\omega(G) = \omega(G_2)$ [120]. Now if our 2-join is an extreme 2-join, we may assume that block G_1 is undecomposable and hence basic in the sense that the maximum weighted clique problem can be solved on that block efficiently. In particular, all of the weights needed to be computed for modifying the weights of G_2 as above can be computed efficiently. We note that this method of computing a maximum clique in the case of even-hole-free graphs (with no star cutset) is not so interesting since the algorithm described in Section 9.1 is more efficient.

Using 2-joins to compute a maximum stable set is more difficult since stable sets can completely overlap both sides of the 2-join. In [120] a simple class of graphs $\mathcal{C}$ decomposable along extreme 2-joins into bipartite graphs and line graphs of cycles with one chord is given for which computing a maximum stable set is NP-hard. Here is how $\mathcal{C}$ is constructed. A *gem-wheel* is a graph made of an induced cycle of length at least 5 together with a vertex adjacent to exactly four consecutive vertices of the cycle. Note that a gem-wheel is a line-graph of a cycle with one chord. A *flat path* of a graph G is a path of length at least 2, whose interior vertices all have degree 2 in G, and whose ends have no common neighbors outside the path. *Extending a flat path* $P = p_1, \ldots, p_k$ of a graph means deleting the interior vertices of P and adding three vertices x, y, z and the following edges: $p_1 x$, xy, yp_k, zp_1, zx, zy, zp_k. *Extending a graph* G means extending all paths of $\mathcal{M}$, where $\mathcal{M}$ is a set of flat paths of length at least 3 of G. Class $\mathcal{C}$ is the class of all graphs obtained by extending 2-connected bipartite graphs. From the definition, it is clear that all graphs of $\mathcal{C}$ are decomposable along extreme 2-joins. One leaf of the decomposition tree is the underlying bipartite graph, and all the others leaves are gem-wheels. The following is shown by Naves [100], and the proof of it can be found in [120].

Theorem 10.3 (Naves [100, 120]) *The problem whose instance is a graph G from $\mathcal{C}$ and an integer k, and whose question is "Does G contain a stable set of size at least k" is NP-complete.*

Let $\mathcal{C}^{PARITY}$ be the class of graphs in which all holes have the same parity. In [120] it is shown how to use 2-joins to compute a maximum stable set in $\mathcal{C}^{PARITY}$.

Let G be a graph with a weight function w on the vertices and $(X_1, X_2, A_1, B_1, \emptyset, A_2, B_2, \emptyset)$ a 2-join of G. For $i = 1, 2$, $D_i = X_i \setminus (A_i \cup B_i)$. For any graph H, $\alpha(H)$ denotes the weight of a maximum weighted stable set of H. Let $a = \alpha(G[A_1 \cup D_1])$, $b = \alpha(G[B_1 \cup D_1])$, $c = \alpha(G[D_1])$ and $d = \alpha(G[X_1])$. The blocks of decomposition by a 2-join that would be useful for computing a largest stable set can be done as follows.

A *flat claw* of a weighted graph G is any set $\{q_1, q_2, q_3, q_4\}$ of vertices such that:

- the only edges between the q_i's are $q_1 q_2$, $q_2 q_3$ and $q_4 q_2$;

- q_1 and q_3 have no common neighbor in $V(G) \setminus \{q_2\}$;

- q_4 has degree 1 in G and q_2 has degree 3 in G.

Define the *even block* G_2 with respect to a 2-join $X_1|X_2$ in the following way. Keep X_2 and replace X_1 by a flat claw on $q_1, \ldots, q_4$ where q_1 is complete to A_2 and q_3 is complete to B_2. Give the following weights: $w(q_1) = d - b$, $w(q_2) = c$, $w(q_3) = d - a$, $w(q_4) = a + b - d$. It can be shown that all weights are in fact non-negative.

A *flat vault* of graph G is any set $\{r_1, r_2, r_3, r_4, r_5, r_6\}$ of vertices such that:

- the only edges between the r_i's are such that r_3, r_4, r_5, r_6, r_3 is a 4-hole;

- $N(r_1) = N(r_5) \setminus \{r_4, r_6\}$;

- $N(r_2) = N(r_6) \setminus \{r_3, r_5\}$;

- r_1 and r_2 have no common neighbors;

- r_3 and r_4 have degree 2 in G.

Define the *odd block* G_2 with respect to a 2-join in the following way. Replace X_1 by a flat vault on $r_1, \ldots, r_6$. Moreover r_1, r_5 are complete to A_2 and r_2, r_6 are complete to B_2. Give the following weights: $w(r_1) = d - b$, $w(r_2) = d - a$, $w(r_3) = w(r_4) = c$, $w(r_5) = w(r_6) = a + b - c - d$. It can be shown that all weights are non-negative, if $c + d \leq a + b$ holds.

By adequately choosing when to use even or odd blocks, it can be shown that for a 2-join in a graph G in $\mathcal{C}^{PARITY}$, $\alpha(G_2) = \alpha(G)$.

We observe that such construction of blocks is not class-preserving, so it would not allow for inductive use of the decomposition theorems. This problem is avoided in [120] by building the decomposition tree in two stages. First using blocks of decomposition constructed as we discussed at the beginning of this section (that are class-preserving). In the second stage the decomposition tree is reprocessed to replace marker paths by gadgets designed for even and odd blocks. This results in the leaves of the decomposition tree that are not basic as in the decomposition theorems used, but some extensions of these basic classes, for which it is shown that the weighted stable set problem can be computed efficiently.

Recently this work was extended by Chudnovsky, Trotignon, Trunck and Vušković [37] to obtain an $\mathcal{O}(n^7)$ coloring algorithm for perfect graphs with no balanced skew-partition, by focusing on decompositions by 2-joins, their complements and homogeneous pairs. Here the notion of trigraphs was quite helpful in obtaining the desired extreme decompositions.

Acknowledgements

We thank Maria Chudnovsky and Nicolas Trotignon for helpful discussions about the presentation of some of the material. This work was partially supported by EPSRC grant EP/H021426/1 and Serbian Ministry of Education and Science projects 174033 and III44006.

References

[1] L. Addario-Berry, M. Chudnovsky, F. Havet, B. Reed, and P. Seymour, Bisimplicial vertices in even-hole-free graphs, *Journal of Combinatorial Theory Series B* **98** (2008), 1119–1164.

[2] P. Aboulker, P. Charbit, M. Chudnovsky, N. Trotignon, and K. Vušković, LexBFS, structure and algorithms, preprint (2012), submitted.

[3] P. Aboulker, M. Radovanović, N. Trotignon, and K. Vušković, Graphs that do not contain a cycle with a node that has at least two neighbors on it, to appear in *SIAM Journal on Discrete Mathematics*.

[4] P. Aboulker, M. Radovanović, N. Trotignon, T. Trunck, and K. Vušković, Linear balanceable and subcubic balanceable graphs, preprint (2012), submitted.

[5] C. Berge, Färbung von Graphen, deren sämtliche bzw. deren ungerade Kreise starr sind (Zusammenfassung), *Technical report, Wiss. Z. Martin Luther Univ. Math.-Natur. Reihe (Halle-Wittenberg)* (1961).

[6] C. Berge, Sur Certains Hypergraphes Généralisant les Graphes Bipartis, in *Combinatorial Theory and its Applications* I (eds. P. Erdös, A. Rényi and V. Sós), *Colloquia Mathematica Societatis János Bolyai* **4**, North-holland, Amsterdam (1970), 119–133.

[7] M. Burlet and J. Fonlupt, Polynomial algorithm to recognize a Meyniel graph, *Discrete Math* **21** (1984), 225-252.

[8] K. Cameron, Antichain sequences, *Order* **2** (1985), 249–255.

[9] K. Cameron, B. Lévêque, and F. Maffray, Coloring vertices of a graph or finding a Meyniel Obstruction, *Theoretical Computer Science* **428** (2012), 10–17.

[10] P. Camion, Characterization of totally unimodular matrices, *Proceedings of the American Mathematical Society* **16** (1965), 1068–1073.

[11] H.-C. Chang and H.-I. Lu, A faster algorithm to recognize even-hole-free graphs, *Proceedings of 23rd Annual ACM-SIAM Symposium on Discrete Algorithms* (2012), 1286–1297.

[12] P. Charbit, M. Habib, N. Trotignon, and K. Vušković, Detecting 2-joins faster, to appear in *Journal of Discrete Algorithms*.

[13] P. Charbit, F. de Montgolfier, and M. Raffinot, Linear time split decomposition revisited, *SIAM Journal on Discrete Mathematics* **26** **(2)** (2012), 499–514.

[14] V. Chvátal and N. Sbihi, Bull-free Berge graphs are perfect, *Graphs and Combinatorics* **3** (1987), 127–139.

[15] V. Chvátal and N. Sbihi, Recognizing claw-free Berge graphs, *Journal of Combinatorial Theory Series B* **44** (1988), 154–176.

[16] M. Chudnovsky, Berge trigraphs and their applications, PhD thesis, Princeton University (2003).

[17] M. Chudnovsky, Berge trigraphs, *Journal of Graph Theory* **53** **(1)** (2006), 1–55.

[18] M. Chudnovsky, The structure of bull-free graphs I: Three-edge paths with centers and anticenters, *Journal of Combinatorial Theory Series B* **102** **(1)** (2012), 233–251.

[19] M. Chudnovsky, The structure of bull-free graphs II and III: A summary, *Journal of Combinatorial Theory Series B* **102** **(1)** (2012), 252–282.

[20] M. Chudnovsky, The structure of bull-free graphs II: elementary trigraphs, manuscript.

[21] M. Chudnovsky, The structure of bull-free graphs III: global structure, manuscript.

[22] M. Chudnovsky, G. Cornuéjols, X. Liu, P. Seymour, and K. Vušković, Recognizing Berge graphs, *Combinatorica* **25** **(2)** (2005), 143–186.

[23] M. Chudnovsky and R. Kapadia, Detecting a theta or a prism, *SIAM Journal on Discrete Math* **22** (2008), 1164–1186.

[24] M. Chudnovsky, K. Kawarabayashi, and P. Seymour, Detecting even holes, *Journal of Graph Theory* **48** (2005), 85–111.

[25] M. Chudnovsky and I. Penev, The structure of bull-free perfect graphs, to appear in *Journal of Graph Theory*.

[26] M. Chudnovsky, N. Robertson, P. Seymour, and R. Thomas, The strong perfect graph theorem, *Annals of Mathematics* **164** (1) (2006), 51–229.

[27] M. Chudnovsky and S. Safra, The Erdős-Hajnal conjecture for bull-free graphs, *Journal of Combinatorial Theory Series B* **98** (2008), 1301–1310.

[28] M. Chudnovsky and P. Seymour, Solution of three problems of Cornuéjols, *Journal of Combinatorial Theory Series B* **98** (2008), 116–135.

[29] M. Chudnovsky and P. Seymour, The three-in-a-tree problem, *Combinatorica* **30** (**4**) (2010), 387–417.

[30] M. Chudnovsky and P. Seymour, Claw-free Graphs I: Orientable prismatic graphs, *Journal of Combinatorial Theory Series B* **97** (2007), 867–903.

[31] M. Chudnovsky and P. Seymour, Claw-free Graphs II: Non-orientable prismatic graphs, *Journal of Combinatorial Theory Series B* **98** (2008), 249–290.

[32] M. Chudnovsky and P. Seymour, Claw-free Graphs III: Circular interval graphs, *Journal of Combinatorial Theory Series B* **98** (2008), 812–834.

[33] M. Chudnovsky and P. Seymour, Claw-free Graphs IV: Decomposition theorem, *Journal of Combinatorial Theory Series B* **98** (2008), 839–938.

[34] M. Chudnovsky and P. Seymour, Claw-free Graphs V: Global structure, *Journal of Combinatorial Theory Series B* **98** (2008), 1373–1410.

[35] M. Chudnovsky and P. Seymour, Claw-free Graphs VI: Colouring, *Journal of Combinatorial Theory Series B* **100** (2010), 560–572.

[36] M. Chudnovsky and P. Seymour, Claw-free Graphs VII: Quasi-line graphs, *Journal of Combinatorial Theory Series B* **102** (2012), 1267–1294.

[37] M. Chudnovsky, N. Trotignon, T. Trunck, and K. Vušković, Coloring perfect graphs with no balanced skew-partitions, preprint (2012), submitted.

[38] V. Chvátal, Star cutsets and perfect graphs, *Journal of Combinatorial Theory Series B* **39** (1985), 189–199.

[39] V. Chvátal and N. Sbihi, Bull-free Berge graphs are perfect, *Graphs and Combinatorics* **3** (1987), 127–139.

[40] S. Cicerone and D. Di Stefano, On the extension of bipartite graphs to parity graphs, *Discrete Applied Mathematics* **95** (1999), 181–195.

[41] M. Conforti, G. Cornuéjols, G. Gasparyan, and K. Vušković, Perfect graphs, partitionable graphs and cutsets, *Combinatorica* **22 (1)** (2002), 19–33.

[42] M. Conforti, G. Cornuéjols, A. Kapoor, and K. Vušković, Universally signable graphs, *Combinatorica* **17 (1)** (1997), 67–77.

[43] M. Conforti, G. Cornuéjols, A. Kapoor, and K. Vušković, Even and odd holes in cap-free graphs, *Journal of Graph Theory* **30** (1999), 289–308.

[44] M. Conforti, G. Cornuéjols, A. Kapoor, and K. Vušković, Balanced $0, \pm 1$ matrices, Part I: Decomposition theorem, and Part II: Recognition algorithm, *Journal of Combinatorial Theory Series B* **81** (2001), 243–306.

[45] M. Conforti, G. Cornuéjols, A. Kapoor, and K. Vušković, Even-hole-free graphs, Part I: Decomposition theorem, *Journal of Graph Theory* **39** (2002), 6–49.

[46] M. Conforti, G. Cornuéjols, A. Kapoor, and K. Vušković, Even-hole-free graphs, Part II: Recognition algorithm, *Journal of Graph Theory* **40** (2002), 238–266.

[47] M. Conforti, G. Cornuéjols, and M. R. Rao, Decomposition of balanced matrices, *Journal of Combinatorial Theory Series B* **77** (1999), 292–406.

[48] M. Conforti, G. Cornuéjols, and K. Vušković, Square-free perfect graphs, *Journal of Combinatorial Theory Series B* **90** (2004), 257–307.

[49] M. Conforti, G. Cornuéjols, and K. Vušković, Decomposition of odd-hole-free graphs by double star cutsets and 2-joins, *Discrete Applied Mathematics* **141** (2004), 41–91.

[50] M. Conforti, G. Cornuéjols, and K. Vušković, Balanced matrices, *Discrete Mathematics* **306** (2006), 2411–2437.

[51] M. Conforti and A. M. H. Gerards, *Stable sets and graphs with no even holes*, preprint (2003), unpublished.

[52] M. Conforti, B. Gerards, and A. Kapoor, A theorem of Truemper, *Combinatorica* **20** (**1**) (2000), 15–26.

[53] M. Conforti and M. R. Rao, Structural properties and decomposition of linear balanced matrices, *Mathematical Programming* **55** (1992), 129–168.

[54] M. Conforti and M. R. Rao, Testing balancedness and perfection of linear matrices, *Mathematical Programming* **61** (1993), 1–18.

[55] G. Cornuéjols, Combinatorial Optimization: Packing and Covering, *SIAM*, Philadelphia, PA (2001). CBMS-NSF Regional Conference Series in Applied Mathematics 74.

[56] G. Cornuéjols and W. H. Cunningham, Composition for perfect graphs, *Discrete Mathematics* **55** (1985), 245–254.

[57] B. Courcelle, The monadic second order logic of graphs. I. Recognizable sets of finite graphs, *Information and Computation* **85** (**1**) (1990), 12–75.

[58] W. H. Cunningham, Decomposition of directed graphs, *SIAM Journal on Algebraic and Discrete Methods* **3** (**2**) (1982), 214–228.

[59] W. H. Cunningham and J. Edmonds, A combinatorial decomposition theory, *Canadian Journal of Mathematics* **32** (**3**) (1980), 734–765.

[60] E. Dahlhaus, Parallel algorithms for hierarchical clustering and applications to split decomposition and parity graph recognition, *Journal of Algorithms* **36** (**2**) (2000), 205–240.

[61] M. V. G. da Silva and K. Vušković, Triangulated neighborhoods in even-hole-free graphs, *Discrete Mathematics* **307** (2007), 1065–1073.

[62] M. V. G da Silva and K. Vušković, Decomposition of even-hole-free graphs with star cutsets and 2-joins, to appear in *Journal of Combinatorial Theory Series B*.

[63] C. M. H. de Figueiredo and F. Maffray, Optimizing bull-free perfect graphs, *SIAM Journal on Discrete Mathematics* **18** (2004), 226–240.

[64] C. M. H. de Figueiredo and K. Vušković, A class of β-perfect graphs, *Discrete Mathematics* **216** (2000), 169–193.

[65] G. A. Dirac, On rigid circuit graphs, *Abh. Math. Sem. Univ. Hamburg* **25** (1961), 71–76.

[66] G. A. Dirac, Minimally 2-connected graphs, *Journal für die Reine und Angewandte Mathematik* **228** (1967), 204–216.

[67] J. Edmonds, Maximum matching and a polytope with 0,1-vertices, *J. Res. Nat. Bur. Standards* **69B** (1965), 125–130.

[68] J. Edmonds, Paths, trees and flowers, *Canad. J. Math.* **17** (1965), 449–467.

[69] P. Erdős and A. Hajnal, Ramsey-type theorems, *Discrete Applied Mathematics* **25** (1989), 37–52.

[70] M. Farber, On diameters and radii of bridged graphs, *Discrete Mathematics* **73** (1989), 249–260.

[71] C. P. Gabor, W. L. Hsu, and K. J. Supowit, Recognizing circle graphs in polynomial time, *Journal of the ACM* **36 (3)** (1989), 435–473.

[72] T. Gallai, Graphen mit triangulierbaren ungeraden Vielecken, *Magyar Tud. Akad. Mat. Kutado Int. Közl.* **7** (1962), 3–36. English translation given by F. Maffray and M. Preissmann in *Perfect Graphs*, J. L. Ramírez-Alfonsín and B. A. Reed (editors), Willey (2001), 25–66.

[73] M. Garey and D. S. Johnson, *Computers and Intractability: A Guide to the Theory of NP-completeness*, W. H. Freeman, San Francisco (1979).

[74] J. Geelen, B. Gerards, and G. Whittle, Towards a matroid-minor structure theory, *Combinatorics, Complexity and Chance*, A tribute to Dominic Welsh, (eds. G. Grimmett and C. McDiarmid), Oxford University Press (2007), 72–82.

[75] R. Giles, L. E. Trotter, and A. Tucker, The strong perfect graph theorem for a class of partition able graphs, in *Topics on Perfect Graphs* (C. Berge and V. Chvátal, eds.), North Holland, Amsterdam (1984), 161–167.

[76] M. Grötschel, L. Lovász, and A. Schrijver, The ellipsoid method and its consequences in combinatorial optimization, *Combinatorica* **1** (1981), 169–197.

[77] M. Grötschel, L. Lovász, and A. Schrijver, *Geometric algorithms and combinatorial optimization*, Springer Verlag (1988).

[78] R. B. Hayward, Weakly triangulated graphs, *Journal of Combinatorial Theory Series B* **39** (1985), 200–209.

[79] P. J. Heawood, Map colour theorem, *Quart J Pure Appl Math* **24** (1890), 332–338.

[80] A. Hertz, A fast algorithm for colouring Meyniel graphs, *Journal of Combinatorial Theory Series B* **50** (1990), 231–240.

[81] R. Hayward, C. T. Hoàng and F. Maffray, Optimizing weakly triangulated graphs, *Graphs and Combinatorics* **5** (1989), 339–349. See erratum in vol. 6 (1990) 33–35.

[82] C. T. Hoàng, Efficient algorithms for minimum weighted colouring of some classes of perfect graphs, *Discrete Applied Mathematics* **55** (1994), 133–143.

[83] W. L. Hsu, Efficient algorithms for some packing and covering problems on graphs, Ph.D. Dissertation, School of Operations Research and Industrial Engineering, Cornell University, Ithaca, New York, (1980).

[84] W. L. Hsu and G. L. Nemhauser, Algorithms for maximum weighted cliques, minimum weighted clique covers, and minimum colorings of claw-free perfect graphs, in *Topics on Perfect Graphs* (eds. C. Berge and V. Chvátal), North-Holland, Amsterdam (1984), 357–369.

[85] T. Kloks, H. Müller, and K. Vušković, Even-hole-free graphs that do not contain diamonds: a structure theorem and its consequences, *Journal of Combinatorial Theory Series B* **99** (2009), 733–800.

[86] B. Lévêque, D. Y. Lin, F. Maffray, and N. Trotignon, Detecting induced subgraphs, *Discrete Applied Mathematics* **157** (2009), 3540–3551.

322   K. Vušković

[87] B. Lévêque, F. Maffray, and N. Trotignon, On graphs with no induced subdivision of K_4, *Journal of Combionatorial Theory Series B* **102** (2012), 924–947.

[88] X. Li and W. Zang, A combinatorial algorithm for minimum weighted coloring of claw-free perfect graphs, *Journal of Combinatorial Optimization* **9 (4)** (2005), 331–347.

[89] L. Lovász, Normal hypergraphs and the perfect graph conjecture, *Discrete Mathematics* **2** (1972), 253–267.

[90] L. Lovász, On the Shannon capacity of a graph, *IEEE Trans. Inform. Theory* **25** (1979), 1–7.

[91] T.-H. Ma and J. Spinrad, An $\mathcal{O}(n^2)$ algorithm for undirected split decomposition, *Journal of Algorithms* **16 (1)** (1994), 154–160.

[92] F. Maffray and M. Preissmann, On the NP-completeness of the k-colorability problem for triangle-free graphs, *Discrete Math.* **162** (1996), 313–317.

[93] F. Maffray and B. Reed, A description of claw-free perfect graphs, *Journal of Combinatorial Theory Series B* **75** (1999), 134–156.

[94] F. Maffray and N. Trotignon, Algorithms for perfectly contractile graphs, *SIAM Journal on Discrete Mathematics* **19 (3)** (2005), 553–574.

[95] F. Maffray, N. Trotignon, and K. Vušković, Algorithms for square-$3PC(\cdot,\cdot)$-free graphs, *SIAM Journal on Discrete Math* **22 (1)** (2008), 51–71.

[96] S. E. Markossian, G. S. Gasparian, and B. A. Reed, β-perfect graphs, *Journal of Combinatorial Theory Series B* **67** (1996), 1–11.

[97] T. McKee, Independent separator graphs, *Utilitas Mathematica* **73** (2007), 217–224.

[98] G. J. Minty, On maximal independent sets of vertices in claw-free graphs, *Journal of Combinatorial Theory Series B* **28** (1980), 284–304.

[99] D. Nakamura and A. Tamura, A revision of Minty's algorithm for finding a maximum weighted stable set of a claw-free graph, *Journal of Operations Research Society of Japan* **44 (2)** (2001), 194–204.

[100] G. Naves, personal communication (2009).

[101] K. R. Parthasarathy and G. Ravindra, The strong perfect graph conjecture is true for $K_{1,3}$-free graphs, *Journal of Combinatorial Theory Series B* **21** (1976), 212–223.

[102] I. Penev, Forbidden substructures in graphs and trigraphs, and related coloring problems, PhD thesis, Columbia University (2012).

[103] J. Petersen, Sur le théorèm de Tait, *L'Intermédiaire Math* **5** (1898), 225–227.

[104] M. D. Plummer, On minimal blocks, *Transactions of the American Mathematical Society* **134** (1968), 85–94.

[105] S. Poljak, A note on the stable sets and coloring of graphs, *Comment. Math. Univ. Carolin.* **15** (1974), 307–309.

[106] M. Rao, Solving some NP-complete problems using split decomposition, *Discrete Applied Mathematics* **156 (14)** (2008), 2768–2780.

[107] B. Reed and N. Sbihi, Recognizing bull-free perfect graphs, *Graphs and Combinatorics* **11** (1995), 171–178.

[108] N. Robertson and P. D. Seymour, Graph minors III. Planar treewidth, *Journal of Combinatorial Theory Series B* **36** (1984), 49–64.

[109] N. Robertson and P. D. Seymour, Graph minors V. Excluding a planar graph, *Journal of Combinatorial Theory Series B* **41** (1986), 92–114.

[110] N. Robertson and P. D. Seymour, Graph minors XVI. Excluding a non-planar graph, *Journal of Combinatorial Theory Series B* **89** (2003), 43–76.

[111] N. Robertson and P. D. Seymour, Graph minors XX. Wagner's Conjecture, *Journal of Combinatorial Theory Series B* **92 (2)** (2004), 325–357.

[112] D. J. Rose, R. E. Tarjan and G. S. Leuker, Algorithmic aspects of vertex elimination on graphs, *SIAM J. Comput.* **5** (1976), 266–283.

[113] F. Roussel and I. Rusu, Holes and dominoes in Meyniel graphs, *Internat. J. Found. Comput. Sci.* **10** (1999), 127–146.

[114] F. Roussel and I. Rusu, An $\mathcal{O}(n^2)$ algorithm to color Meyniel graphs, *Discrete Mathematics* **235 (1–3)** (2001), 107–123.

[115] P. Seymour, Decomposition of regular matroids, *Journal of Combinatorial Theory Series B* **28** (1980), 305–359.

[116] J. Spinrad, Recognition of circle graphs, *Journal of Algorithms* **16 (2)** (1994), 264–282.

[117] R. E. Tarjan, Decomposition by clique separators, *Discrete Mathematics* **55** (1985), 221–232.

[118] N. Trotignon, Decomposing Berge graphs and detecting balanced skew partitions, *Journal of Combinatorial Theory Series B* **98** (2008), 173–225.

[119] N. Trotignon and K. Vušković, A structure theorem for graphs with no cycle with a unique chord and its consequences, *Journal of Graph Theory* **63 (1)** (2010), 31–67.

[120] N. Trotignon and K. Vušković, Combinatorial optimization with 2-joins, *Journal of Combinatorial Theory Series B* **102** (2012), 153–185.

[121] K. Truemper, Alpha-balanced graphs and matrices and GF(3)-representability of matroids, *Journal of Combinatorial Theory Series B* **32** (1982), 112–139.

[122] K. Truemper, A decomposition theory for matroids. V. Testing of matrix total unimodularity, *Journal of Combinatorial Theory Series B* **49** (1990), 241–281.

[123] K. Truemper, *Matroid Decomposition*, Academic Press, Boston (1992).

[124] W. T. Tutte, A homotopy theorem for matroids I,II, *Transactions of the American Mathematical Society* **88** (1958), 144–174.

[125] K. Vušković, Even-hole-free graphs: a survey, *Applicable Analysis and Discrete Mathematics* **4 (2)** (2010), 219–240.

[126] S. H. Whitesides, An algorithm for finding clique cut-sets, *Information Processing Letters* **12** (1981), 31–32.

[127] G. Zambelli, On perfect graphs and balanced matrices, PhD thesis, Carnegie Mellon University (2004).

School of Computing
University of Leeds
Leeds LS2 9JT, UK
and
Faculty of Computer Science (RAF)
Union University
Knez Mihajlova 6/VI
11000 Belgrade, Serbia
k.vuskovic@leeds.ac.uk

Structure in minor-closed classes of matroids

Jim Geelen, Bert Gerards and Geoff Whittle

Abstract

This paper gives an informal introduction to structure theory for minor-closed classes of matroids representable over a fixed finite field. The early sections describe some historical results that give evidence that well-defined structure exists for members of such classes. In later sections we describe the fundamental classes and other features that necessarily appear in structure theory for minor-closed classes of matroids. We conclude with an informal statement of the structure theorem itself. This theorem generalises the Graph Minors Structure Theorem of Robertson and Seymour.

1 Introduction

For the last thirteen years we have been involved in a collaborative project to generalise the results of the Graph Minors Project of Robertson and Seymour to matroids representable over finite fields. The banner theorems of the Graph Minors Project are that graphs are well-quasi-ordered under the minor order [34] (that is, in any infinite set of graphs there is one that is isomorphic to a minor of another) and that for each minor-closed class of graphs there is a polynomial-time algorithm for recognising membership of the class [32]. We are well on track to extend these theorems to the class of $\mathbb{F}$-representable matroids for any finite field $\mathbb{F}$.

It is important to point out here that day-to-day work along this track does not concern well-quasi-ordering or minor testing. The actual task and true challenge is to gain insight into the structure of members of proper minor-closed classes of graphs or matroids. The well-quasi-ordering and minor-testing results are consequences – not necessarily easy ones – of the structure that is uncovered. Ironically, while one may begin studying structure with the purpose of obtaining marketable results, in the end it is probably the structural theorems themselves that are the most satisfying aspect of a project like this. To acquire that structural insight is the bulk of the work and the theorems that in the end describe the entire structure are the main deliveries of a project like this.

The point is related to topics discussed in two interesting articles by Tim Gowers [15, 16]. The latter article deals, amongst other things, with the role played by "rough structure theorems" in combinatorics. From this perspective Szemeredi's Regularity Lemma can be regarded as a rough structure theorem for all graphs. The Regularity Lemma has become an indispensable tool and has been used to prove profound theorems. Of course there is no such thing as an all-purpose tool. As Gowers points out it tells us nothing that would enable us to find better bounds for Ramsey's Theorem. There is currently no tool, analogous to the regularity lemma, available for finding such bounds. He remarks in [15] that

> A better bound seems to demand a more global argument, involving the whole graph, and there is simply no adequate model for such an argument in graph theory.

A little further on he is led to

> fantasise that there might be a sort of classification of red-blue colourings (or equivalently graphs) which would enable one to solve the problem ...

On reading both papers it seems clear that – stated in a language suited to the polemical purpose of this introduction – Gowers would like a rough structure theorem that gives the *global* structure of members of the class of graphs that have no fixed clique as an induced subgraph. We know from Ramsey's theorem that such graphs must have arbitrarily large holes, but it is an understanding of the global structure that would potentially give improved bounds.

Recall that a *minor* of a graph is obtained by deleting and contracting edges. If, instead of asking for the global structure of a graph with no induced K_n for some fixed n, we ask for the global structure of a graph with no K_n *minor*, then such a structure theorem already exists. That theorem is the Graph Minors Structure Theorem of Robertson and Seymour [33]. This theorem tells us that members of the class of graphs with no fixed K_n minor are graphs that have a tree-like decomposition into graphs that are "essentially planar". Here essentially planar allows departure from planarity in three distinct but *bounded* ways where the bounds are a function of the size of the clique we are excluding. The Graph Minors Structure Theorem is truly the workhorse of the Graph Minors Project and it is hard to see how the other goals of the project could be achieved without it.

Early on Robertson and Seymour realised that their results were likely to be special cases of more general results for matroids represented over finite fields. If you are not familiar with matroids, then it suffices at this stage to be aware of the following. Any finite set of vectors in a vector space over a field $\mathbb{F}$ has an associated matroid M that captures the key combinatorial properties of the collection. The matroids associated with such collections of vectors are the $\mathbb{F}$-representable matroids. Any graph G

has an associated matroid that is $\mathbb{F}$-representable for any field $\mathbb{F}$. There is a natural notion of substructure in matroids that we call "minor". Minors in graphs and matroids correspond in that the matroid associated with a minor of a graph G is a minor of the matroid associated with G.

Let q be a prime power. Robertson and Seymour conjectured – although apparently not in print – that the class of $GF(q)$-representable matroids is well-quasi-ordered, that is, that in any infinite set of such matroids there is one that is isomorphic to a minor. As said above we are well on track to prove this conjecture and also the conjecture that membership testing for minor-closed classes of $GF(q)$-representable matroids can be performed in polynomial time. Not that we have already laid out full proofs of these two conjectures, but we have achieved a structure theorem for minor-closed classes of $GF(q)$-representable matroids and we know how to apply that to well-quasi-ordering and membership testing. But, actually carrying out these applications, does require a full write-up of the proof of the structure theorem. That write-up will be laborious and so will take some time: the structure alone is already more involved for matroids than for graphs, and proving the Graph Minor Structure Theorem required two thirds of the 750 pages of the 23 Graph Minors papers. We add that, while it would be inconceivable to prove a structure theorem for matroids without having the Graph Minors Structure Theorem as a guide, the structural results that we have obtained are very far from being an incremental extension of the Graph Minors Project.

We expect our structure theorem to have numerous applications other than well-quasi-ordering and minor testing. We mention one here. Rota [36] conjectured that for any finite field $\mathbb{F}$ there are a finite number of minor-minimal obstructions to representability over $\mathbb{F}$. Rota's Conjecture is probably the most famous conjecture in matroid theory. We expect that the structure theorem will prove an essential ingredient in a future proof of Rota's Conjecture.

Having said that, from now on we forget about applications and focus simply on the structure theory itself. The goal of this paper is not to give an explicit formal statement of a matroid minors structure theorem. That would involve technicalities and complications that would cloud the overall picture. Instead we adopt the following strategy. Sections 3, 4 and 5 present background material that gives evidence for the existence of structure in minor-closed classes of matroids. In Sections 6–10 we discuss topics that are necessary to an understanding of the structure theory. Structure is outside our control; it is what it is. The goal of these sections is primarily to make it clear why certain features necessarily play a role in a structure theorem. In Section 11, we present a very informal statement of the structure theorem itself. If, as a reader, you go away feeling that you have gained some insight into what this branch of structural mathematics is about we would be absolutely delighted.

2   A matroid primer

This paper does not require an extensive background in matroid theory, although it is unlikely that a complete matroid novice would get much from reading it. Terminology and notation generally follows Oxley's comprehensive text [28]. Oxley has also written a concise introduction to matroids [29]; an updated version of this paper is available on his website. In this section we cover some basic topics giving a slant that is relevant to a reading of this paper.

If A is a matrix over a field $\mathbb{F}$ with columns indexed by a finite set E, then the *column matroid* of A is the pair $M(A) = (E, r)$, where, for a subset X of E, we have $r(X)$ is the rank of the set of columns indexed by X. It is easy to see that $M(A)$ satisfies the following:

(i) $0 \le r(X) \le |X|$ for each $X \subseteq E$;

(ii) $r(X) \le r(Y)$ whenever $X \subseteq Y \subseteq E$;

(iii) $r(X \cup Y) + r(X \cap Y) \le r(X) + r(Y)$ for all $X, Y \subseteq E$.

More generally, if E is a finite set and $r : 2^E \to \mathbb{Z}$ is a function, then the pair $M = (E, r)$ is a *matroid* with *rank function* r if the above properties hold. The set E is the *ground set* of M and the members of E are the *elements* of M. The column matroids of matrices over $\mathbb{F}$ are called $\mathbb{F}$-*representable* or *representable over* $\mathbb{F}$.

It is almost certainly a rare property of a matroid to be representable over some field, yet, amazingly enough, it seems that this has yet to be proved [27]. But that is an aside, our interest here is primarily in representable matroids. The rank of a set of columns in a matrix is preserved under row operations and column scaling. It follows that column matroids are preserved under such operations. Readers of this paper with little prior knowledge of matroid theory could do worse than think of representable matroids as column labelled matrices where it is understood that two matrices are equivalent if one can be obtained from the other by row operations and column scaling. While it is certainly possible for two matroids that are not equivalent in the above sense to have identical column matroids it turns out that, for the problems that we discuss here, the distinction is not significant. (Nonetheless, in other contexts, so-called *inequivalent* representations of matroids are genuinely problematic and have been well studied in recent years [14].) Modulo this subtlety one can regard properties of a matrix that are invariant under row operations as being its matroidal properties. Note that row operations are just about the first thing that students learn to do in linear algebra so there are no surprises in the observation that many natural properties of matrices are matroidal.

Terminology for matroids is frequently inherited from matrices and vectors. For example, a set is *independent* if its rank is equal to its cardinality; otherwise it is *dependent*. A *basis* of a matroid M is a maximal independent set. Readers with some familiarity with matroids may have seen them defined differently from here; most likely from their independent sets. It is a feature of the subject that matroids can be defined in many different, but equivalent, ways.

Graphic matroids

Other terminology is inherited from graph theory, for example a minimal dependent set is a *circuit*. The reason for this is that graphs also provide us with a fundamental class of matroids. Let $G = (V, E)$ be a graph and $\mathbb{F}$ be a field. A *signed incidence matrix* of G is a matrix $A \in \mathbb{F}^{V \times E}$ where each column has exactly two nonzero entries, a 1 and a -1 in the rows labelled by its incident vertices (if the edge is a loop, then the column has only zeros). The matroid obtained is independent of the sign pattern and the field. It is called the *cycle matroid* of the graph G and is denoted by $M(G)$. Note that the elements of the cycle matroid are the *edges* of the graph. A set of edges is independent in $M(G)$ if and only if it is the edge set of a forest in G. All matroids obtained in this way are called *graphic*.

At one extreme it may seem that cycle matroids do not capture much of the graph. Indeed all forests with edge set E have the same cycle matroid, a rather trivial one where every set is independent. But with mild extra connectivity the picture changes quickly: a 3-connected graph can, up to the labels of the vertices, be recovered completely from its cycle matroid.

The fact that the elements of a graphic matroid correspond to edges of an associated graph creates a small wrinkle for the uninitiated. The vector space perspective makes it natural to think of the elements of a matroid as points, not edges. It's not hard to reconcile the perspectives of course. Let A be a matrix obtained from a graph G as described above. Let B be a $|V| \times |V|$ identity matrix and consider the matrix $[B|A]$. Each element of the graphic matroid $M(G)$ lies in the span of a pair of elements of B. Thus B plays a role akin to the vertices of G and the elements of the cycle matroid are seen as naturally corresponding to edges of a graph. More generally we can see that any matroid with a basis B having the property that all other elements are in the span of pairs of elements of this basis will have some sort of graphic structure. We'll return to this topic when we discuss frame matroids.

Minors and duality

For a subset D of E, the *deletion* of D from M is the matroid $M \backslash D = (E - D, r')$ where $r'(X) = r(X)$ for all subsets X of $E - D$. In the case that M is the column matroid of a matrix A, then $M \backslash D$ is the column matroid of the matrix obtained by removing the columns labelled by D from A. For a subset C of E, the *contraction* of C from M is the matroid $M/C = (E - C, r')$, where $r'(X) = r(X \cup C) - r(C)$ for all $X \subseteq E - C$. Geometrically contraction corresponds to projecting the vectors labelled by $E - C$ from the subspace spanned by C onto a subspace that is skew to C. From a matrix perspective, if $M = M(A)$ for a matrix A, then we can obtain a representation of M/C by first performing row operations so that the columns labelled by C are in echelon form, then removing the rows that have nonzero entries in a column labelled by C, and finally removing the columns labelled by C. A *minor* of M is a matroid of the form $M \backslash D / C$ for some disjoint subsets C and D of E. Minors form a natural partial order on matroids. A class of matroids is *minor closed* if every minor of a member of the class also belongs to the class. Evidently the class of $\mathbb{F}$-representable matroids is minor closed.

Given a matroid M with ground set E, there is a unique matroid M^* with ground set E having the property that B^* is a basis of M^* if and only if $B^* = E - B$ for some basis B of M. We say that M^* is the *dual* of M. Evidently $(M^*)^* = M$. Moreover duality interchanges deletion and contraction in that $(M \backslash X)^* = M^*/X$ for any set X of elements of M.

Matroid duality turns out to be perfectly natural in our fundamental examples. For example, if M is represented over $\mathbb{F}$ by the matrix A, then M^* is also $\mathbb{F}$-representable. Indeed, modulo appropriate labelling of the columns, a matrix A' represents M^* if and only if the row spaces of A and A' are orthogonal subspaces. It follows from this, that if A is regarded as the generator matrix of a linear code, then the parity-check matrix of that code is a matrix representing M^*.

More surprising, and somehow truly fundamental, is the following fact. If G is a graph, then the dual of the cycle matroid of G is graphic if and only if G is planar. Moreover, in the case that G is a planar graph, then the dual of the cycle matroid of G is the cycle matroid of G^* where G^* is any planar dual of G. Thus a matroid property defined in a purely combinatorial or algebraic way is seen to have a fundamental connection with a topological property. We will see more of this phenomenon. It is one of the remarkable aspects of the theory of matroid minors and, indeed, of graph minors.

A matroid is called *cographic* if it has the form $M(G)^*$ for some graph G. Cographic matroids are sometimes called *bond* or *cocycle* matroids as the circuits of $M(G)^*$ are the bonds of G.

Connectivity

The *connectivity function* λ_M of a matroid M is defined, for all partitions (A, B) of $E(M)$ by

$$\lambda_M(A, B) = r(A) + r(B) - r(E(M)) + 1.$$

We usually abbreviate $\lambda_M(A, B)$ to $\lambda_M(A)$. The $+1$ is there solely to make graph and matroid connectivity align in a certain way and is often omitted. We say that (A, B) is a *k-separation* of M if $\lambda(A, B) \leq k$. If (A, B) is a k-separation, then we say that A is *k-separating*.

What is λ_M measuring? Say that M is represented as a collection of vectors in a vector space, and (A, B) is a partition of $E(M)$, then $r(A)$, $r(B)$ and $r(E(M))$ measure the dimension of the subspaces $\langle A \rangle$, $\langle B \rangle$ and $\langle E(M) \rangle = \langle A \cup B \rangle$ spanned by A, B and $A \cup B$ respectively. In this case $\lambda(A)$ is one more than the dimension of the subspace $\langle A \rangle \cap \langle B \rangle$. The intuition is that, the higher the value of $\lambda(A, B)$, the more potential for communication there is between A and B. At the bottom extreme we have $\lambda(A, B) = 1$. In the represented cases, this means that A and B span skew subspaces so that the potential for mutual interaction is nonexistent.

We now consider the connection between graph and matroid connectivity. Let $G = (V, E)$ be a graph. For a set X of edges of G, let $V(X)$ denote the set of vertices incident with edges in X. The *connectivity function* λ_G of G is defined for all partitions (A, B) of E by $\lambda_G(A, B) = |V(A) \cap V(B)|$. Evidently $V(A) \cap V(B)$ is a vertex cut that separates A from B. Moreover, if A and B induce connected subgraphs, then $\lambda_G(A, B) = \lambda_{M(G)}(A, B)$. In the case that A and B do not induce connected subgraphs these values can differ significantly. Nonetheless it is still true to say that matroid connectivity generalises vertex connectivity. Of course this is another wrinkle for the uninitiated as, having got one's head around the fact that the elements of a graphic matroid are the edges of the graph, one might expect matroid connectivity to generalise edge connectivity, but that's not the case. It's definitely vertex connectivity that we capture most naturally in matroids.

Some important matroids

Certain matroids will turn out to be of considerable importance for us. Before we discuss these we need to acquaint ourselves with some elementary structures. An element e of a matroid is a *loop* if $r(\{e\}) = 0$, equivalently e is a loop if $\{e\}$ is dependent. A pair of elements $\{f, g\}$ of a matroid is a *parallel pair* if neither are loops and $\{f, g\}$ is dependent. This generalises graph terminology in that, if M is the cycle matroid of a graph G, then the loops and parallel pairs of M are precisely the loops and parallel pairs of G. A matroid is *simple* if it has no loops or parallel pairs. As in graph theory, for many purposes we can focus on simple matroids. Nonetheless, loops and parallel pairs are created by contraction so we allow their presence in general.

Let q be a prime power. The r-dimensional vector space over $GF(q)$ is denoted $V(r, q)$. Evidently $V(r, q)$ is a matroid. From our point of view it contains some degeneracies in that the zero vector is a loop and two nonzero vectors on the same 1-dimensional subspace are a parallel pair. We obtain a canonical simple matroid by removing all but one element of each of the 1-dimensional subspaces. Via this, or one of a number of equivalent constructions, we obtain the projective geometry $PG(r - 1, q)$. Note that, while $PG(r - 1, q)$ has dimension $r - 1$, regarded as a matroid it has rank r. If M is a simple matroid representable over $GF(q)$ of rank at most r, then M is isomorphic to a *submatroid* of $PG(r - 1, q)$, that is, a matroid isomorphic to M can be obtained by deleting points from $PG(r - 1, q)$.

The role played by projective geometries in representable matroids is analogous to that played by cliques in graphs. Every simple rank-r graphic matroid is a submatroid of the matroid $M(K_{r+1})$. Note that a spanning tree in K_{r+1} has r edges so that $M(K_{r+1})$ has rank r. Cliques and projective geometries control the number of elements in simple rank-r graphic and $GF(q)$-representable matroids. In other words they control the *growth rates* of these classes. We'll return to this topic soon.

Let M be a matroid obtained by randomly choosing an n-element set of vectors from a rank-r vector space over an infinite field. The probability that M has a dependent set of size less than $r + 1$ is 0, that is, every r-element subset of $E(M)$ is a basis. Put in matroid language, with a probability of 1 the matroid M is the *uniform matroid* $U_{r,n}$. Of particular importance to us is the case $r = 2$. Here the uniform matroid $U_{2,n}$ is just an n-point line. Note that $PG(1, q) \cong U_{2,q+1}$ so that no $GF(q)$-representable matroid can contain $U_{2,q+2}$ as a minor. A matroid is *binary* if it is representable over $GF(2)$. Evidently $U_{2,4}$ is not binary. It is a seminal – but not difficult – theorem of Tutte that the converse holds in that a matroid with no $U_{2,4}$-minor is binary. This is a striking example of the central theme of this paper. Matroids in general are wild, but the simple fact of excluding $U_{2,4}$ as a minor has put us into the highly structured class of binary matroids.

We don't always win this way. Let G be a graph consisting of three edges, one a loop, the others forming a path of length 2. If excluding $U_{2,4}$ imposes structure, surely excluding $M(G)$ must as well; if anything it's an even simpler matroid. But it's probably not so. It is widely believed that having an $M(G)$ minor is rare in that the proportion of n-element matroids with an $M(G)$-minor is asymptotically zero as n tends to infinity. Notwithstanding this, Tutte's theorem is encouraging.

3   Growth rates of minor-closed classes

The *growth rate* of a class $\mathcal{M}$ of matroids is the function f, where $f(r)$ is the maximum number of elements in a simple rank-r member of $\mathcal{M}$, if that maximum exists. The class of all matroids does not have a well-defined growth rate, for example long lines give arbitrarily large simple rank-2 matroids. However many natural minor-closed classes do.

The projective geometry $PG(r-1,q)$ has $\frac{q^r-1}{q-1}$ elements, so this is precisely the growth rate of $GF(q)$-representable matroids. We say that this class has a growth rate that is *base-q exponential*.

The clique K_{r+1} has $\frac{r^2+r}{2}$ edges, so graphic matroids have polynomial, indeed quadratic, growth rates. There is a big jump between exponential and polynomial, let alone between exponential and quadratic. Kung [24] conjectured that there are no minor-closed classes with growth rates that are intermediate between exponential and quadratic. That's a remarkably large gap, but it turns out that this conjecture is true as we shall see later in this section. The picture that emerges when growth rates are studied provides clear evidence for structure in members of minor-closed classes.

In this section we consider growth rates in somewhat more detail. We know from graphs that significant jumps in growth rates can occur. We cannot resist giving the proof of the following theorem of Mader [25].

Theorem 3.1 (Mader's Theorem) *Let m be an integer greater than 2. There is a constant c_m such that, if $G = (V, E)$ is a graph with no K_m-minor, then $|E| \leq c_m |V|$.*

Proof Let $c_m = 2^{m-3}$. We use induction on m. The case $m = 3$ is easy. Assume $m \geq 4$. Let u be a vertex of a minor-minimal graph $G = (V, E)$ satisfying $|E| \geq 2^{m-3}|V|$. By minimality there are more than 2^{m-3} edges incident with u and every such edge is in at least 2^{m-3} triangles. Hence the graph G_u induced by the vertices adjacent to u has minimum degree at least 2^{m-3} and hence $|E(G_u)| \geq 2^{m-4}|B(G_u)|$. By induction G_u contains a K_{m-1}-minor. The corresponding minor induced by G_u and v is a K_m-minor in G. $\qquad\square$

If $\mathcal{G}$ is any proper minor-closed class of graphs, then $\mathcal{G}$ avoids some graph H and hence avoids a clique with as many vertices as H. It follows from Mader's Theorem that the growth rate of $\mathcal{G}$ is bounded by a linear function. An immediate consequence is that the growth rate of any proper minor-closed class of graphic matroids is bounded by a linear function.

The situation is in striking contrast to the case where we are avoiding K_m as a subgraph. Evidently $(m-1)$-partite graphs cannot have a K_m subgraph and these can have quadratically many elements. As is well known, Turán's Theorem [39] identifies the extremal examples and growth rate exactly in this case.

What is going on in the graph minors case? The sharp drop in growth rate shows that it is difficult for a graph to avoid a fixed graph H as a minor. The graph must have low density and its minors must have low density also. The conclusion that avoiding H as a minor imposes structure on a graph is inescapable. (This is with the benefit of hindsight, and hindsight gives 20-20 vision.) Of course, one way to avoid having H as a minor is to be embeddable on a surface on which H cannot be embedded. This is certainly explicit structure. Readers already familiar with the Graph Minors Structure Theorem of Robertson and Seymour [33] will be aware that this is not the full story, but it is the heart of the story. We'll give a fuller picture later. However note that, just as topology entered into the picture in a surprising way via the connection with planar graph duality, it has, once again, emerged as part of the answer after asking a purely combinatorial question.

Being embeddable on an appropriate surface is one way to avoid H. Here is another. Consider $K_{3,n}$. This is a graph that does not have K_5 as a minor and by making n sufficiently large we can avoid embeddability on any surface that we like. Here we achieve the trick of avoiding K_5 by having too thin connectivity. This is another structural reason and we'll come back to that later as well.

Back to more general matroids. As noted earlier, a minor-closed class of matroids need not have a defined growth rate. But it turns out that once we exclude a line as a minor, then we do. Moreover, in the prime power case, we can characterise the extremal matroids. Kung [24] proved the following theorem. Recall that $U_{2,n}$ is the line with n points.

Theorem 3.2 (Kung's Theorem) *Let q be a positive integer exceeding 1. If M is a simple rank-r matroid with no $U_{2,q+2}$-minor, then $|E(M)| \leq \frac{q^r-1}{q-1}$. Moreover if equality holds and $r \geq 4$, then q is a prime power and M is the projective geometry $PG(r-1,q)$.*

Here is part of the proof. A *parallel class* is a maximal set whose 2-element subsets are parallel pairs. Assume that M has no $U_{2,q+2}$-minor. Choose an element $e \in E(M)$. The parallel classes of M/e are easily seen to be the lines of M containing e with e removed. By induction, there are at most $\frac{q^{r-1}-1}{q-1}$ of them. But each has at most q elements, as otherwise M has a $(q+2)$-point line as a restriction. Therefore M has at most $q\frac{q^{r-1}-1}{q-1} + 1 = \frac{q^r-1}{q-1}$ elements as required. Identifying the extremal structures in the case that equality holds is also quite straightforward and given by Kung in [24].

Earlier we mentioned that Kung had conjectured that there were no growth rates for minor-closed classes in between quadratic and exponential. This was part of his Growth Rates Conjecture. Another part of his conjecture was that an analogue of Mader's Theorem held for matroids. This part was proved by Geelen and Whittle [13]. In that paper it is proved that, if $\mathcal{M}$ is a minor-closed class of matroids that does not contain $U_{2,n}$ for all $n \geq 2$ and does not contain all graphic matroids, then the growth rate of $\mathcal{M}$ is bounded by a linear function. Further progress on Kung's conjecture was made by Geelen and Kabell [11]. Eventually the full conjecture was proved by Geelen, Kung and Whittle [12].

Theorem 3.3 (Growth Rate Theorem) *Let $\mathcal{M}$ be a minor-closed class of matroids. Then either,*

(i) *There exists an integer c such that $|E(M)| \leq c(r(M))$ for all simple matroids $M \in \mathcal{M}$,*

(ii) *$\mathcal{M}$ contains all graphic matroids and there exists an integer c such that $|E(M)| \leq c(r(M))^2$ for all simple matroids $M \in \mathcal{M}$,*

(iii) *there is a prime-power q and an integer c such that $\mathcal{M}$ contains all $GF(q)$-representable matroids and $|E(M)| \leq cq^{r(M)}$ for all simple matroids $M \in \mathcal{M}$, or*

(iv) *$\mathcal{M}$ contains $U_{2,n}$ for all $n \geq 2$.*

Let $\mathcal{U}(q)$ denote the class of matroid with no $U_{2,q+2}$-minor. We've already seen that $\mathcal{U}(2)$ is precisely the class of binary matroids. It follows from the Growth Rate Theorem that in general $\mathcal{U}(q)$ has essentially the same growth rate as the class of $GF(q')$-representable matroids where q' is the largest prime power less than or equal to q and $\mathcal{U}(q)$ contains all matroids representable over a field of order at most q. But as q grows, the $\mathcal{U}(q)$ zoo is seen to rapidly contain some pretty wild beasts. A notoriously wild class of matroids is the class of *spikes*; see for example [4, 20]. Spikes have no $U_{2,6}$-minor but have linear linear growth rate so have no impact on growth rates.

Incidentally, there are natural ways in which one can associate a matroid with a hypergraph. Doing this with, for example, 3-regular hypergraphs where each edge is incident with three vertices gives classes of matroids with cubic growth rates. It follows from the Growth Rate Theorem that these classes are not minor closed. Depending on the precise way that the associated matroid is defined, one inevitably builds either arbitrarily long lines or large projective geometries as minors of members of the class.

4   Frame matroids and varieties

Kung gained the intuition for his Growth Rate Conjecture from joint work with Kahn [23] where they studied so-called *varieties* of matroids. We will describe this seminal work in this section as it gives insight into why certain classes of matroids necessarily play a role in matroid structure theory. Part (i) of the Growth Rate Theorem indicates that graphic matroids play a fundamental role, and that is clearly true, but there are certain somewhat richer classes with quadratic growth rates that necessarily play a role in structure theory. We begin by describing these classes.

We say that a matroid M is a *frame* matroid if there exists a matroid M' with a basis B having the property that $M = M' \backslash B$ and that every element of M' is in the span of a pair of elements of B. Note that, if M is a frame matroid, then so is the matroid M' described above; to see that simply add and delete a basis consisting of elements parallel to the elements of B. The point is that we allow, but do not require, the elements of B to be part of our matroid. We have seen that graphic matroids are frame matroids. It is not hard to see that minors of frame matroids are frame matroids so that frame matroids form a minor-closed class of matroids.

Let q be a prime power and r a positive integer. We now construct an important class of frame matroids over $GF(q)$. We define them as submatroids of $PG(r-1,q)$. Let $B = (b_1, b_2, \ldots, b_r)$ be a basis of $PG(r-1,q)$. We lose no generality in saying that the elements of B label the columns of an identity matrix. For each pair i,j with $i < j$ add all distinct vectors having 1 in row i, a non-zero element in row j and zeros in all other coordinates. There are $q-1$ such vectors for each choice of i and j. Let $D(r,q)$ denote the submatroid of $PG(r-1,q)$ consisting of the elements we have described. It turns out that dependencies in $D(r,q)$ are determined purely by the structure of the multiplicative group $GF(q)^*$ of $GF(q)$. These matroids were studied by Dowling [3] who showed that one can begin with an arbitrary finite group Γ and define a frame matroid $D(r, \Gamma)$ whose dependencies are determined by group multiplication. In the case that $\Gamma = GF(q)^*$ we have $D(r, \Gamma) = D(r, q)$.

These matroids can also be defined graphically. We can obtain a directed group-labelled graph from our original construction as follows. Let B label the vertices of our graph. For each pair b_i and b_j of vertices where $i < j$ we add $q - 1$ edges directed from b_i to b_j and label these edges by the elements of $GF(q)^*$. Take a cycle in the graph and traverse it. Multiply the group values of the edges that have been traversed in the direction of the edge with the inverses of the group values of the edges that have been traversed against the direction of the edge. It is not hard to see that the cycle is a circuit in $D(r, q)$ if and only if the product is 1. We can also see that the assignment of directions to edges was essentially arbitrary. If we reverse the direction of an edge we simply invert the group value. The procedure can be reversed of course and, beginning with a group-labelled directed graph, one can describe rules to obtain an associated matroid. If the graph is the one defined above, then the associated matroid is $D(r, q)$. Note that there are other matroids that can be associated with group-labelled graphs. The whole area is the subject of an extensive study by Zaslavsky; see for example [42].

Let Γ be a finite group. We call a matroid that is a minor of $D(r, \Gamma)$ a *Dowling matroid over* Γ. Certainly Dowling matroids over Γ are frame matroids, but we also have the stronger property that a rank-r simple Dowling matroid over Γ is a submatroid of $D(r, \Gamma)$. Thus, just as we have seen with graphic and $GF(q)$-representable matroids, each simple rank-r member of the class is a submatroid of a canonical rank-r member of the class. Observe too that all members of the class of Dowling matroids over a group Γ are representable over a field $\mathbb{F}$ if and only if Γ is a subgroup of $\mathbb{F}^*$. Note also that $D(r, \Gamma)$ has $\binom{r}{2}|\Gamma| + r$ elements so that the class of Dowling matroids has, as predicted by the Growth Rates Theorem, quadratic growth rate.

The multiplicative group of $GF(2)$ is the trivial group. In this case the associated class of Dowling matroids is precisely the class of graphic matroids. This is perhaps surprising since we have defined graphic matroids to be matroids on the edges of a graph, whereas $D(r, \Gamma)$ is a matroid on the edges and vertices of an appropriate graph. Indeed, if Γ is the trivial group, then $D(r, \Gamma) \cong M(K_{r+1})$. To see this, take the representation of $M(K_{r+1})$ consisting of all columns with exactly two nonzero entries. The set of all edges incident with a fixed vertex of K_{r+1} is a basis B of $M(K_{r+1})$. We may perform row operations to convert the columns labelled by B into an identity matrix. Observe that every other edge is in a triangle with an edge in B, so that, in the new matrix, all other edges have exactly two nonzero entries, and the matrix we have is precisely the one obtained in the construction of $D(r,2)$.

We now consider Kahn and Kung's theorem [23]. Let M_1 and M_2 be matroids on disjoint ground sets E_1 and E_2. Then the *direct sum* $M_1 \oplus M_2$ is defined to be the matroid on $E_1 \cup E_2$ with rank function $r_{M_1 \oplus M_2}(X) = r_{M_1}(X \cap E_1) + r_{M_2}(X \cap E_2)$ for all $X \subseteq E_1 \cup E_2$. It is elementary that $M_1 \oplus M_2$ is in fact a matroid. If X is a set of elements of the matroid M, then the *restriction* of M to X, denoted $M|X$ is the matroid obtained by deleting the elements of M not in X, that is, $M|X = M\backslash(E - X)$. It is easily seen that if (A, B) is a partition of the ground set of the matroid M, then $M = M|A \oplus M|B$ if and only if $\lambda_M(A, B) = 1$.

Let $\mathcal{M}$ be a minor-closed class of matroids that is closed under isomorphism. A *sequence of universal models* for $\mathcal{M}$ is a sequence $U_1, U_2, \ldots$ of matroids such that for all integers $r > 0$, the matroid U_r has rank-r and such that each simple rank-r matroid in $\mathcal{M}$ is a isomorphic to submatroid of U_r. We say that $\mathcal{M}$ is a *variety* if the following hold:

(V1) $\mathcal{M}$ is closed under direct sums, that is, if M_1 and M_2 are in $\mathcal{M}$, then so too is $M_1 \oplus M_2$.

(V2) $\mathcal{M}$ has a sequence of universal models.

It is clear that if G is a finite group, then the Dowling matroids over G form a variety whose universal models are the Dowling geometries $D(r, G)$ and if q is a prime power, then the class of $GF(q)$-representable matroids forms a variety whose universal models are the projective geometries over $GF(q)$. Kahn and Kung [23, Theorem 14] prove that, apart from three classes that are degenerate due to extremely low connectivity, these are the only varieties.

Kahn and Kung's theorem is certainly a strong theorem. It gives insight into why one would expect projective geometries and Dowling geometries to play a role in structure theory. Nonetheless, one may ask why the condition (V1) is imposed? The condition that we are closed under direct sums seems to be relatively mild but it is vital as otherwise the theorem does not hold. If we remove that condition we uncover many more minor-closed classes with sequences of universal models. Some of these have degeneracies due to low connectivity. It is natural to believe that the others are in some sense closely related to varieties. Indeed one can conjecture that they are obtained by "perturbing" varieties in ways related to perturbations discussed later in this paper. There are many interesting open problems in this area.

5 Regular matroids

Growth rates and varieties give us evidence for structure from a very general perspective. Evidence comes from quite a different direction when we see explicit structural characterisations of specific minor-closed classes of matroids. Most strikingly, Seymour [37] gave such a characterisation for regular matroids. In this section we discuss that characterisation.

A matrix A over the rationals is *totally unimodular* if the determinant of every square submatrix of A is in $\{0, 1, -1\}$. A matroid is *regular* if it can be represented by a totally unimodular matrix. For a variety of reasons regular matroids are one of the most well-studied classes of matroids. Tutte [41] proved that a matroid is regular if and only if it is representable over all fields and gave an excluded-minor characterisation of the class. Graphic and cographic matroids are regular. Moreover, Heller [17] proved that the growth rate of regular matroids is precisely that of the class of graphic matroids. Apart from that the class remained poorly understood and many questions remained open for totally-unimodular matrices and regular matroids until Seymour proved his Regular Matroid Decomposition Theorem [37], to date probably the most famous theorem in matroid theory. This theorem gives a precise structural characterisation of regular matroids. We'll begin by stating a consequence of it.

Let (A, B) be a partition of $E(M)$ for some matroid M. Recall that (A, B) is a 3-separation if $\lambda(A, B) \leq 3$. In general a regular matroid need not be either graphic or cographic. But, as an immediate consequence of Seymour's theorem we obtain

Corollary 5.1 *Let M be a 3-connected regular matroid with at least* 11 *elements having the property that, whenever (A, B) is a 3-separation of M, then either $|A| \leq 5$ or $|B| \leq 5$. Then M is either graphic or cographic.*

Thus, whenever we impose a reasonably mild connectivity condition we fall into one of two fundamental subclasses. What happens when we fail that connectivity test? Then there must exist a 3-separation that splits the matroid in some relatively serious way. But, in this case, we can perform a standard matroid decomposition operation to obtain two smaller matroids, both of which are minors of the original matroid. This process can be repeated until we obtain pieces that are either graphic, cographic or copies of a certain 10-element matroid. Moreover, the original matroid can be canonically built from the pieces. The decomposition and composition operations are not difficult to describe – they are generalisations of the clique-sum operations for graphs – but the details need not concern us here.

Seymour's Theorem is a remarkable result. It is an exact structural description of regular matroids. It provides a road map for proving theorems about regular matroids; if a property holds for graphs, cographs and is preserved under certain elementary sums, then it holds for regular matroids. Moreover it leads to a polynomial-time algorithm for recognising regular matroids [40]. We could not ask for more from a structure theorem.

Seymour's theorem is extremely suggestive. Regular matroids form a proper minor-closed class of binary matroids. Highly connected regular matroids must either be graphic or cographic. We can fantasise that regular matroid decomposition is indicative of structure for any proper minor-closed class of binary matroids; or $GF(q)$-representable matroids for that matter. What hope is there for such a fantasy?

Life rapidly becomes more complicated in more general minor-closed classes. It's not hard to see that regular matroids are, in essence, the matroids obtained by excluding $PG(2,2)$. What happens if we are in the class obtained by excluding a larger projective geometry, say $PG(10,2)$? This class contains all graphic matroids of course and, by the Growth Rates Theorem, it has quadratic growth rate. If structure is there, graphic matroids should play a fundamental role. Is this the case? Consider an example. Construct a matroid as follows. Take the cycle matroid of a large clique represented by columns with exactly two nonzero entries. Now add eight arbitrary "junk" rows to the matrix and consider the matroid M represented by the columns of the matrix A we have constructed. Removing a row from A corresponds to first adding a vector v with a 1 in that row and zeros elsewhere, and then contracting v from the extended matroid. We see that we can make M graphic after a sequence of eight such operations. This is clearly a property preserved under minors. But it is not possible to make $PG(10,2)$ graphic after a sequence of eight such operations. We have shown that our matroid M does not have a $PG(10,2)$-minor. The junk rows ensure that M is unlikely to be graphic and it is clearly not possible to decompose M across small separations.

This will come as no surprise to readers familiar with the Graph Minors Structure Theorem. The role played by the junk rows is analogous to the role played by apex vertices in that theorem. Just as we can add only a bounded number of apex vertices we would expect to be able to add only a bounded number of junk rows. The upshot is clear. In general we of course have to let go of having graphic or cographic as the core objects of our structure. Instead the best we could hope for are objects that are in some sense close to being graphic or cographic.

6 Modularity

What about the role played by 3-separations in regular matroid decomposition? In fact there really is something special about 3-separations that persists for more general classes of binary matroids. Consider an example. Choose a binary matroid with a partition (A, B) such that $M|A \cong M(G)$ for some large graph G. Say that (A, B) is a t-separation. Recall that this means that $\lambda(A, B) = r(A) + r(B) - r(M) + 1 \leq t$. Assume that $\langle B \rangle \cap A \cong M(K_t)$. We hope this makes sense. The picture to have in mind is that of M consisting of the cycle matroid of a large graph where the space of interaction with the rest of the matroid B is spanned by the relatively small subclique $M(K_t)$ of A. The question we ask is "Can we build a minor containing A that is incompatible with the graphic structure we now see on A?" Effectively we are allowed to contract elements from B so long as we always contract elements outside the span of A.

What if (A, B) is a 3-separation, so that $\lambda(A, B) \in \{1, 2, 3\}$? In the case that $\lambda(A, B) = 1$, we have a direct sum and nothing we do to B affects the structure of A. Say $\lambda(A, B) = 2$, then we have a single edge e in K_2 and all that we can do from B is to produce other edges in parallel with e. Say $\lambda(A, B) = 3$. Then the matroid $M(K_3)$ is a 3-point line and, in binary space, this is a full line. Again, operations from B can only put points in parallel with existing points. Thus, for 3-separations, nothing we do from B can perturb the graphic structure that we see in A.

What if $\lambda(A, B) = 4$? Here $M(K_4)$ has rank-3 and spans a plane in the underlying projective space. The points on this plane form a copy of $PG(2, 2)$, the well-known Fano plane. This plane has seven points, but K_4 has only six edges, so there is a point p not occupied by the points of $M(K_4)$. If it is possible to project an element of M onto p then we have obtained a minor that conflicts, albeit in a small way, with the graphic structure we have for G. Now imagine that B is not the only such set "poking out" of A. Then maybe it would be possible to accumulate the evidence of such incompatibility and build a serious conflict with the graphical structure of G. How one accumulates such evidence is precisely what one needs to think about in proving structure theorems.

What if it is not possible to project an element of B onto p? Then our copy of $M(K_4)$ is a so-called "modular flat". In this case we have no conflict with the graphical structure of A. It turns out to be a consequence of the modularity of $M(K_4)$ that the graphical structure we have found for G can be enlarged and extended into B – not necessarily all the way as there may be 3-separations within B. So a 4-separation will either give us evidence for lack of graphic structure or an opportunity to enrich our graphic structure.

Now assume that we have extended the graphic structure of G as far as possible into the rest of the matroid. It follows from the above discussion that we may have arbitrarily many 3-separating sets glued onto A, but each higher-order k-separating set glued onto A will be providing us with evidence for lack of graphic structure, and this cannot happen in an uncontrolled way without building a significantly richer non-graphic minor.

The crucial observation was that the modularity of $M(K_4)$ controlled the structure of life in B. The fact that modularity has this effect is further evidence for the existence of structure and we now consider the topic of modularity in more detail. A flat F of a matroid M is *modular* if $r(X \cap F) + r(X \cup F) = r(X) + r(F)$ for all flats X of M. Recall that contraction of an element in a matroid corresponds to projection from that element. Let F be a modular flat of a matroid and x be an element outside of F. It may be that F is no longer a flat in M/x as some element z is captured in the span of F in M/x. It follows from the definition of rank in the contraction that, in this case $r(F \cup \{x, z\}) = r(F) + 1$. Let Z be the flat spanned by $\{x, z\}$. Then $r(Z) = 2$ and, by the modularity of F we have $r(Z \cap F) = 1$. Thus there is a point $z' \in Z \cap F$ and the effect of contracting x is to do nothing more than add z in parallel with z'; not a particularly exciting event. The same argument applies when we contract any set X of elements that is *skew* to F, that is $r(F \cup X) = r(F) + r(X)$. Hence modular flats are precisely those flats that cannot be enriched by contraction of elements outside their span.

As observed above, modularity imposes significant structure on a matroid modulo reasonable connectivity assumptions. Assume that the matroid M is 3-connected and M has a modular flat F. It is not difficult to prove that if $M|F$ is isomorphic to the 3-point line $U_{2,3}$, then M must be binary. Assume that $M(K_4)$ is modular in M. Again we easily see that M must be binary. Moreover, if we raise the connectivity a little and assume that 3-separations have the property that one side has at most three elements, then we obtain the very strong consequence that M must be graphic. This follows from a theorem of Seymour characterising when there is no circuit containing a given 3-element set in a binary matroid [38]. As observed above, this result gives us a powerful tool for demonstrating whether we have a matroid that is close to being graphic or whether we can produce something richer as a minor.

In binary matroids we would expect from the Growth Rates Theorem that significant extra richness implies that we can build increasingly large projective geometries as minors. For other fields that is not the case as we have the intermediate classes of Dowling matroids to consider. These have quadratic growth rates, but are significantly richer than graphic matroids.

We now consider the same question as that considered at the start of this section, only this time we may assume that $M|A$ is a Dowling, rather than a graphic matroid. Consider a specific case. Assume that we are in the class of $GF(3)$-representable matroids. Here the class of Dowling matroids over the 2-element group is a class of $GF(3)$-representable matroids. As before we would like to be able to extend our Dowling structure into the set B or get evidence from B that there is richer structure in our matroid that is incompatible with the Dowling structure we have obtained. Our analogues of cliques are the Dowling geometries $D(r,3)$. If $r = 2$, then $D(r,3)$ is the 4-point line and this is modular in any $GF(3)$-representable matroid, so, as before, we have to live with 3-separations. What about $r = 3$? Here a genuine unpleasantness arises. The matroid $D(3,3)$ has representation.

$$D = \begin{bmatrix} 1 & 0 & 0 & 1 & 1 & 0 & 0 & 1 & 1 \\ 0 & 1 & 0 & 1 & -1 & 1 & 1 & 0 & 0 \\ 0 & 0 & 1 & 0 & 0 & 1 & -1 & 1 & -1 \end{bmatrix}$$

It turns out that are several copies of $M(K_4)$ sitting inside $D(3,3)$. Indeed the matroid represented by the last six columns of D is isomorphic to $M(K_4)$. Now imagine that the partition (A, B) of our matroid M has the properties that $\lambda(A, B) = 4$, that $M|A$ is a ternary Dowling matroid with a restriction isomorphic to $D(3,3)$, and that the $M(K_4)$ restriction of $D(3,3)$ we have just described is extended to the cycle matroid of a large clique given by the set A. In other words, on one side of (A, B) we see graphic structure, on the other side we see Dowling structure. It is a straightforward fact about Dowling matroids that, while $M(K_4)$ can be embedded in ternary Dowling matroids in a curious way, higher order cliques cannot. It follows that we cannot extend the Dowling structure of B into the larger matroid. We have produced a matroid in which $D(3,2)$ is modular, but in which the Dowling structure of $D(3,2)$ is not extendable.

In summary we see that for binary-matroid structure there really is something special about 3-separations, just as with regular matroids. But for larger fields that specialness extends – at least – to certain types of 4-separations.

7 Branch width and tangles

We gain insight into structure from considering growth rates and varieties partly because we slip past degeneracies caused by low connectivity. We saw a similar situation for regular matroids in Corollary 5.1 where, by insisting that matroids be sufficiently connected, we were able to obtain an extremely simple outcome for the structure theorem. There are two ways in which low connectivity affects structure. It can happen that we have many low order separations and it is possible to decompose the matroid in a tree-like way across these separations. If this does not occur, then we have regions of high connectivity separated from each other by lower order separations. We are able to identify such regions of high connectivity through the notion of a "tangle".

We say that a tree is *cubic* if all of its internal vertices have degree 3. A *branch decomposition* of a matroid M is a cubic tree T whose leaves are labelled by the elements of M. Each edge of this tree induces a partition of the ground set of M, that is, it induces a separation of a certain order. The *width* of the branch decomposition is the maximum order of a separation induced by an edge of T. The *branch width* of M is the minimum width over all branch decompositions of M.

One can use the connectivity function of the graph to produce an analogous definition of the branch width of a graph. It turns out that, while there are differences between the connectivity function of a graph and a matroid, these differences are unimportant in that, apart from certain trivial exceptions, the branch width of a graph and its cycle matroid are the same [18]. Some readers may be more familiar with the related concept of tree width. While tree width does generalise to matroids [21], branch width is the more natural concept for matroids. Moreover, as we now explain, from a qualitative point of view, the two concepts are effectively equivalent.

A class of matroids has bounded branch width if there is a k such that all members of the class have branch width at most k. It is known that a class of matroids has bounded branch width if and only if it has bounded tree width. If $\mathcal{M}$ is a class of matroids of bounded branch width, then we have an entirely satisfactory structural outcome for members $\mathcal{M}$. All members decompose across small separations into tree-like structures. There is now a huge literature on bounded tree width for graphs and we will not delve into that. The main point here is that, for matroids, most of the key wins that bounded tree width gives for graphs extend to classes of $GF(q)$-representable matroids of bounded branch width. For example such classes are well-quasi-ordered [7]. Also Hliněný [19] has extended Courcelle's Theorem [1] to such classes. Hliněný has shown that problems that can be stated in a certain logic have a polynomial-time algorithm for $GF(q)$-representable matroids of bounded tree width. The precise logic is not important here; the point is that, just as with graphs, many natural problems that are NP-complete for $GF(q)$-representable matroids in general can be stated in this logic. It follows that, in the case of $GF(q)$-representable matroids, the property of having a tree-like decomposition across small separations is an extremely powerful controller of the complexity of the underlying matroid. We also know that there are only a finite number of excluded minors for $GF(q)$-representability that have branch width at most k for any positive integer k.

The upshot is that, from the point of view of matroid minors, for matroids representable over finite fields, classes of bounded branch width are easily dealt with. Note that this is *not* the case if we step outside that cosy world. For more general matroids bounding branch width does not seem to give us any gains. This is an appropriate time to take stock and refocus. It is conceivable that deep structural theorems exist for very general classes of matroids, but these theorems will not be constructive and will not give us algorithmic consequences. To illustrate the point consider the class of matroids representable over the rationals. One may perhaps expect this class to be reasonably well behaved. We are after all dealing with finite structures coordinatisable over a familiar field. But from the perspective of the problems we are interested in this is an intractable class. For example it's easy to construct infinite antichains of Q-representable matroids of branch width 3. See also [20, 26] for other examples that illustrate just how radically things change here. That transition is a fact we have to accept.

We can now focus on classes of matroids of unbounded branch width. A matroid with very high branch width need not be highly connected in any strict sense of the word as it may have many low order separations; rather it must have regions of high connectivity. In [31] Robertson and Seymour make the notion of a region of high connectivity precise through the concept of a "tangle". Tangles are generalised to matroids in [10]. A basic problem of identifying an area of high connectivity in a graph or matroid is one of erosion. It's not possible to see such a region as a subgraph or submatroid without damaging valuable information. As an example from graphs, consider the road map of the United Kingdom. One highly connected region will correspond to Manchester. It's certainly not possible to find low order separations that split Manchester in any serious way. But if we attempt to isolate Manchester as a subgraph we will lose vertices and edges outside the city limits that the good citizens will tell you are valuable for gaining connectivity. How often do we hear comments like "I would never drive through town to get to my Aunt's place, rather I drive via X so that I can bypass the congestion on Y St." We might think that being able to take minors helps, but the problem persists.

Instead of trying to isolate a highly connected region as a substructure Robertson and Seymour had the following idea. If we claim that a region of the structure has connectivity at least k, then we are making the claim that it is not possible to split that region in any serious way with a separation of

order less than k. Given a separation (A, B) of order less than k one should
be able to unambiguously say whether the region lies mainly in A or B. For
example, if (A, B) is a 50-separation of our road map, then it's unlikely
that this separation could split Manchester in any serious way, so that
most of Manchester is, say, unambiguously in B. It might be that most
of London is in A, but, from the point of view of capturing Manchester
we would think of B as the "big" side. This ability to unambiguously
decide on an ordering into big and small sides relative to a region of high
connectivity for each small separation is precisely what we need and we
can describe it axiomatically as follows.

A *tangle* $\mathcal{T}$ of *order* k in a matroid M is a collection of separations of
order at most $k - 1$ having the following properties.

(i) For each separation (X, Y) of M of order less than k, exactly one of
(X, Y) or (Y, X) is in $\mathcal{T}$.

(ii) If (A_1, B_1), (A_2, B_2) and (A_3, B_3) are in $\mathcal{T}$, then $A_1 \cup A_2 \cup A_3 \neq E(M)$.

(iii) $(\{e\}, E(M) - \{e\}) \notin \mathcal{T}$ for each element e of M.

If $(A, B) \in \mathcal{T}$, then we think of B as the "big" side. Property (iii) is a
non-triviality condition. Property (ii) says that $E(M)$ is not the union of
three small sides and is a strengthening of the condition that we have an
unambiguous small side. Assume that we fail that condition. Construct
a cubic tree with one vertex of degree 3 and leaves labelled by A_1, A_2
and A_3. We have started to build a branch decomposition of order $k - 1$.
Evidently we have not found an obstruction to branch width k. But with
(ii) we have indeed found such an obstruction and it turns out that tangles
are precisely dual to branch width in that a matroid has branch width at
most k if and only if it has no tangles of order k. This is proved for graphs
in [31] and extended to matroids in [10].

If $\mathcal{T}$ is a tangle of order k in M and $k' < k$, then it is easily seen that
we obtain a tangle $\mathcal{T}'$ of order k' by simply restricting attention to the
separations of order at most $k' - 1$ in $\mathcal{T}$. We say that $\mathcal{T}'$ is a *truncation* of
$\mathcal{T}$. A tangle is *maximal* if it is not a truncation of any other tangle. The
question arises as to how the maximal tangles can interact. Anyone who
has considered a modern conurbation will be aware of the issue. But it
turns out that the answer is remarkably simple and the maximal tangles
can indeed be disentangled. If $\mathcal{T}$ and $\mathcal{T}'$ are distinct maximal tangles
of order k and k' respectively then there exists a separation (X, Y) of
order less than $\min\{k, k'\}$ such that $(X, Y) \in \mathcal{T}$ and $(Y, X) \in \mathcal{T}'$. Such a
separation is a *distinguishing separation*. A *tree decomposition* of $E(M)$ is

a tree T whose vertices label a partition of $E(M)$. Note that each edge of a tree decomposition of $E(M)$ induces a separation of M.

Theorem 7.1 *Let* $\{\mathcal{T}_1, \mathcal{T}_2, \ldots, \mathcal{T}_n\}$ *be the maximal tangles of a matroid* M. *Then there is a tree decomposition of* M *with vertex set* $\{1, 2, \ldots, n\}$ *such that the separation induced by and edge* $\{i, j\}$ *is a minimal-order distinguishing separation for* $\mathcal{T}_i$ *and* $\mathcal{T}_j$.

Thus we can think of the matroid as being built from its maximal tangles in a tree-like way. In fact Theorem 7.1 holds for any set of tangles of M that are incomparable in the truncation order. We discussed regular-matroid decomposition as a possible model for more general structure in minor closed classes. Note that regular-matroid decomposition is, essentially, decomposition relative to a maximal set of incomparable tangles of order at most 3.

From the tree of tangles we learn that the real task is to understand the structure of a matroid relative to a fixed high order tangle. The presence of other tangles certainly complicates the analysis and leads to a number of technical issues for proofs and so on, but, to gain insight into what is going on for the purposes of this paper we are well served by focussing on the case where M is highly connected in the sense that it has a single high-order maximal tangle.

8 Grids

From now on, we may assume that matroids we are dealing with have arbitrarily high branch width, in other words we are dealing with matroids that contain arbitrarily high order tangles. The question is "What does high branch width give us?" and the answer is "It gives us grid minors".

An $n \times n$ *grid* is a planar graph G_n with vertex set $\{(i, j) : 1 \leq i, j \leq n\}$ where vertices (i, j) and (i', j') are adjacent if and only if $|i-i'|+|j-j'| = 1$. The branch width of a grid is an increasing function of n; indeed $M(G_n)$ contains a tangle of order n. It is also straightforward to prove that if H is a planar graph, then H is a minor of a sufficiently large grid. To see this, imagine drawing H with slightly thick edges on fine graph paper. Robertson and Seymour prove that a graph with sufficiently large tree width has a large grid minor [30]. This theorem is extended to matroids in [9]. The matroid $U_{q,q+2}$ is the dual of the $U_{2,q+2}$, the $(q+2)$-point line.

Theorem 8.1 (Grid Theorem for Matroids) *There is a function* $f(n, q)$ *such that, if* $k \geq f(n, q)$ *and* M *is a matroid of branch width at least* k, *then* M *has either* $M(G_n)$, $U_{2,q+2}$ *or* $U_{q,q+2}$ *as a minor.*

We have seen that $U_{2,q+2}$ is not $GF(q)$-representable. Since representability is closed under duality, neither is $U_{q,q+2}$. We immediately obtain the next corollary.

Corollary 8.2 *Let q be a prime power. Then, for all n there exists k such that, if M is a $GF(q)$-representable matroid of branch width at least k, then M has an $M(G_n)$-minor.*

In other words, for $GF(q)$-representable matroids we can force arbitrary grids as minors by making the branch width sufficiently large. Using the observation that any given planar graph is a minor of a sufficiently large grid, the contrapositive of Corollary 8.2 says

Corollary 8.3 *Let q be a prime power and H be a fixed planar graph. Then the class of $GF(q)$-representable matroids with no H-minor is a class of bounded branch width.*

This is the structure theorem for excluding the matroid of a planar graph. To avoid a planar graph we must decompose in a tree-like way across bounded size separations. As planar graphs themselves form a class of unbounded branch width, the converse of Corollary 8.3 also holds. It follows from Corollary 8.3 and the fact that minor-closed classes of $GF(q)$-representable matroids of bounded branch width are well-quasi-ordered, that any minor-closed class of $GF(q)$-representable matroids that does not contain all planar graphs is well-quasi-ordered.

We mentioned earlier that topological properties can arise in unexpected ways. Even within graphs it's surprising the way that topology asserts itself; for matroids even more so. Imagine an intelligent life form based on the operating system of a computer. One could believe that their natural geometry was binary geometry and the natural objects of study binary matroids. Our creatures would discover branch width as a controller of complexity and would discover cycle matroids of planar graphs as the unique minimal minor-closed class of binary matroids having unbounded branch width. Planarity is an unavoidable part of the picture. Not only that, as we shall soon see, they will also be forced to eventually discover more general surface embeddings.

The Grid Theorem for Matroids is quite general in that it is not just a theorem for representable matroids. But having said that, it is not completely satisfactory. Ideally all the outcomes should be matroids of increasing branch width and lines and their duals do not have this property. Let G be a graph. Embed the vertices of G as a simplex in real space. Regard each edge of G as a point placed freely on the line joining its incident vertices. The real-represented matroid formed by these points is the *bicircular matroid* of G, denoted $B(G)$. Cycle matroids of grids are essentially self dual. This is not the case with bicircular matroids of grids. The next conjecture is due to Robertson and Seymour, although not in print.

Conjecture 8.4 *For all n there exists k such that if M is a matroid with branch width k, then M has a minor isomorphic to one of $U_{n,2n}$, $M(G_n)$, $B(G_n)$, or $B(G_n)^*$.*

Apart from very small values of n the minors excluded by this conjecture form an antichain. Moreover, as n grows the branch width of these matroids increases. Hence, if true, the conjecture is best possible. We believe Conjecture 8.4, but it is likely to be difficult to prove.

Finally we note that, as well as the original proof of the grid theorem for graphs given in [30], there are two other published proofs [2, 35]. The proof given in [2] is particularly intuitive. However none of the proof techniques extend to matroids in any natural way. The proof for matroids given in [9] uses quite a different strategy.

9   Core classes and deviations

Let $q = p^k$ be a prime power and let $\mathcal{M}$ be a minor-closed class of $GF(q)$-representable matroids. We are gradually converging towards a statement of the essential features of a structure theorem for members of $\mathcal{M}$. Inevitably there will be layers to the structure depending on what we exclude. The Grid Theorem has already shown us what is, essentially, the bottom layer. If $\mathcal{M}$ does not contain all planar graphs, then $\mathcal{M}$ is a class of bounded branch width. Thus we may assume that $\mathcal{M}$ contains all planar graphs.

If our interest is in graphs, there is only one more layer to the structure. A proper minor-closed class $\mathcal{G}$ of graphs cannot contain all cliques so that there is an integer n such that K_n does not belong to $\mathcal{G}$. This situation is handled by the Graph Minors Structure Theorem. In this case, as observed in the introduction, graphs in the class have a tree-like decomposition into parts that are "essentially planar", where essentially planar allows bounded deviations from planarity in three specific ways. Thus we can say that the "core class" for the Graph Minors Structure Theorem is the class of planar graphs. In this section we first describe the core classes that arise in structure theory for our minor-closed class $\mathcal{M}$. We'll then discuss the issue of the deviations from the core classes.

Core classes   Consider the analogous question for the structure theorem for our class $\mathcal{M}$ of $GF(q)$-representable matroids. What are the core classes we obtain here? The existence of Dowling geometries and projective geometries guarantee that there will be more layers to the hierarchy for matroids. We know from the Growth Rate Theorem that if $\mathcal{M}$ is a class of matroids that does not contain the cycle matroids of all cliques, then, just as in the graph case, $\mathcal{M}$ has linear growth rate. But life is not that simple. Consider the class of cographic matroids. Members of this class have no K_5 minor and we have a class that is very far from being graphic, let alone embeddable on a surface. Assume then, that there is an n such that $\mathcal{M}$ does not contain either $M(K_n)$ or $M(K_n)^*$. Then we have a clear answer. There is a subgroup Γ of the multiplicative group of G such that the core class is the class of Dowling matroids associated with group-labelled planar graphs where the group labels come from Γ.

We may now assume that we are dealing with a class of matroids that either contains all graphic matroids or all cographic matroids. Up to duality we may assume that $\mathcal{M}$ contains all graphic matroids. As we are in a proper minor-closed class there is some r such that $PG(r-1,q)$ is not in $\mathcal{M}$. But if $k \neq 1$, so that q is not prime, then the class of all $GF(q)$-representable matroids has no $PG(2,q)$-minor. It is apparent that we have to consider the subfields of $GF(q)$. For the next stage of the hierarchy we assume that there is an r such that $\mathcal{M}$ does not contain all projective geometries over the base field; that is, $\mathcal{M}$ does not contain all projective geometries over $GF(p)$ where $q = p^k$ and p is prime. In other words, there is an r such that $PG(r-1,p)$ is not in $\mathcal{M}$. By the Growth Rate Theorem we are now in a class of quadratic growth rate. Moreover, it turns out that the dual of a large projective geometry has a reasonably large projective geometry as a minor. This means that the earlier complication caused by cographic matroids has disappeared. But those caused by subgroups of $GF(q)^*$ certainly have not. The answer now is that there is a subgroup Γ of $GF(q)^*$ such that our core objects are Dowling matroids over Γ.

Finally, we may assume that we contain all $GF(p)$-representable matroids. In this case there is a subfield $\mathbb{F}$ of $GF(q)$ such that our core class is the class of $\mathbb{F}$-representable matroids. In other words, for some integer l that divides k, the core class is the class of $GF(p^l)$-representable matroids.

Deviations   The Graph Minors Structure Theorem allows three deviations from planarity. For the first we allow graphs to be embedded on a surface in which K_n does not embed. That is, we allow embeddings in surfaces of bounded genus. For the second we allow a bounded number of *vortices* of bounded *vortex width*. We have no intention of discussing vortices in detail apart from observing that one can think of a vortex as a graph of bounded path width attached to a face of the graph in a way that respects the facial ordering of the vertices of attachment. Finally, we allow a bounded number of *apex vertices*. These are vertices attached in an arbitrary way to the surface structure. It is clear that there is a need for such vertices. Consider a very large grid with a single extra vertex attached to every vertex of the grid. This graph has no K_6 minor, but it is not embeddable over a surface of small genus, nor can it be explained through vortices. Put another way, the class of apexed planar graphs is a class with no K_6 minor that demonstrates the need for apex vertices in a structure theorem.

Since we've mentioned apexed planar graphs we'll take the opportunity to deviate for a moment from our discussion of deviations. We say that an *apexed planar graph* is a graph with a single apex vertex attached to a planar graph. The class of apexed planar graphs is an interesting class and arises naturally in a number of situations. Recall from the discussion on Dowling matroids, that when we take the Dowling matroid associated with a graph, we allow matroid elements on the vertices as well as the edges. If we do this with a planar graph G and we perform the construction in the standard way, we obtain a matrix labelled by the vertices and edges of G with the following properties. The columns labelled by the vertices form an identity matrix wile the remaining columns labelled by the edges of G have two nonzero entries and give the cycle matroid of G in the usual way. It turns out that this matroid is isomorphic to the cycle matroid of the graph obtained by adding a single apex vertex connected to every vertex of G. Thus, from a Dowling matroid perspective – which we necessarily have in matroid structure theory – the class of apexed planar graphs is not, in itself, a deviation from planarity. Of course, if we have more than one apex vertex the situation changes.

If $\mathcal{M}$ is a class of $GF(q)$-representable matroids where we have excluded $M(K_n)$ and $M(K_n)^*$, then similar deviations arise from our core class. We allow group labelled graphs on surfaces of bounded genus. Vortices also inevitably arise, but we do not have graphic structure on the elements in the vortices. Nonetheless, very loosely speaking, vortices resolve as matroids of bounded path width glued to a face of the graph where the path decomposition respects the ordering of the vertices given by the face. What about apex vertices? In graphs we can think of apex vertices as adding a bounded amount of uncontrolled junk. It's clearly necessary to allow this. For matroids there is a somewhat more general way of adding a bounded amount of junk which we capture via the notion of a "perturbation". Thus, for $\mathcal{M}$, we replace adding apex vertices by performing a "bounded rank perturbation". We'll discuss perturbations more carefully in the next section.

Moving up to the higher levels in the matroid hierarchy things change. There is no road map from graphs to follow. Fortunately, as we move up, at least this aspect of structure simplifies. While we necessarily allow bounded rank perturbations, there is no higher-order analogue of surface structure or vortices. We have a single type of deviation from our core structures of Dowling matroids and matroids representable over subfields.

10   Bounded rank perturbations

Let A and B be matrices over a field $\mathbb{F}$. Then B is a *rank-t perturbation* of A if there are matrices A' and B' that are row equivalent to A and B respectively such that the matrix $B' - A'$ has rank at most t. Evidently, if B is a rank-t perturbation of A, then A is a rank-t perturbation of B. The *perturbation distance* between A and B is the minimum value t for which B is a rank-t perturbation of A. Note that, for the perturbation distance to be defined it suffices for A and B to have the same number of columns as we may add rows of zeros to obtain matrices A' and B' for which $B' - A'$ is defined. The case we are interested in is when A and B represent matroids $M[A]$ and $M[B]$ on a common ground set. In this case it is straightforward to prove that a bound on the perturbation distance between A and B also gives a bound on the difference between the sizes of a maximum clique minor or maximum projective geometry minor of $M[A]$ and $M[B]$. In other words, bounded-rank perturbations cannot arbitrarily increase the size of such minors. Thus bounded-rank perturbations are a necessary feature of structure theory.

Let M_1 and M_2 be matroids on a common ground set. Then M_2 is a *projection* of M_1 if there is a matroid M with a set X of elements such that $M_1 = M \backslash X$ and $M_2 = M/X$. If M_2 is a projection of M_1, then M_1 is a *lift* of M_2. Moreover the *rank* of the projection or lift is $r(M_1) - r(M_2)$. Note that projections are often called quotients in the matroid literature; see, for example Oxley [28]. In the represented case projections are geometrically quite natural. In this case $E(M)$ labels a set of vectors. Now add a new vector **p** and project the points labelled by $E(M)$ from **p** onto a hyperplane that avoids **p**. The resulting matroid M' on E that labels the projected points is a rank-1 projection of M. From a matrix perspective we have added a new column representing **p** to the matrix representing M and contracted **p** in the usual way, that is, we pivot on an entry in **p** to replace it by a column with a single non-zero entry. We then delete this column and the row containing the nonzero entry. Evidently higher-rank projections can be described as a sequence of such rank-1 projections so that we are adding a set of vectors and projecting the elements of M from that set.

The inverse of the above procedure describes lifts in the represented case. It probably says something about the wiring of the human brain that our geometric intuition seems more comfortable with projections than lifts. Having said that, from a matrix perspective, lifts are easily described. A rank-1 lift is obtained by adding a new row to the matrix representing M. Consider an example. Take a matrix over $GF(q)$ representing $M(G)$ for some graph G, where we have two nonzero entries in each column in the usual way. Now add an arbitrary row; the resulting matroid M' is an elementary lift of $M(G)$. It is easy to see that any perturbation can be resolved as a sequence of lifts and projections. Indeed, if M_1 and M_2 are any two represented matroids on the same ground set then they can be so resolved; simply project M_1 to a matroid consisting only of loops, encoded by a matrix of zeros, and then perform the lifts by adding the rows that we need to get a representation of M_2. We'll say that the *lift-project distance* between M_1 and M_2 is the minimum of the sum of the ranks of projections and lifts that transform the representation of M_1 to a representation of M_2.

We can now describe the relationship between perturbations, and lifts and projections more precisely. First observe that if the matrix A represents a matroid M, then we may add a row to A that is in the row space of A and still have a representation of M as we have not changed the linear independence or otherwise of sets of columns. Indeed, we may as well take the whole row space $R(A)$. Note that $R(A)$ encodes all the representations equivalent to A; to obtain such a representation we simply choose a set of rows containing a basis of $R(A)$.

Proposition 10.1 *Let M_1 and M_2 be $\mathbb{F}$-represented matroids on a common ground set represented by A_1 and A_2 respectively. Let R_1 and R_2 denote the row space of A_1 and A_2 respectively. Let $k = \dim(R_1 \cap R_2)$, $k_1 = \dim(R_1) - k$, and $k_2 = \dim(R_2) - k$. Then the following hold.*

(i) *The perturbation distance between A_1 and A_2 is the maximum of $\{k_1, k_2\}$.*

(ii) *The lift-project distance between A_1 and A_2 is $k_1 + k_2$.*

Finally we consider the relationship between apex vertices and perturbations. Say that the graph G_2 is obtained from the graph G_1 by adding a set A of apex vertices to $V(G_1)$. Now add a set T of edges that forms a tree that spans the vertices in A. Let G_3 be the graph obtained by contracting T. Then $M(G_3)$ is a projection of $M(G_2)$ and G_3 can be obtained from G_1 by adding a single apex vertex. But we observed in the previous section that, in this case, $M(G_3)$ is a Dowling matroid that can be associated with the graph G_1.

More on group-labelled graphs We can think of Dowling matroids as certain matroids associated with group-labelled graphs. We have already seen that these are fundamental in structure theory. The study of perturbations leads to other ways of associating matroids with group-labelled graphs that also play an important role. We will not give the full story here, but focus on the special case of binary lifts of graphic matroids. If we perform a sequence of k such lifts then the resulting matroid can be represented by a matrix that is graphic in the usual way with k extra rows. Each column can now be interpreted as a graph edge labelled by a k-tuple from the additive group of the vector space $V(k, 2)$. Our matroid now has an interpretation as a matroid associated with a group-labelled graph. This is quite distinct from the Dowling matroid that would be associated with such a group-labelled graph.

The full picture is more complex, but, associated with a perturbed Dowling matroid, there is a group-labelled graph that captures much of the crucial information of the matroid. This is important and extremely useful. On the one hand, one can generalise results for graph minors and develop structure theory for minor-closed classes of group labelled graphs, see for example [5, 22]. On the other hand, via the correspondences discussed above, it is often possible to reduce problems for matroids to questions about group-labelled graphs.

11   Back to tangles

The Graph Minors Structure Theorem is frequently stated in a way that makes no reference to tangles. Rather, it states that a graph can be decomposed using the operation of clique sum into parts that are almost planar. While there is an analogue of clique sum for matroids representable over finite fields, it seems unlikely that this will lead to a useful form of a structure theorem for such matroids. Rather, we focus on structure relative to a tangle in a matroid. The global structure is then understood via the structure relative to the maximal tangles together with the tree of tangles. In fact the same is true for graphs in that the strategy of the proof of the structure theorem is to focus on structure relative to a given tangle.

A reader familiar only with the definition may find a tangle a somewhat intangible thing. But there is a variety of techniques that enable one to deal with tangles. For example, there is a matroid associated with a tangle $\mathcal{T}$. But the most important fact is that grids can be connected with tangles. The grid theorem tells us that in a matroid with very high branch width, there is a large grid minor, but there is no use in having that minor in some part of the matroid that is not highly connected to the tangle – a grid in London does not give us structure in Manchester. There is a natural tangle of order n that can be associated with the cycle matroid $M(G_n)$ of the $n \times n$ grid. Any tangle in a minor N of a matroid M induces a tangle in M in a natural way. Thus each grid minor of M induces a tangle in M. We say that a tangle $\mathcal{T}$ in M *dominates* an $M(G_n)$-minor of M, if this tangle is a truncation of $\mathcal{T}$. The next theorem is proved in [10].

Theorem 11.1 *For each finite field* $\mathbb{F}$ *and positive integer* k, *there exists an integer* θ *such that, if* M *is an* $\mathbb{F}$-*representable matroid and* $\mathcal{T}$ *is a tangle in* M *of order* θ, *then* $\mathcal{T}$ *dominates an* $M(G_k)$-*minor.*

This is a qualitative converse to the fact that grids have large tangles. Through it, structure relative to high order tangles becomes more tangible as it is, in essence, structure relative to a large grid minor. We are now in a position to give an informal statement of our structure theorem. Recall from Section 6 that certain low-rank separations are unavoidable. For graphs 3-separations are certainly unavoidable, but we saw that for matroids certain 4-separations necessarily arise. But that is as bad as it gets. Indeed, when we get to projective geometries, it turns out that 2-separations are as bad as it gets. At times the mathematical universe is unreasonably kind to us. Let $\mathbb{F}$ be a finite field, let $\mathbb{F}_0$ be a subfield of $\mathbb{F}$, let M be an $\mathbb{F}$-representable matroids, and let $\mathcal{T}$ be a tangle in M.

- For any integer n, there exists θ such that, if $\mathcal{T}$ has order θ in M and does not control an $M(K_n)$- or $M(K_n)^*$-minor, them M or M^* is essentially a frame matroid over a subgroup of $\mathbb{F}^*$ whose associated graph is embeddable in a surface of low genus. Here "essentially" allows low-rank perturbations, vortices and suppressing small sides of 4-separations.

- For any integer n, there exists m such that, if $\mathcal{T}$ is a tangle in M that controls an $M(K_m)$-minor, but does not control a $PG(n-1, \mathbb{F}_0)$-minor, then M or M^* is essentially a frame matroid. Here "essentially" allows low-rank perturbations and suppressing small sides of 4-separations.

- For any integer n, there exists m such that, if $\mathcal{T}$ controls a $PG(m-1, \mathbb{F}_0)$-minor but does not control a $PG(n-1, \mathbb{F})$-minor, then M is essentially representable over a subfield of $\mathbb{F}$. Here "essentially" allows low-rank perturbations and suppressing small sides of 2-separations.

12   Progress towards Rota's Conjecture

An *excluded minor* for minor-closed class $\mathcal{M}$ is a matroid M that is not in $\mathcal{M}$ having the property that all of its proper minors belong to $\mathcal{M}$. As noted in the introduction, probably the most famous conjecture in matroid theory is

Conjecture 12.1 (Rota's Conjecture) *For any finite field $GF(q)$, there are only a finite number of excluded minors for the class of $GF(q)$-representable matroids.*

Applying structure theory to well-quasi-ordering and minor testing essentially follows the model established by Robertson and Seymour for graph minors. However we also expect structure theory to be a valuable tool to help resolve Rota's Conjecture. This has a somewhat different flavour and we conclude by briefly discussing how we expect structure theory to be applied to this conjecture. The following two complementary results add considerable support to Rota's Conjecture. The first is a theorem of Geelen, Gerards and Whittle [8].

Theorem 12.2 *For any prime power q, there are only finitely many excluded minors for the class of $GF(q)$-representable matroids that contain a $PG(q+6, q)$-minor.*

This is significant since it opens Rota's Conjecture to an attack via structural results on the class of $GF(q)$-representable matroids with no $PG(q+6, q)$-minor. The second is a result of Geelen and Whittle [6].

Theorem 12.3 *For any prime power q and fixed integer k there are only a finite number of excluded minors for $GF(q)$-representability of branch width k.*

This is significant since we may combine it with the grid theorem for matroids to obtain

Theorem 12.4 *For any planar graph H and prime power q, there are only a finite number of excluded minors for $GF(q)$-representability that do not contain a minor isomorphic to $M(H)$.*

Thus, if Rota's Conjecture is false, then for any given planar graph G, there is an excluded minor containing G. Equivalently, there are excluded minors containing arbitrarily large grids as minors.

Even with the structure theory there are still major conceptual obstacles to overcome to be able to resolve Rota's Conjecture completely; these obstacles are largely caused by inequivalent representations of matroids and we will not go into details here. Nonetheless recent work has been promising and we are confident of a resolution of Rota's Conjecture in the not-too-distant future.

References

[1] B. Courcelle, The monadic second order logic of graphs. I. Recognizable sets of finite graphs, *Information and Computation* **85** (1990), 12–75.

[2] R. Diestel, T. R. Jensen, K. Yu. Gorbanov and C. Thomassen, Highly connected sets and the excluded grid theorem, *J. Combin. Theory Ser. B* **75** (1999), 61–73.

[3] T. A. Dowling, A class of geometric lattices based on finite groups, *J. Combin. Theory Ser. B* **14** (1973), 61–83.

[4] J. Geelen, Some open problems on excluding a uniform minor, *Adv. in Appl. Math.* **41** (2008), 628–637.

[5] J. Geelen and B. Gerards, Excluding a group labelled graph, *J. Combin. Theory Ser. B* **99** (2009), 247–253.

[6] J. Geelen and G. Whittle, Branch width and Rota's Conjecture, *J. Combin. Theory Ser. B* **86** (2002), 315–330.

[7] J. Geelen, B. Gerards and G. Whittle, Branch-width and well-quasi-ordering in matroids and graphs, *J. Combin. Theory Ser. B* **84** (2002), 270–290.

[8] J. Geelen, B. Gerards and G. Whittle, On Rota's Conjecture and excluded minors containing large projective geometries, *J. Combin. Theory Ser. B* **96** (2006), 405–425.

[9] J. Geelen, B. Gerards and G. Whittle, Excluding a planar graph from $GF(q)$-representable matroids, *J. Combin. Theory Ser. B* **97** (2007), 971–998.

[10] J. Geelen, B. Gerards and G. Whittle, Tangles tree-decompositions and grids in matroids, *J. Combin. Theory Ser. B* **99** (2009), 657–667.

[11] J. Geelen and K. Kabell, Projective geometries in dense matroids, *J. Combin. Theory Ser. B* **99** (2009), 1–8.

[12] J. Geelen, J. Kung and G. Whittle, Growth rates of minor-closed classes of matroids, *J. Combin. Theory Ser. B* **99** (2009), 420–427.

[13] J. Geelen and G. Whittle, Cliques in dense $GF(q)$-representable matroids, *J. Combin. Theory Ser. B* **87** (2003), 264–269.

[14] J. Geelen and G. Whittle, Inequivalent representations of matroids over prime fields, submitted.

[15] T. Gowers, The two cultures of mathematics, in *Mathematics, Frontiers and Perspectives* (ed. V. Arnold, M. Atiyah, P. Lax, B. Mazur), American Mathematical Society, Rhode Island (2000).

[16] T. Gowers, Rough structure and classification, in *Geom. Funct. Anal., Special Volume, Part I* (2000), 79–117.

[17] I. Heller, On linear systems with integral valued solutions, *Pacific J. Math.* **7** (1957), 1351–1364.

[18] I. V. Hicks and N. B. McMurray, The branchwidth of graphs and their cycle matroids, *J. Combin. Theory Ser. B* **97** (2007), 681–692.

[19] P. Hliněný, Branch-width, parse trees and monadic second-order logic for matroids, *J. Combin. Theory Ser. B* **96** (2006), 325–351.

[20] P. Hliněný, Some hard problems on matroid spikes, *Theory of Computing Systems* **41** (2007), 551–562.

[21] P. Hliněný and G. Whittle, Matroid tree width, *European J. Combin.* **27** (2006), 1117–1128.

[22] T. Huynh, 2009. The linkage problem for group-labelled graphs, PhD Thesis, University of Waterloo,

[23] J. Kahn and J. P. S. Kung, Varieties of combinatorial geometries, *Trans. Amer. Math. Soc.* **271** (1982), 485–499.

[24] J. P. S. Kung, Extremal matroid theory, *Contemporary Mathematics* **147** (1993), 21–61.

[25] W. Mader, Homomorphieeigenschaften und mittlere Kantendichte von Graphen, *Math. An.* **174** (1967), 265–268.

[26] D. Mayhew, M. Newman and G. Whittle, On excluded minors for real representability, *J. Combin. Theory Ser. B* **99** (2009), 685–689.

[27] D. Mayhew, M. Newman, D. Welsh and G. Whittle, On the asymptotic proportion of connected matroids, *European Journal of Combinatorics* **32** (2011), 882–890.

[28] J. G. Oxley, *Matroid Theory*, Oxford University Press, Oxford (2011).

[29] J. G. Oxley, What is a matroid?, *Cubo* **5** (2003), 179–218.

[30] N. Robertson and P. D. Seymour, Graph Minors. V. Excluding a planar graph, *J. Combin. Theory Ser. B* **41** (1986), 92–114.

[31] N. Robertson and P. D. Seymour, Graph Minors. X. Obstructions to tree decomposition, *J. Combin. Theory Ser. B* **59** (1991), 153–190.

[32] N. Robertson and P. D. Seymour, Graph Minors. XIII. The disjoint paths problem, *J. Combin. Theory Ser. B* **63** (1995), 65–110.

[33] N. Robertson and P. D. Seymour, Graph Minors. XVI. Excluding a non-planar graph, *J. Combin. Theory Ser. B* **89** (2003), 43–76.

[34] N. Robertson and P. D. Seymour, Graph Minors. XX. Wagner's Conjecture, *J. Combin. Theory Ser. B* **92** (2004), 325–357.

[35] N. Robertson, P. D. Seymour and R. Thomas, Quickly excluding a planar graph, *J. Combin. Theory Ser. B* **62** (1994), 323–348.

[36] G. -C. Rota, Combinatorial theory, new and old, in *Actes du Congres International des Mathematiciens (Nice, Sept. 1970)* Tome 3, pp. 229–233, Gaulthier-Villars, Paris (1971).

[37] P. D. Seymour, Decomposition of regular matroids, *J. Combin. Theory Ser. B* **28** (1980), 305–354.

[38] P. D. Seymour, Triples in matroid circuits, *European J. Combin.* **7** (1986), 177–185.

[39] P. Turán, On an extremal problem in graph theory, *Matematikai és Fizikai Lapok* **48** (1941), 436–452.

[40] K. Truemper, On the efficiency of representability tests for matroids, *European J. Combin.* **3** (1982), 275–291.

[41] W. T. Tutte, A homotopy theorem for matroids, I, II, *Trans. Amer. Math. Soc.* **88** (1958), 144–174.

[42] T. Zaslavsky, Biased graphs. II. The three matroids, *J. Combin. Theory Ser. B* **51** (1991), 46–72.

Department of Combinatorics and Optimization
University of Waterloo, Waterloo, Canada
jfgeelen@math.uwaterloo.ca

Centrum Wiskunde & Informatica
Amsterdam, The Netherlands
and
Maastricht University School of Business and Economics
Maastricht, The Netherlands

School of Mathematics Statistics and Operations Research
Victoria University, Wellington, New Zealand
geoff.whittle@vuw.ac.nz

Automatic counting of tilings of skinny plane regions

Shalosh B. Ekhad and Doron Zeilberger

Abstract

The deductive method ruled mathematics for the last 2500 years, now it is the turn of the inductive method. We make a start by using the C-finite ansatz to enumerate tilings of skinny place regions, inspired by a Mathematics Magazine Problem proposed by Donald Knuth.

1 Very important

This article is accompanied by the Maple packages

- http://www.math.rutgers.edu/~zeilberg/tokhniot/RITSUF,

- http://www.math.rutgers.edu/~zeilberg/tokhniot/RITSUFwt,

- http://www.math.rutgers.edu/~zeilberg/tokhniot/ARGF,

to be described below. In fact, more accurately, this article accompanies these packages, written by DZ and the many output files, discovering and proving deep enumeration theorems, done by SBE, that are linked to from the webpage of this article: http://www.math.rutgers.edu/~zeilberg/mamarim/mamarimhtml/ritsuf.html.

2 How it all started: April 5, 2012

During one of the Rutgers University Experimental Mathematics Seminar dinners, the name of Don Knuth came up, and two of the participants, David Nacin, who was on sabbatical from William Patterson University, and first-year graduate student Patrick Devlin, mentioned that they recently solved a problem that Knuth proposed in *Mathematics Magazine* [5]. The problem was:

1868. *Proposed by Donald E. Knuth, Stanford University, Stanford, California.*

Let $n \geq 2$ be an integer. Remove the central $(n-2)^2$ squares from an $(n+2) \times (n+2)$ array of squares. In how many ways can the remaining squares be covered with $4n$ dominoes?

As remarked in the published solution in Math. Magazine, the problem was already solved in the literature by Roberto Tauraso [7]. The answer turned out to be very elegant: $4(2F_n^2 + (-1)^n)^2$.

David and Patrick, as well as the solution published in *Math. Magazine*, used human ingenuity and *mathematical deduction*. But, as already preached in [8] and [9], the following is a **fully rigorous proof**:

> "By direct counting of tilings, the first 10 terms (starting at $n = 2$) of the enumerating sequence are
>
> $$36, 196, 1444, 9604, 66564, 454276, 3118756, 21362884,$$
> $$146458404, 1003749124,$$
>
> but so are the first 10 terms of the sequence $\{4(2F_n^2+(-1)^n)^2\}$. Since the statement is true for the first 10 terms, it must be true for **all** $n \geq 2$. **QED!**"

In order to *justify* this "empirical" proof, all we need to say is that both sides are obviously *C-finite* sequences whose minimal recurrences have order ≤ 5, hence their difference is a C-finite sequence of order ≤ 10 and hence if it vanishes for the first 10 terms, it *always* vanishes. See below for a justification.

But first let's remind ourselves what are *C*-finite sequences, and recall how to find upper bounds for the orders of the recurrences.

3   *C*-finite sequences

Recall that a *C*-finite sequence $\{a(n)\}_{n=0}^{\infty}$ is a sequence that satisfies a **homogeneous linear-recurrence equation with constant coefficients**. It is known (but not as well-known as it should be!) and easy to see (e.g. [9], [4]) that the set of *C*-finite sequences is an *algebra*. Even though a *C*-finite sequence is an "infinite" sequence, it is in fact, **like everything else in mathematics** (and elsewhere!) a **finite** object. An order-L *C*-finite sequence $\{a(n)\}_{n=0}^{\infty}$ is completely specified by the coefficients $c_1, c_2, \ldots, c_L$ of the recurrence

$$a(n) = c_1 a(n - 1) + c_2 a(n - 2) + \cdots + c_L a(n - L), \qquad \text{(Cfinite)}$$

and the **initial conditions**

$$a(0) = d_1, \quad \ldots, \quad a(L - 1) = d_L.$$

So a *C*-finite sequence can be **coded** in terms of the $2L$ "bits" of information

$$[[d_1, \ldots, d_L], [c_1, \ldots, c_L]].$$

For example, the Fibonacci sequence is written:

$$[[0, 1], [1, 1]].$$

Since this ansatz (see [8]) is fully decidable, it is possible to decide equality, and evaluate *ab initio*, wide classes of sums, and things are easier than the *holonomic ansatz* (see, e.g., [9]). The wonderful new book by Manuel Kauers and Peter Paule [4] also presents a convincing case.

4 The art of guessing

If you know *a priori*, or suspect, that a given sequence $\{a(n)\}$, is C-finite of order $\leq L$, all you need is to compute the first $2L$ terms of $a(n)$, and then plug-in $n = L, L + 1, \ldots 2L - 1$ into the *ansatz* (Cfinite), get a system of L linear equations for the L unknowns $c_1, \ldots, c_L$, and let the computer solve them. Having found them, you should test the recurrence for quite a few larger n, just to make sure that it is indeed C-finite.

5 Orders of recurrences

We need a

Crucial Lemma. If $a(n)$ is C-finite of order L then, for every $i \geq 0$, $a(n + i)$ is a *linear combination* of $\{a(n), \ldots, a(n + L - 1)\}$.

Proof For $i < L$ this is utterly trivial. For $i = L$ it follows from Equation (Cfinite), with $n \to n + L$, and for $i > L$ it follows by induction on i, by substituting $n \to n + i$ into (Cfinite), and then noting, that by the inductive hypothesis, $a(n + j)$ for $j < i$ are all linear combinations of $\{a(n), \ldots, a(n + L - 1)\}$, and a linear combination of linear combinations is a linear combination. □

We are now ready to recall the following facts (see [4]):

Fact 0. If $\{a(n)\}$ and $\{b(n)\}$ are C-finite of orders L_1 and L_2 respectively, then $\{a(n) + b(n)\}$ is C-finite of order $L_1 + L_2$.

Proof By the Crucial Lemma, for all $i \geq 0$, $a(n + i)$ is a linear combination of $a(n), \ldots, a(n + L_1 - 1)$ and $b(n + i)$ is a linear combination of $b(n), \ldots, b(n + L_2 - 1)$, hence putting $c(n) := a(n) + b(n)$, $c(n + i)$ is a linear combination of

$$a(n), \ldots, a(n + L_1 - 1), b(n), \ldots, b(n + L_2 - 1),$$

hence the $L_1 + L_2 + 1$ sequences $c(n), \ldots, c(n + L_1 + L_2)$ are linearly dependent, hence the sequence $c(n)$ is C-finite of order $L_1 + L_2$. □

Fact 1. If two sequences satisfy the same homogeneous linear recurrence with constant coefficients (but with different initial conditions) so does their sum.

Fact 2. If two sequences $a(n)$ and $b(n)$ satisfy the same homogeneous linear recurrence with constant coefficients of order L, then their product $a(n)b(n)$ satisfies another such recurrence of order $\binom{L+1}{2}$.

Proof By the Crucial Lemma, for all $i \geq 0$, $a(n+i)$ ia a linear combinations of $\{a(n),\ldots,a(n+L-1)\}$. Now note that $b(n)$ is a linear combination of $a(n),\ldots,a(n+L-1)$, (since the latter form a *basis* for the L-dimensional vector space of all solutions of the recurrence). It follows that also $b(n+i)$, for $i \geq 0$, being a linear combinations of $\{b(n),\ldots,b(n+L-1)\}$ (by the Crucial Lemma) is a linear combination of $\{a(n),\ldots,a(n+L-1)\}$.

So $a(n+i)b(n+i)$ (for any $i \geq 0$) is a certain linear combination of

$$\{a(n+\alpha)a(n+\beta)\,;\, 0 \leq \alpha \leq \beta \leq L-1\}.$$

There are $M := \binom{L+1}{2}$ such sequences, so the $M+1$ sequences

$$\{a(n)b(n), a(n+1)b(n+1)\ldots,a(n+M)b(n+M)\}$$

must be linearly dependent, giving an order-$\binom{L+1}{2}$ recurrence. □

More generally, with an analogous proof:

Fact 3. If r sequences $a_1(n), a_2(n), \ldots a_r(n)$ satisfy the same homogeneous linear recurrence with constant coefficients of order L, then their product $a_1(n)a_2(n) \cdots a_r(n)$ satisfies another such recurrence of order $\binom{L+r-1}{r}$. In particular, if $a(n)$ is a C-finite sequence of order L, then $a(n)^r$ is a C-finite sequence of order $\binom{L+r-1}{r}$.

Fact 4. If two sequences $a(n)$ and $b(n)$ satisfy different homogeneous linear recurrences with constant coefficients of orders L_1 and L_2, respectively, then their product satisfies such a recurrence of order $L_1 L_2$.

Proof By the Crucial Lemma, $a(n+i)$ $(i \geq 0)$ is a linear combination of $\{a(n),\ldots,a(n+L_1-1)\}$. Likewise, $b(n+i)$ $(i \geq 0)$ is a linear combination of $\{b(n),\ldots,b(n+L_2-1)\}$, So $a(n+i)b(n+i)$ $(i \geq 0)$ is a linear combination of the $L_1 L_2$ sequences

$$\{a(n+\alpha)b(n+\beta) : 0 \leq \alpha < L_1\,,\, 0 \leq \beta < L_2\},$$

hence, putting $M = L_1 L_2$, we see that the $M+1$ sequences

$$\{a(n)b(n),\ldots,a(n+M)b(n+M)\}$$

must be linearly dependent. □

6   Rational generating functions

Equivalently, a C-finite sequence is a sequence $\{a(n)\}$ whose **ordinary generating function**, $\sum_{n=0}^{\infty} a(n)z^n$, is rational and where the degree of the denominator is more than the degree of the numerator. These come up a lot in combinatorics and elsewhere (e.g. formal languages). See the *old testament* [6], Chapter 4, and the *new testament* [4], Chapter 4.

7   Why is the number of tilings of the Knuth square-ring C-finite?

Each of the four "corners" (2×2 squares) is tiled in a certain way, where either the tiles covering it *only* cover it, or some tiles also cover neighboring cells *not* in the corner-square. There are finitely many such scenarios for each corner-square, hence for the Cartesian product of these scenarios. Having decided how to cover these four corner squares, one must decide how to tile the remaining four sides, each of which is either a 2 by n rectangle, or one with a few bites taken from one or both ends. By the *transfer matrix method* ([6], 4.7), it follows *a priori* that each of these enumerating sequences is C-finite. In fact each individual side, for each scenario, satisfies a certain recurrence (that is easily seen to be second-order), hence each product-of-four scenarios satisfies a certain order-5 (Fact 3 with $r = 4$ and $L = 2$) recurrence, and hence so does their sum (Fact 1). By the same reasoning $\{4(2F_n^2 + (-1)^n)^2\}$ also satisfies an order-5 recurrence, so their difference satisfies an order 10 recurrence, by Fact 0.

But we still need to be able to compute the first 10 terms. If we had a very large and very fast computer, we could have actually constructed *all* the tilings, and then counted them, but 1003749124 is a pretty big number, so we need a *more efficient way*.

8   A more efficient way

Suppose that you are given a set of unit-squares (let's call them cells) and you want your computer to find the number of ways of tiling it with dominoes. You pick any cell (for the sake of convenience the left-most bottom-most cell), and look at all the ways in which to cover it with a domino piece. For each of these ways, removing that tile leaves a smaller region, thereby getting an obvious *dynamical programming* recurrence.

Procedure NT(R) in the Maple package RITSUF computes the number of domino tilings of any set of cells.

Using this method, the first-named author solved Knuth's problem in 15 seconds, by typing (in a Maple session, in a directory where `RITSUF` has been downloaded to)

```
read RITSUF: SeqFrameCsqDirect(2,2,2,2,20,t);.
```

We will soon try to solve analogous problems for fatter frames, but then we would need to be *even* more efficient, and using this more efficient method, the same calculation would take less than half a second, typing:

```
read RITSUF: SeqFrameCsq(2,2,2,2,20,t);.
```

9 An even more efficient way

Our general problem is to find an efficient automatic way to compute the C-finite description, and/or its generating function (that *must* always be a rational function), for the sequence enumerating domino tilings of the region, that we denote, in `RITSUF`, by

```
Frame(a1,a2,b1,b2,n,n)
```

that consists of an $(a_1 + n + a_2) \times (b_1 + n + b_2)$ rectangle with the "central" $n \times n$ square removed.

Before describing the algorithm, let us mention that this is accomplished by the procedure

```
SeqFrameC(a1,a2,b1,b2,N,t).
```

For example, `SeqFrameC(1,3,3,1,30,t);` yields

$$-4\,\frac{-9 + 8\,t + 29\,t^2 - 10\,t^3 - 7\,t^4 + 2\,t^5}{(t^2 - 4\,t + 1)\,(t^4 - 4\,t^2 + 1)}.$$

The fifth argument, N is a parameter for "guessing" the C-finite description, indicating how many data points to gather before one tries to guess the C-finite description. It is easy to find a priori upper bounds, but it is more fun to let the user take a guess, and increasing it, if necessary.

10 Mihai Ciucu's amazing theorem

The sequence of positive integers, demanded by Knuth, enumerating the domino tilings of `Frame(2,2,2,2,n,n)`, turned out to be all *perfect squares*. This is not a coincidence! A beautiful theorem of Mihai Ciucu [1] tells us that whenever there is a reflective symmetry, the number of domino tilings is either a perfect square or twice a perfect square. Since we know that (and even if we didn't, we could have discovered it empirically for each case that we are computing), we have to gather far less data points. For example, according to the first-named author of this article, the number of domino tilings of `Frame(3,3,3,3,n,n)` is $2B(n)^2$, where

$$\sum_{n=0}^{\infty} B(n)t^n = -2\,\frac{-29 + 19\,t + 102\,t^2 - 32\,t^3 - 25\,t^4 + 7\,t^5}{(t^2 - 4\,t + 1)\,(t^4 - 4\,t^2 + 1)},$$

while the number of domino tilings of `Frame(4,4,4,4,n,n)` is $C(n)^2$ where

$$\sum_{n=0}^{\infty} C(n)t^n = \frac{P(t)}{Q(t)}$$

and

$$
\begin{aligned}
P(t) = -4\big(&901 + 2517\,t - 17574\,t^2 - 46322\,t^3 + 112903\,t^4 + 291045\,t^5 \\
&- 269376\,t^6 - 741508\,t^7 + 215233\,t^8 + 786069\,t^9 - 21836\,t^{10} \\
&- 352896\,t^{11} - 24137\,t^{12} + 67487\,t^{13} + 5874\,t^{14} - 5056\,t^{15} \\
&- 359\,t^{16} + 97\,t^{17}\big) \\
Q(t) = (t &- 1)\,(t + 1)\,(t^4 + t^3 - 5\,t^2 + t + 1)\,(t^4 - 11\,t^3 + 25\,t^2 - 11\,t + 1) \\
&(t^4 + 7\,t^3 + 13\,t^2 + 7\,t + 1)\,(t^4 - t^3 - 5\,t^2 - t + 1).
\end{aligned}
$$

If you want to see the analogous expressions for `Frame(5,5,5,5,n,n)` and `Frame(6,6,6,6,n,n)`, then you are welcome to look at the output file

http://www.math.rutgers.edu/~zeilberg/tokhniot/oRITSUF2

that you can generate yourself, *ab initio* by running (once you uploaded RITSUF onto a Maple session)

http://www.math.rutgers.edu/~zeilberg/tokhniot/inRITSUF2.

Our method is *pure guessing*, but in order to guess, we need to *efficiently* generate sufficiently many terms of the counting sequence. We must start with *rectangles* of fixed width.

11   The number of domino tilings of a rectangle of a fixed width

Let m be a fixed positive integer. We are interested in a C-finite description, as a function of n, of the sequence $A_m(n)$, the number of domino tilings of an $m \times n$ rectangle. In fact, for this *specific* problem there is an "explicit" solution, famously found by Kasteleyn [3] and Fisher & Temperly [2], but their solution only applies to domino tiling, and we want to illustrate the *general* method.

Also the general approach, using the transfer-matrix method, is not *new* as such (see, e.g. [6]), but since we need it for counting more elaborate things, let's review it.

Consider the task of tiling the n columns of an $m \times n$ rectangle. Let's label the cells of a given column from bottom to top by $\{1, \ldots, m\}$. When we start, all the m cells of the leftmost column are available, so we start with the *state* $\{1, \ldots, m\}$. As we keep on going, not all cells of the current column are available, since some of them have already been tiled by the previous column. The other extreme is the empty set, where nothing is available, and one must go immediately to the next column, where now everything is available, i.e. the only follower of $\emptyset$ is the universal set $\{1, \ldots, m\}$.

Let's take $m = 4$ and see what states may follow the state $\{1, 2, 3, 4\}$. We may have two vertical tiles $\{1, 2\}$ and $\{3, 4\}$, leaving all the cells of the next column available, yielding the state $\{1, 2, 3, 4\}$. We may decide instead to *only* use horizontal tiles, leaving nothing available for the next column, resulting in the state $\emptyset$. If we decide to have one vertical tile in the current column, then:

- If it is $\{1, 2\}$ then both 3 and 4 are parts of horizontal tiles that go to the next column, leaving the set of cells $\{1, 2\}$ available;

- If it is $\{2, 3\}$ then both 1 and 4 are parts of horizontal tiles that go to the next column, leaving the set of cells $\{2, 3\}$ available;

- If it is $\{3, 4\}$ then both 1 and 2 are parts of horizontal tiles that go to the next column, leaving the set of cells $\{3, 4\}$ available.

It follows that the "followers" of the state $\{1, 2, 3, 4\}$ are the five states

$$\emptyset, \{1, 2, 3, 4\}, \{1, 2\}, \{2, 3\}, \{3, 4\}.$$

Who can follow the state $\{1, 4\}$? Since only cell 1 and cell 4 are available there can't be any vertical tiles, and both must be parts of horizontal tiles, occupying $\{1, 4\}$ of the next column, and leaving $\{2, 3\}$ available, so the state $\{1, 4\}$ has only one follower, the state $\{2, 3\}$.

Check out procedure `Followers(S,m)` in RITSUF.

This way we can view any tiling of an $m \times n$ rectangle as a walk of length n in a directed graph whose vertices are labeled by subsets of

$\{1, \ldots, m\}$. The transfer matrix of this graph is the $2^m \times 2^m$ matrix whose rows and columns naturally correspond to subsets (in RITSUF we made the natural convention that the indices of the rows and columns correspond to the binary representations of the subsets plus 1). Let's call $SN(i)$ the set of positive integers corresponding to the positive integer i. For example, $SN(1)$ is the empty set, $SN(10)$ is $\{1, 4\}$ etc.

Check out procedure TM(m) in RITSUF for the transfer matrix.

Calling this transfer matrix A_m, the number of domino tilings of an $m \times n$ rectangle is the $(2^m, 2^m)$ entry of the matrix A_m^n; since we have to completely tile it, the starting state must be $\{1, \ldots, m\}$, of course, but so is the ending state, since every thing must be covered, leaving the next column completely available.

Check out procedure SeqRect(m,N) in RITSUF for the counting sequence for the number of domino tilings of an m by n rectangle for $n = 0, 1, \ldots, N$.

But not just the $(2^m, 2^m)$ entry of A_m^n is informative. Each and every one of the $(2^m)^2$ entries contains information! A typical (i, j) entry of A_m^n tells you the number of ways of tiling an $m \times n$ rectangle where the leftmost column only has the cells in $SN(i)$ available for use, while the rightmost column has some tiles that stick out, leaving available for the $(n+1)$-th column the cells of $SN(j)$.

12 Counting the number of domino tilings of a holey rectangle

We want to figure out how to use matrix-multiplication to determine the number of tilings of the region

$$\text{Frame}(a_1, a_2, b_1, b_2, m, n)$$

that consists of an $(a_1 + m + a_2) \times (b_1 + n + b_2)$ rectangle with the central $m \times n$ rectangle removed.

There are four corner rectangles in our frame:

- the left-bottom (SW) corner consisting of an $a_1 \times b_1$ rectangle,
- the right-bottom (SE) corner consisting of an $a_1 \times b_2$ rectangle,
- the left-top (NW) corner consisting of an $a_2 \times b_1$ rectangle,
- the right-top (NE) corner consisting of an $a_2 \times b_2$ rectangle.

If we look at a typical tiling of the region $\mathrm{Frame}(a_1, a_2, b_1, b_2, m, n)$, and focus on the induced tilings of the four corner-rectangles, we get a tiling with (usually) some of the tiles intersecting the adjacent non-corner rectangles.

Indeed, for the left-bottom $a_1 \times b_1$ corner-rectangle (usually) some of the tiles covering its very top row intersect the very bottom row of the left (West) $m \times b_1$ rectangle, and (usually) some of the tiles covering its very right column intersect the leftmost column of the bottom (South) $a_1 \times n$ rectangle, and similarly for the other three corner rectangles. So suppose that the West $m \times b_1$ rectangle has already been tiled, with some of its bottom tiles overlapping with our above-mentioned left-bottom $a_1 \times b_1$ corner-rectangle, leaving only some of the cells in the top row available, and after we complete the tiling of that left-bottom $a_1 \times b_1$ corner-rectangle, we may use-up some of the cells of the left column of the bottom (South) $a_1 \times n$ rectangle, and the complement of that occupied set is only available for tiling.

This leads naturally to a transfer matrix between the "states" of one side of a corner-rectangle to the states of another side of that corner-rectangle. So let's define $R(a, b, S_1, S_2)$, for positive integers a and b, and distinct $S_1, S_2 \in \{1, 2, 3, 4\}$ where we make the convention

$$1 = \text{Up Side}, \quad 2 = \text{Left Side}, \quad 3 = \text{Down Side}, \quad 4 = \text{Right Side},$$

that tells you the number of ways of tiling the $a \times b$ rectangle where the two sides that are not in $\{S_1, S_2\}$ are "smooth" (i.e. nothing sticks out) and the two sides S_1, S_2 may (and usually do) have some of their tiles "sticking out".

Going counterclockwise starting at the left-bottom (SW) corner, we need to find

- For the left-bottom (SW) corner consisting of an $a_1 \times b_1$ rectangle we need $R(a_1, b_1, 1, 4)$,

- For the right-bottom (SE) corner consisting of an $a_1 \times b_2$ rectangle we need $R(a_1, b_2, 2, 1)$,

- For the right-top (NE) corner consisting of an $a_2 \times b_2$ rectangle we need $R(a_2, b_2, 3, 2)$,

- For the left-top (NW) corner consisting of an $a_2 \times b_1$ rectangle we need $R(a_2, b_1, 4, 3)$.

Like the matrices A_m that for each needed m, we only compute **once** and then record it, (using `option remember`), we compute $R(a, b, S_1, S_2)$ only once for each needed a, b, S_1, S_2 and remember it.

But how to compute $R(a, b, S_1, S_2)$? We first construct, *literally*, all the domino tilings that completely cover the cells of the $a \times b$ rectangle where nothing sticks out of the sides that are not labelled S_1 or S_2, but that may (and usually do) stick out from the sides labelled S_1 and S_2. Then we look at all the pairs of states, and form a matrix whose (i, j) entry is the number of stuck-out tilings of the $a \times b$ rectangle where the "stick-out" state of side S_1 corresponds to the set labelled i, and the "stick-out" state of side S_2 corresponds to the set labelled j.

See procedure RTM(a,b,S1,S2) for its implementation in RITSUF.

It follows that, in terms of the matrices A_m and $R(a, b, S_1, S_2)$, the quantity of interest, the number of tilings of the rectangular picture-frame Frame$(a_1, a_2, b_1, b_2, m, n)$, is

$$\operatorname{Tr}\big(R(a_1, b_1, 1, 4) A_{a_1}^n R(a_1, b_2, 2, 1) A_{b_2}^m R(a_2, b_2, 3, 2) A_{a_2}^n R(a_2, b_1, 4, 3) A_{b_1}^m\big).$$

Since matrix power-raising is very fast, and so is matrix-multiplication, we can quickly crank-out sufficiently many terms in the enumerating sequence, and since we know *a priori* that it is C-finite (by the Cayley-Hamilton theorem each entry of A^n is C-finite (for *any* matrix A), and the rest follows from Facts 0-4 above), we can guess a C-finite description (and/or a rational generating function), that is proved rigorously *a posteriori* by checking that the order bounds are right.

13   The bivariate generating function

If you are interested in the discrete function of the *two* discrete variables m and n, for the number of domino tilings of Frame$(a_1, a_2, b_1, b_2, m, n)$, then it is *doubly* C-finite, meaning that its bivariate generating function has the form

$$P(x, y)/(Q_1(x)Q_2(y)),$$

for some polynomials $P(x, y), Q_1(x), Q_2(y)$. Using an analogous method for guessing, after cranking-out enough data, we can get these generating functions easily, using procedure

GFframeDouble(a1,a2,b1,b2,x,y,N)

in RITSUF. For example, if $D(m, n)$ is the number of domino tilings of Frame$(2, 2, 2, 2, m, n)$, then

$$\sum_{m=0}^{\infty} \sum_{n=0}^{\infty} D(m, n) x^m y^n = \frac{P(x, y)}{Q(x, y)},$$

where

$$P(x,y) = 4\,x^3 y^3 - 7\,x^3 y^2 - 7\,x^2 y^3 - 14\,x^3 y + 10\,x^2 y^2 - 14\,xy^3 + 13\,x^3$$
$$+ 35\,x^2 y + 35\,xy^2 + 13\,y^3 - 30\,x^2 + 10\,xy - 30\,y^2 - 23\,x$$
$$- 23\,y + 36$$

and

$$Q(x,y) = (x-1)\,(x+1)\,\left(x^2 - 3\,x + 1\right)(y-1)\,(y+1)\,\left(y^2 - 3\,y + 1\right).$$

14   Tiling crosses

In how many ways can we tile a cross whose center is a 2×2 square and each of the four arms has length n? The answer is obtained by typing, in RITSUF,

    SeqCrossCsq(2,2,20,t);,

getting (in 0.072 seconds!) that the number is $2B_2(n)^2$, where

$$\sum_{n=0}^{\infty} B_2(n)t^n = \frac{1}{(t+1)\,(t^2 - 3\,t + 1)}.$$

Incidentally, $\{B_2(n)\}$ is http://oeis.org/A001654, the "Golden rect-angle number" $F_n F_{n+1}$, so the number of tilings of this cross is $2F_n^2 F_{n+1}^2$, and it is a fairly simple exercise for humans to prove this fact. But we doubt that any human can derive, by hand, the answer to the analogous question for the cross whose center is a 4×4 square. The answer turns out to be $B_4(n)^2$, where

$$\sum_{n=0}^{\infty} B_4(n)t^n$$
$$= -2\,\frac{3 - t - 5\,t^2 + 13\,t^3 - 11\,t^4 - 2\,t^5 + 2\,t^6}{(t-1)\,(t^4 - 11\,t^3 + 25\,t^2 - 11\,t + 1)\,(t^4 + 7\,t^3 + 13\,t^2 + 7\,t + 1)}.$$

Even SBE needed 4.528 seconds to derive this formula after DZ typed:

    SeqCrossCsq(4,4,30,t);.

Once again, the approach is *purely empirical*. The computer finds all the tilings that completely cover the central square (or rectangle), possibly (and usually) with some tiles extending beyond. For each such scenario we have the state for each of the four arms of the cross. Then we use the previously computed (and remembered!) matrices A_m, find the corresponding entry in $(A_m)^n$, and multiply all these four numbers. Finally we (meaning our computers) add up all these scenarios, getting the desired number. See the source code of `SeqCross(a,b,N)`.

15 Monomer-dimer tilings

The beauty of programming is that, once we have finished writing a program, it is easy to modify it in order to solve more general, or analogous, problems. By typing `ezraMD();` in RITSUF the readers can find the list of analogous procedures for monomer-dimer tilings, where one can use either a 1×2 a 2×1 or 1×1 tile, or equivalently, tiling with dominoes where one is not required to cover all the cells.

For example, if $A(n)$ is the number of ways of tiling the Knuth region (obtained by removing the central n^2 squares from an $(n + 4) \times (n + 4)$ array of squares) by dimers *and* monomers, the answer is much messier. It took SBE 15 seconds to discover that

$$\sum_{n=0}^{\infty} A(n)t^n = \frac{P(t)}{Q(t)},$$

where

$$P(t) = -94\,t^{30} + 1361\,t^{29} + 43975\,t^{28} - 494267\,t^{27} - 5787443\,t^{26}$$
$$+ 61186056\,t^{25} + 266911158\,t^{24} - 3200500450\,t^{23} - 3505671568\,t^{22}$$
$$+ 74767156291\,t^{21} - 29007687275\,t^{20} - 796609853769\,t^{19}$$
$$+ 823823428983\,t^{18} + 3924729557742\,t^{17} - 4977782472712\,t^{16}$$
$$- 9040256915004\,t^{15} + 11643454084810\,t^{14} + 9751493606823\,t^{13}$$
$$- 11693567793807\,t^{12} - 4837640809485\,t^{11} + 5123918478955\,t^{10}$$
$$+ 1059903067708\,t^{9} - 944330286322\,t^{8} - 87120095554\,t^{7}$$
$$+ 67451928324\,t^{6} + 657867045\,t^{5} - 1679236205\,t^{4}$$
$$+ 73176689\,t^{3} + 6962033\,t^{2} - 226706\,t - 10012,$$

and

$$Q(t) = (t - 1)\left(t^3 - 7t^2 + 11t - 1\right)\left(t^3 + 7t^2 - 33t - 1\right)$$
$$\left(t^3 - 11t^2 + 7t - 1\right)\left(t^3 + t^2 - 3t - 1\right)\left(t^3 - 27t^2 + 107t - 1\right)$$
$$\left(t^3 + 3t^2 - t - 1\right)\left(t^6 + 20t^5 + 55t^4 - 304t^3 - 337t^2 + 8t + 1\right)$$
$$\left(t^6 - 37t^4 - 76t^3 - 37t^2 + 1\right).$$

16 Weighted counting

What if instead of *straight counting* one wants to do *weighted counting?*. The weight of a domino tiling is defined to be

$$h^{\#HorizontalTiles}v^{\#VerticalTiles}$$

and the weight of a monomer-dimer tiling is defined to be

$$h^{\#HorizontalTiles}v^{\#VerticalTiles}m^{\#MonomerTiles}.$$

For this we have the Maple package RITSUFwt, freely downloadable from

http://www.math.rutgers.edu/~zeilberg/tokhniot/RITSUFwt,

that also contains all the procedures of RITSUF. To get a list of the procedures for weighted counting, type:

ezraWt();

For example, to get the weighted analog of the Knuth problem, type:

SeqFrameCwt(2,2,2,2,30,t,h,v);

To see the output, go to:

http://www.math.rutgers.edu/~zeilberg/tokhniot/oRITSUFwt1

where one can also see statistical analysis.

17 The Maple package ARGF

The Maple package ARGF (short for "Analysis of Rational Generating Functions") downloadable from:

http://www.math.rutgers.edu/~zeilberg/tokhniot/ARGF,

does automatic statistical analysis of random variables whose weight-enumerators are given by rational functions, like the one outputted by RITSUFwt, and whose procedures are also included in the latter. We urge the readers to look at

http://www.math.rutgers.edu/~zeilberg/tokhniot/oRITSUFwt4

and the other output files in the webpage of this article

http://www.math.rutgers.edu/~zeilberg
/mamarim/mamarimhtml/ritsuf.html

for examples. Here we use the methodology of [10], [11].

18 Conclusion

Sometimes the naive, empirical, *inductive* approach, of "guessing", is superior to the traditional *deductive* method, that ruled mathematics for the last 2500 years.

Acknowledgements

The authors wish to thank the anonymous referee and also the non-anonymous editor (Mark Wildon) for helpful suggestions and minor corrections. Supported in part by the USA National Science Foundation.

References

[1] Mihai Ciucu, Enumeration of perfect matchings in graphs with reflective symmetry, *J. Combin. Theory Ser. A* **77** (1997), 67–97.

[2] M. Fisher and H. Temperley, Dimer Problems in Statistical Mechanics-an exact result, *Philos. Mag.* **6** (1961), 1061–1063.

[3] P. W. Kasteleyn, The statistics of dimers on a lattice: I. The number of dimer arrangements in a quadratic lattice, *Physica* **27** (1961), 1209–1225.

[4] Manuel Kauers and Peter Paule, *The Concrete Tetrahedron*, Springer, 2011.

[5] Donald E. Knuth, Tiling a square ring with dominoes, proposed problem #1868, Math. Magazine **84(2)** (April 2011), and Solution by John Bonomo and David Offner, Math. Magazine **85(2)** (April 2012), 154–155.

[6] Richard Stanley, *Enumerative Combinatorics*, volume 1, Wadsworth and Brooks/Cole, Pacific Grove, CA, 1986, second printing, Cambridge University Press, Cambridge, 1996.

[7] Roberto Tauraso, A New Domino Tiling Sequence, *Journal of Integer Sequences*, **7** (2004), Article 04.2.3. Available on-line: http://www.cs.uwaterloo.ca/journals/JIS/VOL7/Tauraso/tauraso3.pdf.

[8] Doron Zeilberger, The C-finite Ansatz, a "sanitized" version is to appear in the Ramanujan Journal, the original, uncensored version is available on-line: http://www.math.rutgers.edu/~zeilberg/mamarim/mamarimhtml/cfinite.html.

[9] Doron Zeilberger, An Enquiry Concerning Human (and Computer!) [Mathematical] Understanding, in C. S. Calude (editor), *Randomness & Complexity, from Leibniz to Chaitin*, World Scientific, Singapore, 2007. Available on-line: `http://www.math.rutgers.edu/~zeilberg/mamarim/mamarimhtml/enquiry.html`.

[10] Doron Zeilberger, The Automatic Central Limit Theorems Generator (and Much More!), in *Advances in Combinatorial Mathematics*, Proceedings of the Waterloo Workshop in Computer Algebra 2008 in honor of Georgy P. Egorychev, chapter 8, pp. 165–174, (I. Kotsireas, E. Zima, editors), Springer Verlag, 2009. Available on-line: `http://www.math.rutgers.edu/~zeilberg/mamarim/mamarimhtml/georgy.html`.

[11] Doron Zeilberger, HISTABRUT: A Maple Package for Symbol-Crunching in Probability theory, *Personal Journal of Shalosh B. Ekhad and Doron Zeilberger*, Aug. 25, 2010, `http://www.math.rutgers.edu/~zeilberg/mamarim/mamarimhtml/histabrut.html`.

Department of Mathematics, Rutgers University (New Brunswick)
Hill Center-Busch Campus, 110 Frelinghuysen Rd.
Piscataway, NJ 08854-8019, USA
zeilberg@math.rutgers.edu